The Influence of Technology on Engineering Education

Edited by
J. Bourne
A. Brodersen
M. Dawant
Vanderbilt University
Nashville, TN

CRC

CRC Press
Boca Raton New York London Tokyo

Library of Congress Cataloging-in-Publication Data

Catalog record is available from the Library of Congress.

This material is based upon work supported by the National Science Foundation under a study entitled: "Paradigm Shifts in Engineering Education: The Influence of Technology" SED-9253002. The U.S. Government has certain rights to this material.

Any opinions, findings, and conclusions or recommendations are those of the author(s) and do not necessarily reflect the views of the National Science Foundation.

Direct all inquiries to CRC Press, Inc., 2000 Corporate Blvd., N.W., Boca Raton, Florida 33431.

International Standard Book Number 0-8493-2639-7
Printed in the United States of America 2 3 4 5 6 7 8 9 0
Printed on acid-free paper

EDITOR'S NOTE

This book is the outcome of a National Science Foundation study entitled: "Paradigm Shifts in Engineering Education: The Influence of Technology," SED-9253002. The overall objective of this study was to forecast which of various possible futures in engineering education were most promising to pursue. The first part of the book contains a series of critical review papers that survey the state-of-the-art in various aspects of engineering education and attempts to look into the future to determine directions for future directions for engineering education. The second part of the book contains data and summaries from meetings held by focus groups convened to discuss possible alternative forecasts.

The editors of this book wish to thank our consultants for their effort in creating the review papers and for debating issues over the course of the study. The consultants were: James Avery (University of Colorado), Bruce Black (Rose-Hulman Institute of Technology), James Dally (University of Maryland), Christopher Dede (George Mason University), Francis Kulacki (University of Minnesota, formerly of Colorado State University), Randy Katz (University of California at Berkeley), Roy Mattson (National Technological University), Joseph Tront (Virginia Polytechnic Institute), and Robert Thomas (Cornell University).

The editors wish to thank the National Science Foundation for its support and a special thanks goes to Jim Dietz, our project monitor from the Foundation for his encouragement throughout the project. Also acknowledged are Vanderbilt University for support of faculty time, Jackie Lusk, who organized focus groups and meetings, and Ellen Bourne and Jackie Lusk, who organized the materials found in this book. A special thanks goes to CRC Press who has made it possible to publish and distribute this book.

EXECUTIVE SUMMARY:
VISION FOR THE FUTURE IN ENGINEERING EDUCATION AND FUTURE DIRECTIONS

Computer and communications technology will enable engineering education to adapt to new and varied curricular structures and learning models that will be far more student-centered than present teaching and learning paradigms

For either evolutionary or revolutionary change in the engineering curriculum, computer and communications technology offers the potential to enrich the learning experience for students via constructivist and anchored instructional modalities. Faculty effectiveness and the faculty's ability to deal with complex, trans-disciplinary problems will also be enhanced through the use of these technologies.

Computer and communications technology will especially influence the engineering curriculum by facilitating the teaching of engineering design. Visualization and simulation tools will allow students to move beyond a focus on calculation methods toward the development of engineering judgment through the rapid explanation of alternative solutions and problem formulation.

Computer and engineering technology will make possible the use of digital libraries, multimedia, hypertext, and distance learning modalities that will shift both the place and time when learning takes place. Flexible, highly responsive, and multidimensional learning environments will result from the application of such technologies for both institution-centered and dispersed groups of students.

Computer and communications technology will ultimately enable engineering schools to provide education to everyone, anywhere and at anytime. The development of highly connected electronic "knowledge highways" will offer students, faculty, and practitioners the capability to freely access knowledge they need. This knowledge will be available via taxonomies that permit the use of common materials, browsable expert lectures, and simulations. Linkages to multimedia-facilitated tutorials will dramatically enhance the breadth and richness of information available to students and faculty. The faculty will evolve new roles as learning facilitators and knowledge arbitrators within the context provided by these technology and techniques derived from them.

The recommendationsfrom this study can be summarized in three categories as shown in the Table below. This table shows categories and suggested methods for facilitating achieving progress in these areas.

Summary Recommendations by Category

Category	Recommendations
1. Capturing expertise and making it easily available	Create Electronic Digital Libraries: study access mechanisms, knowledge structures, software, standards
	Collect, store, disseminate expert knowledge
2. Research on how we teach and learn in engineering education	Study cooperative learning environments, team learning methodologies.
	Investigate student centered learning including anchored instruction (= learning design and problem solving).
	Determine how visualization/simulation teaching and learning methods can assist in engineering education.
3. Training/dissemination	Assist faculty in learning about use of technology in engineering education.
	Outreach

PART I: TABLE OF CONTENTS

PART II: TABLE OF CONTENTS

Part I

THE INFLUENCE OF TECHNOLOGY ON ENGINEERING EDUCATION: OVERVIEW AND CRITICAL REVIEW PAPERS

SECTION I: INTRODUCTION

Chapter 1

OVERVIEW

John R. Bourne and Arthur J. Brodersen, Vanderbilt University

The basic hypothesis of the work described in this book is that technology will enable fundamental changes in engineering education in the upcoming years. The materials herein examine how technology can facilitate engineering education in multiple dimensions. The perspectives taken focus on how we must change our pedagogy, and how technology can create improved environments to support learning.

This book provides both individual viewpoints from our study group and amalgamated recommendations and visions that were created by the same group in a single meeting during the summer of 1993. Commencing with this chapter, the first section of the book introduces the study, the second section examines educational issues in detail, the third section probes deeply into futures for education technology and finally, the combined material is evaluated and contrasted. To set the stage for the reading of the book, this chapter provides a *reader's guide* to the material.

A. THE READER'S GUIDE

Part I: Section I

The three chapters contained in the first section of Part I are:

- Chapter 1. Overview.
- Chapter 2. Engineering Curriculum: The Contextual Basis for Reform in the Information Age, Francis Kulacki, University of Minnesota.
- Chapter 3. Artificial Realities, Virtual Communities, and Intelligent Artifacts: Implications for Engineering Education, Christopher Dede, George Mason University.

Chapters 2 and 3 set the stage for the remainder of the book by providing high contrast descriptions of our world of engineering education and what this world might become in the future. Kulacki's article examines where we are now and Dede's paper takes us on a brief journey into an educational world of the future.

The premise of Kulacki's article is that engineering education has grown too large, requiring on the average over nine semesters to complete a baccalaureate degree. A historical perspective on where we have been in engineering education during the last two centuries is given that will be useful in providing the reader with a context in which to ruminate about the problems facing engineering education today. A major proposition put forward by Kulacki is that the current curriculum must be transformed into a curriculum that can be offered within a four year period. Kulacki shows that this transformation can be achieved by moving the curriculum to a more science-oriented base. He demonstrates this with a detailed curriculum revision.

In contrast to Kulacki's paper, Dede provides the reader with a detailed look at how technology might enable transformation of engineering education. Moving from a contrast between school and work place models, Dede provides a series of vignettes set in the late 1990s to illustrate how learning methods might change based on learning principles rooted in cognitive science. He provides illustrations that use multimedia, learning by doing, and learning while in cyberspace, and he shows how technology can facilitate a faculty member's interaction with her entire academic existence. Finally, Dede creates a series of recommendations, based on his prior presentation, that describe how change can be created by engineering educators.

2. Part I: Section II

This section of the book contains chapters that focus on educational issues, models and implementation strategies.

- Chapter 4. Educational Issues and Models. Charles Kinzer, Vanderbilt University.
- Chapter 5. The Influence of Technology on Engineering Education: Anchored Instruction and Engineering Education, James Dally, University of Maryland.
- Chapter 6. The Future of Mathematics Education for Engineering Students. Malcolm Getz, Vanderbilt. Computer Algebra Systems at Rose-Hulman Institute of Technology, Bruce Black.
- Chapter 7. The Role of Distance Learning in Engineering Education, Roy Mattson, National Technological University.
- Chapter 8. The Impact of Technology on the Engineering Laboratory, James Avery, University of Colorado.

The above collection of papers steps us through a number of pedagogical ideas that range from setting the pedagogical context to what we might actually do in traditional learning-by-doing engineering laboratories.

In Chapter 4, Kinzer sets the stage by examining how anchored instruction can be useful in engineering education. The paper describes how anchored instruction can assist students in learning in highly contextualized situations in which students engage in problem solving and discovery. This paper provides definitions, a detailed background about anchored instruction, and includes examples relevant to engineering education.

In chapter 5, James Dally extends the cognitive issues discussed by Kinzer directly into engineering education domains. He reviews anchored instruction at a number of institutions, analyzes the current curriculum and process, describes problems with this curriculum, suggests re-engineering the curriculum to include anchored instruction and discusses what this might mean for engineering education. Dally provides a detailed example of the use of anchored instruction in mechanics. This paper includes an emphasis on the need for teaching design through experience, the benefits of learning in teams and the utility of learning-by-doing. The review of similar activities in other countries is worth reading.

In Chapter 6, Malcolm Getz examines the way technology can impact the teaching of mathematics to engineers. In an appendix to this chapter, Bruce Black describes the actual experiences with teaching mathematics with computers at Rose-Hulman. Getz discusses teaching styles, including talking, writing, and group activities as part of the teaching of mathematics with computers that link his discussions back to information presented in earlier chapters. As an economist, Getz brings the cost and time dimensions to the discussion, presenting a compelling case for using computers to teach mathematics. Finally, he ends with a discussion of how we might implement computer mathematics throughout the curriculum. As a counterpoint to optimism, Black recounts experiences in teaching computer math at Rose-Hulman that did not reveal improvements in math skills due to the use of computer math.

Chapter 7 provides a detailed examination of the role of distance education in engineering education by Roy Mattson from the National Technological University, a university that is a pioneer in distance education. The extensive presentation ranges from needs and a history of distance education, through a presentation of historical NTU activities in this area, to a discussion of many other activities in distance education worldwide. Mattson concludes with a discussion of how colleges and government should proceed to implement distance learning ideas.

In Chapter 8, the final chapter in this section, Avery discusses an important aspect of engineering education—the laboratory. He discusses the need for data-acquisition systems and computer-aided design systems. He contrasts virtual and physical laboratories, wondering whether simulation can replay reality. Finally, he discusses progress at the University of Colorado and provides a prospectus for change.

3. Part I: Section III

The final section of this part of the book is devoted to technology. Computer and communications technology, multimedia and the digital library, and, finally, publishing technology are described. The following three chapters are included:

- Chapter 9. The Impact of Computer and Communications Technology on Engineering Education, Bourne and Brodersen, Vanderbilt University.

- Chapter 10. New Paradigms for Engineering Education Using the National Information Infrastructure and Electronically Mediated Instruction, Katz, Thomas and Tront, U.C. Berkeley, Cornell, and Virginia Tech.

In Chapter 9, Bourne and Brodersen attempt to take a long view of the future of technology and how it will impact engineering education. Impacts in different areas in academia are assessed and futures in hardware and software presented.

Chapter 10 first examines prospective paradigm shifts in engineering education, then describes the National Information Infrastructure (NII) and finally discusses how the NII can impact the delivery of engineering education.

4. Part I: Section IV

The final section of this part of the text rolls up the information presented so far into a cohesive discussion. Also included is a discussion of the outcomes presented in Part II of the book.

5. Part II: Results of the Workshop on Paradigm Shifts in Engineering Education, July, 1993.

This part of the book presents the methods and findings from the workshop held by the participants in the study in the summer of 1993. The efforts of the participants in understanding tradeoffs in three different areas in engineering education: distributed learning, curriculum paradigms, and delivery mechanisms are presented. Also included are surveys made at several conferences and on the Internet to provide the group with background information about the state of engineering education. Pressures and opportunities in the three areas were acquired from focus groups in the workshop, a Kepner-Tregoe tradeoff of alternatives in the areas was analyzed, and vision statements were created. At the end of the workshop a road map for the future was produced.

6. Summary Statement.

The editors of this book hope that engineering educators will be able to use the materials that are presented to help guide them toward an improved future. As pointed out throughout the chapters, there are many potential pitfalls. New technologies offer potential as well as danger. Several authors suggest that controlled studies should be commenced and that the government should help in this effort. More than anything, however, we as engineering educators must confront the biggest enemy—ourselves—and be willing to embrace change. It is clear that paradigms can and must shift; we must move to a mechanism in which students learn better, learn more, and learn faster. And we must create systems that permit these improvements with an ever-shrinking resource pool. As we move to the next century, it is clear that change is upon us and we cannot rest comfortably assured that what we have done in the past will work in an uncertain future.

B. CONTRIBUTORS

The individuals who contributed to this book and to the study meetings are listed below.

John R. Bourne, professor of Electrical Engineering and professor of Management of Technology, Vanderbilt University, Nashville, Tennessee. Bourne's research areas include intelligent systems and educational technology. He is the editor of the *CRC Critical Reviews in Biomedical Engineering* and is director of the Center for Innovation in Engineering Education at Vanderbilt University. He has been at Vanderbilt University for twenty-five years and has taken one sabbatical in industry and another sabbatical to Sweden.

Arthur J. Brodersen received the B.S., the M.S. and Ph.D. degrees from the University of California, Berkeley in 1961, 1963, and 1966, respectively, all in electrical engineering. He was on the faculty at the University of Florida, Gainesville, from 1966 to 1974. He has been at Vanderbilt since 1974 and has served as chairman of electrical and biomedical engineering and associate dean of the school of engineering. He is currently professor of electrical engineering, director of the computer engineering program and co-director of the Center for Innovation in Engineering Education .

In recent years, his research interests have been centered on the uses of computer and communications technology to enhance the undergraduate education of engineering students.

Martine M. Dawant has a MS degree in Civil Engineering from the Université Catholique de Louvain (Belgium). She spent the 4 last years with Vanderbilt University working on various projects including creating intelligent tutorial systems for engineering education, building systems for teaching about quality methods. She organized surveys and analyzed data for the NSF-funded study on engineering education published in this book. She also organized the materials found in this book.

Charles K. Kinzer is Associate Professor in the Department of Teaching and Learning and Research Scientist in the Learning Technology Center at Peabody College of Vanderbilt University. His areas of interest include research in teacher cognition and its influences on instruction, vocabulary comprehension processes in reading and writing, and the use of technology in teaching and learning. He has edited and contributed to several books about computer and other technological applications to schools and learning, and has co-authored a major textbook in the field of reading education, now in its third edition. He has also been the technology consultant to several publishers developing software in the areas of reading comprehension and job-related literacy, most recently for Houghton Mifflin's Reading Comprehension K-4 series.

Dr. Kinzer has been the Principal or Co-principal Investigator on a number of federally-funded research projects, from agencies such as OERI, FIPSE, NSF, NIMH, and OSEP; he has had major roles in many other funded projects. His expertise has facilitated collaborative relationships with colleagues from the Learning Technology Center, the Center for Intelligent Systems, the Vanderbilt University Medical School and with teams of researchers in education, special education and psychology.

Dr. Kinzer has published extensively, in journals such as the *Journal of Special Education Technology*, the *Journal of Reading Behavior*, *Reading Research and Instruction*, *Applied Cognitive Psychology*, *Educational Researcher*, the *Journal of Reading*, and others. He is Director of Graduate Studies for the Department of Teaching and Learning at Peabody College of Vanderbilt University, serves on several editorial boards, chairs the International Reading Association Research and Studies Committee: Programs, and is the co-editor of the *NRC Yearbook.*

Roy H. Mattson was born in Chisholm, Minnesota on December 28, 1927. After service in the US Navy as an electronic technician, he attended the University of Minnesota. There he received the BS degree in Electrical Engineering with distinction in 1951 and the MSEE degree in 1952. He was at the Bell Telephone Laboratories from 1952 until 1956. He then attended Iowa State University and received the Ph.D. degree in 1959. He was an Assistant Professor at Iowa State University from 1956 to 1959 and Associate Professor in Electrical Engineering from 1959 until 1961. He then joined the University of Minnesota as Associate Professor. In 1966 he accepted a position as Professor and Head of the Electrical Engineering Department at the University of Arizona. He was Department Head until 1986, and continued as a Professor until 1988. He then joined the National Technological University as the Academic Vice President.

Dr. Mattson has been active in engineering education activities serving as the Editor of IEEE Transactions on Education, the General Chairman of Frontiers in Education Conference, Chairman of electrical Engineering Division for the American Society for Engineering Education, Chairman of the IEEE Validation of Education Achievement, he served on the IEEE Education Activities Board and was Chairman of the Electrical Engineering Department Heads association. He has authored two books, been granted six patents, and published numerous articles.

Malcolm Getz has been a faculty member in economics at Vanderbilt since 1973. From 1984 to 1994, he was Director of the Jean and Alexander Heard Library and from 1985 to 1994, he was Associate Provost for Information Services and Technology. He earned his B.A. in economics from Williams College in 1967, and his Ph.D. in economics from Yale University in 1973. He was the recipient of a Woodrow Wilson Fellowship, a Southern Fellowship, and a National Science Foundation Fellowship. He has published books on urban fire departments, public libraries, and a microeconomics textbook.

Malcolm Getz's recent writing addresses economic issues in libraries, computing in instruction, information storage, and electronic publishing. His latest book (with Clotfelter and others), *Economics Challenges in Higher Education* (Chicago, 1991), examines changing costs per student in higher education through the 1980s. A current essay describes innovation in higher education.

Bruce A. Black received his B. S. degree from Columbia University, the M.S. from Massachusetts Institute of Technology, and his Ph.D. from the University of California at Berkeley, all in electrical engineering. He has held faculty positions at the University of Toronto, Clarkson University, and is currently Professor of Electrical and Computer Engineering at Rose-Hulman Institute of Technology. His areas of interest are communication systems, computer networks, and computer architecture, and he is active in curriculum development. He lives in Terre Haute, Indiana with his wife and two children, accompanied by a dog, a cat, ten gerbils, two guinea pigs, and a mouse.

James P. Avery holds a B.S. in Computer Science from Michigan State University, and a Ph.D. in Analytical Chemistry from Illinois. He spent four years on the chemistry faculty at Illinois. He moved to the Electrical and Computer Engineering Department at the University of Colorado in 1981, and is currently the Associate Chair, with responsibility for undergraduate programs. He is also the Technical Director of the Integrated Teaching Laboratory, a college-wide laboratory devoted to integrating laboratory innovations into engineering curriculum. His research interests include signal processing systems for atmospheric radars, chemical instrumentation, and data visualization.

James W. Dally obtained a bachelor of science degree and a master of science degree, both in mechanical engineering, from the Carnegie Institute of Technology. He obtained a doctoral degree in mechanics from the Illinois Institute of Technology. He has taught at Cornell University, the Illinois Institute of Technology, and the University of Rhode Island. He is currently a professor in the Department of Mechanical Engineering at the University of Maryland. Professor Dally has also worked at the Mesta Machine Company; Armour Research Foundation; and IBM, Federal Systems Division. He is a fellow of the American Society of Mechanical Engineers, Society for Experimental Mechanics, and American Academy of Mechanics. He was elected to the National Academy of Engineering in 1984 and was appointed as an honorary member of the Society for Experimental Mechanics in 1983. Professor Dally has co-authored five books: *Experimental Stress Analysis, Photoelastic Coatings, Instrumentation for Engineering Measurement, Static and Dynamic*

Photoelasticity and Caustics, and *Packaging of Electronic Systems*. He has written over 100 scientific papers and holds five patents.

Christopher Dede is a Full Professor at George Mason University in Fairfax, Virginia, where he has a joint appointment in the Schools of Information Technology & Engineering and of Education. He is Director of Federal Relations and Strategic Alliances for the university and is on the core advisory faculty of GMU's Institute for Public Policy. His research interests span technology forecasting and assessment, artificial intelligence, virtual reality, and strategic planning. He currently has grants from the National Science Foundation to develop virtual reality environments for science education and to improve engineering education via the use of instructional technology.

Francis A. Kulacki is a Professor of Mechanical Engineering and Dean of the Institute of Technology at the University of Minnesota. He received his undergraduate degree in mechanical engineering and the masters degree in gas engineering at the Illinois Institute of Technology. His doctorate was obtained at the University of Minnesota in mechanical engineering with a specialty in thermal sciences. From 1971 to 1979, Dr. Kulacki was on the faculty of the mechanical engineering department at the Ohio State University. He moved to the University of Delaware, where he chaired the Department of Mechanical and Aerospace Engineering from 1980 to 1985. From 1986 to 1993, he was Dean of the College of Engineering at Colorado State University. Dr. Kulacki's research has focused on several fundamental problems on convective heat transfer in heat generating fluids and porous media. As department chair and dean, he has overseen reform in the teaching of engineering design and the development of departmental and college-level computing environments. He has been an active participant in the affairs of the American Society of Mechanical Engineers, and is currently active on the Educational Advisory Group of the National Society of Professional Engineers.

Gearold R. Johnson became Academic Vice President of the National Technological University in 1994, succeeding Dr. Roy Mattson who retired in March, 1994. Dr. Johnson is currently on leave of absence from Colorado State University where he holds the George T. Abell Chair in Engineering. Dr. Johnson joined Colorado State University in 1971; he served as chair of engineering science from 1979 to 1984. He founded the Center for Computer Assisted Engineering in 1981 and designed and implemented the College's computing network. Dr. Johnson's research interests are in engineering education, formal methods in engineering design, and computing environments. He received the B.S.A.E., M.S.E. and Ph.D. from Purdue University.

Evan Vlachos is Professor of Sociology and Civil Engineering, and Associate Director of the International School for Water Resources, at Colorado State University. Dr. Vlachos has over 20 years of experience in academic teaching and administration, applied research and consulting. His work has focused on natural resources management, water resources planning, forecasting and futurism, technology assessment, and demography. Dr. Vlachos holds a LL.B. from the School of Law, Athens, Greece, and both an M.A. and Ph.D. from Indiana University, as well as a Certificate of Russian Studies.

Robert J. Thomas received the Ph.D. degree (Electrical Engineering) from Wayne State University in June, 1973. He currently holds the position of Professor of Electrical

Engineering at Cornell University where he has been on the faculty since September, 1973. During the 1979-1980 academic year he spent his sabbatical leave with the U.S. Department of Energy Office of Electric Energy Systems (EES) in Washington, DC In 1987 and 1988, he was on assignment from Cornell University to the National Science Foundation (NSF) as the first Program Director for the Large-Scale Nonlinear Systems Program in the Engineering Directorate's Division of Electrical, Communications and Systems Engineering (ECSE). He has been a consultant to NASA, to the Aero-space Corporation, to Oak Ridge National Laboratories, to the Solar Energy Research Institute and to the U.S. Department of Energy through the Meridian Corporation on various electric utility system related issues. He has been a consultant to the Acrotech Company on circuit and control system design techniques and to the General Electric Company CRD on control of motor drives. He has been an Electric Power Research Institute (EPRI) ERA Professor and a member of the NSF's Division of Electrical and Communication Systems advisory committee. He was the IEEE Transactions on Circuits and Systems Associate Editor for Power Networks, Graph Theory, and Large-Scale Systems. He is currently an Associate Editor of the Wiley Interscience Journal on Computer Applications in Engineering Education. He is a member of the IEEE United States Activity board's Energy Policy Committee. His current technical research interests are broadly in the areas of analysis and control of nonlinear continuous and discrete time systems with applications to large-scale electric power systems. He is in his third year as the EE faculty advisor on the Cornell Hybrid Vehicle Team. The team won the overall competition in 1993 and took top honors in six categories in the 1994 competition. His educational research interests are concerned with new methods of electronically mediated instruction. He is a member of Tau Beta Pi, Eta Kappa Nu, Sigma Xi, ASEE and a Fellow of the IEEE. He was Cornell's founding Principal Investigator of the SYNTHESIS National Engineering Education Coalition. He is the Director of the Power Systems Engineering Research Center, a consortium of universities (Cornell, Berkeley, Howard, Illinois and Wisconsin) involved with research on issues associated with restructuring the nation's electric utility system.

Joseph G. Tront is the Assistant Dean for Engineering Computing and an Associate Professor of Electrical Engineering at Virginia Polytechnic Institute and State University, where he teaches both graduate and undergraduate courses in computer engineering and electronics. His research interests include: investigating the use of educational technologies in engineering education, VLSI design, parallel processing, digital circuit modeling and simulation, and microprocessor applications. Dr. Tront is the director of the SUCCEED coalition's Center for Technology and Communications wherein groups of researchers explore how to best integrate educational technologies into the teaching/learning process. As Assistant Dean, he has overseen the formation of the college's Multimedia Development Lab and has been instrumental in working with faculty to make use of this new facility and this new paradigm in engineering education. Dr. Tront's work has been presented at a variety of national and international conferences and has been published in the IEEE Trans. on Computers, the IEEE Trans. on Electromagnetic Compatibility, and the IEEE Trans. on Education. His research has been sponsored by NSF, NASA, and the U.S. Navy.

Randy Katz received his A. B. (with distinction in all subjects) in mathematics and computer science from Cornell University in 1976. He received in the M. S. and Ph.D. degrees in computer science from the University of California, Berkeley in 1978 and 1980 respectively. After a year in industry, he joined the University of Wisconsin-Madison as an Assistant Professor of Computer Science. In 1983, he joined the faculty of the University of California, Berkeley where he is now a Full Professor in the Electrical Engineering and Science Department. He has been an IBM Predoctoral Fellow, an IBM Faculty Development Award, and was selected as a National Science Foundation Presidential Young Investigator in 1984.

Katz has won numerous best paper and best presentation awards in conferences. He has authored over 100 papers on computer-aided design, multiprocessor systems, high performance disk array and tertiary storage systems (his research group coined the "RAID"), and more recently, video server architectures and wireless computing systems. He won a Best Teaching Award of the Berkeley Academic Senate in 1992. He has written an introductory textbook in hardware design that is now in use at over 80 college campuses.

In 1993, Katz took a leave of absence from Berkeley to serve as a Program Manager in the Computing Systems Technology Office (CSTO) of the Advanced Research Projects Agency (ARPA) of the Department of Defense. He has been co-chair of the White House Technology Task Force, served on Vice President Al Gore's National Performance Review (Reengineering Government Through Information Technology team), serves on the Federal Wireless Policy Committee, co-chairs the High Performance Computing and Communications (HPCC) Program's committee on Information Infrastructure Technology and Applications (IITA), and supports the Technology Policy Working Group (TPWG) of the Information Infrastructure Task Force (IITF). He is manager for ARPA's National-scale Information Enterprises (NIE) program, part of ARPA's technical response to the challenge of developing the National Information Infrastructure. In addition, he co-manages ARPA's Wireless, Adaptive, Mobile Information Systems (WAMIS) program.

Chapter 2

ENGINEERING CURRICULUM: THE CONTEXTUAL BASIS FOR REFORM IN THE INFORMATION AGE

Francis A. Kulacki, Dean Institute of Technology, University of Minnesota;
Evan Vlachos, Professor, Department of Sociology, Colorado State University;
Gearold R. Johnson, Academic Vice President, National Technological University.

I. INTRODUCTION

Engineers and their system of education have been bound up in the progress of our nation from the development of the textile and chemical industries in the thirteen colonies to the present when the U.S. still enjoys a global economic and military leadership position. A variety of geographic, social, and political factors have contributed to this progress, but perhaps the most important underlying element has been the creativity and entrepreneurship of the work force. Technologies and innovations have arisen over the past 200 years that have broadly transformed industry, the economy, and the standard of living.

Initially, the engineer and the craftsman provided the impetus for the introduction of new products and manufacturing processes. More recently, engineers and scientists have assumed the leadership roles. Table 1 provides a brief historical summary of how the profession has developed along with the transformation of the nation's economy and industry.

While curricular designs and accreditation criteria developed after the publication of the Grinter Report (Grinter, 1955) and Goals Report (Walker, et al., 1968) have served industry and the educational enterprise well, there is growing recognition that the current system of education and the curriculum needs an overhaul. Forces for this come from both within and without the engineering schools. It is reasonable to assume that these forces signal a beginning of a period when the reformulation of undergraduate curricula may well be more than incremental and evolutionary, with a wide range of educational approaches gaining validity and acceptance within the professional and educational systems (Hutchinson and Muler, 1988).

The trend toward completing a four-year program in five (or more) years underscores points of friction with established curricula. Faculties have been unwilling, or at best careful, to reduce content within the current system that generally favors specialization throughout undergraduate studies. Indeed, engineering is the only field of study that has not upgraded its primary educational requirements beyond the four-year baccalaureate as the first designated degree for entry into professional practice. The four-year baccalaureate was well established in 1900 and represented then the most rigorous requirement of all professional fields. Law, medicine, and other fields now require considerably more formal education and practice before full professional status is granted. For engineering students, the unpleasant reality is the aforementioned impacted curriculum owing to expanding scientific and engineering knowledge and the efforts of many programs to duplicate each other with respect to curricular breadth and content[1].

1790-1850	Technology transfer from Europe helps build new American industries. Craftsmen apply technology with little scientific knowledge. Engineering program at the U.S. Military Academy offers formal instruction in civil engineering in 1802. Rensselaer Polytechnic Institute established in 1824. Approximately two self-proclaimed engineers exist in each state by 1816. Most engineering activity classified as either civil or mechanical engineering. Erie Canal completed in 1825. Gunpowder, textile, and chemical industries established. Rapid advance of railroads occurs, driven by introduction of high strength steels.
1840-1890	Emergence of the professional engineer and expansion of engineering schools and curricula. Educational specialization begins within engineering. Colleges introduce applied science and engineering studies: Union, Yale, Brown, Harvard, Dartmouth, Michigan, and Cornell. Institute schools are developed: Worcester Polytechnic Institute, Stevens Institute of Technology, Massachusetts Institute of Technology. Morrill Act establishes the Land Grant College with a mission in agriculture, mechanic arts, and related classical studies. Engineering schools graduate about 100 bachelor's degree engineers per year. Inventions: cotton gin, horse-drawn reaper, long-distance telegraph, typewriter, rotary press, telephone, Westinghouse air brake, and Edison electric light and power system.
1900-1940	Engineers number 45,000. Engineering schools produce about 4,500 graduates per year. Post-baccalaureate study programs are developed in industry and by colleges. Engineering education has a technology and applications emphasis and focus. A large percentage of the curriculum is devoted to current industrial practice and technology.
1900-1940	In 1921, Thorstein Veblen describes revolutionary role of the engineer in society and emergence of a technocracy. Man-made fibers are developed and introduced. National highway system proposed.
1940s	The rise of science and applied science begins to affect approaches to graduate and undergraduate education, methods of instruction, and research. Engineering enrollment explodes due to benefits of GI Bill. World War II demonstrates effectiveness of military R&D conducted on university campuses. Nuclear Age begins. Transistors invented and semiconductors are developed.
1950s	An overhaul of engineering curricula is implemented at the national level. This includes removal of an integrated course sequence, a growing emphasis on engineering science, elimination of many instructional laboratories, and efforts of Engineers Council for Professional Development to improve design, synthesis, and systems content. Soviet Union launches Sputnik in 1957. Digital electronic computers are developed.

Table 1. Historical perspective on engineering education in the United States. From technical craft to scientific professionalism[1].

[1] Many sources have been used to compile this table. In addition to a variety of historical texts and summaries on the development of American industry, science and technology, key source materials related to engineering education are: Cunningham, 1987; National Academy Press, 1985; Perrucci and Gerstl, 1969a,b; Wickenden, 1930/1934; Grinter, 1955; and Walker, *et al.*, 1968.

1960s	Design requirements updated for all ECPD-approved degree programs. Bachelor's degree programs in engineering technology started. Computer science as a separate discipline begins under industry impetus. Integrated circuits developed. U.S. lands man on the moon. Environmental awareness begins. The National Science Foundation supports acquisition of scientific computers by major research universities.
1970s	Rapid increase of student enrollment. Resource problems arise in both public and private colleges. Quality of programs not maintained by universities, i.e., lack of financial support of both programs and facilities/laboratories occurs. The "research college" becomes the expected role for many engineering schools. VLSI technologies developed leading to the development of microprocessors, and memory chips. ARPAnet initiated. First and second energy crises occur. Environmental quality becomes a political and socio-technical issue. U.S. begins to lose consumer electronics industry but is world leader in agricultural exports, commercial and military aviation, and computer hardware.
Mid-1970s	A "crisis" situation is believed to have developed, with desire for a faculty career declining significantly. Engineering enrollment grows. Supercomputer and microcomputer technologies developed. Interactive computer graphics technologies developed. Biotechnology industry develops focus on engineered organisms and small batch processing.
1980s	Enrollment peak reached in 1983-1984. Increase of multidisciplinary research centers on many campuses. Fragmented curricula and specialization continue. New initiatives taken by government and industry to bring university-based research closer to commercial needs, e.g., materials, manufacturing, electronics. New accreditation guidelines assure flexibility in curriculum, i.e., 1.5 years of arts and science and 1.5 years of engineering science. U.S. manufacturing experiences loss of jobs due to international competition in consumer goods. Quality of U.S. products is an issue. Graduate engineering education becomes a national issue with decline of Americans in doctoral programs. Satellite-based distance education initiated. Internet becomes a computer network of computer networks spanning the globe. Large investment in computer hardware in engineering schools.
1990s	Retirement of faculty hired in the 1950s and 1960s starts. Issues of cultural and gender diversity in student body and faculty arise. Global challenges to U.S. leadership and economic productivity arise. Internet characterized as a collection of millions of computers with millions of users. Initiation of the development of the national information infrastructure, i.e., the information superhighway.

Table 1 (continued). Historical perspective on engineering education in the United States. From technical craft to scientific professionalism.

More recently, public debate within industry, professional societies, and academia has focused on the adequacy of the baccalaureate degree as the terminal degree for engineering practice. A traditional debate centers on the skills a new graduate brings to an initial work assignment versus the ability of that same graduate to work in different technical areas and at higher levels of technical and managerial responsibility over an average career span of forty years. New elements in this debate arise from the changing nature of engineering work and the global integration of national and regional economies, particularly as related to manufacturing (Drucker, 1980, 1989; Landis, 1987; Hill, 1991).

It is, perhaps, these last two factors that have the potential for providing the greatest impetus to curricular redesign over the next few years. For example, Landis (1987) suggests a future in which most engineering tasks will involve the development and use of automated design tools, for both engineers and technologists.[2] Hill (1991) argues that the current system of engineering education was established for "old-era" industrialization as the framework for engineering and manufacturing, and that to meet the needs of the new era, such as lean production processes (Womack et al., 1990), a transformation of the curriculum is needed. The coming of the post-industrial society will further increase the demands for a flexible curriculum capable of adapting to quickly changing circumstances and to the requisites of a predominantly information-based culture.

A predominantly information-based culture requires knowledge and skills that heretofore have not been part of the traditional curriculum. Surely it will not be enough to be simply a user of telecomputing (the merger of communication and computation). Engineers and technologists will be the primary agents for the transformation to the post-industrial society, and any revised or revamped curriculum must encompass telecomputing.

II. THE TRANSFORMING CONTEXT

There are thus several aspects to the need for the transformation of engineering curricula at this time. Most of them are professionally based and ask questions as to the nature of educational inputs, outcomes, and processes. Figure 1 illustrates this in a simplistic way. In general, we define "process" to be the curriculum, which comprises both curricular design (plan and structure) and instructional modality. The intersection of educational input (I), process (P), and outcomes (O) emphasizes the context of the transformation. Again, process is broadly interpreted to be the curriculum, although a wider range of action agents and definitions may be applied, e.g., technology, accreditation guidelines, and professional registration requirements.

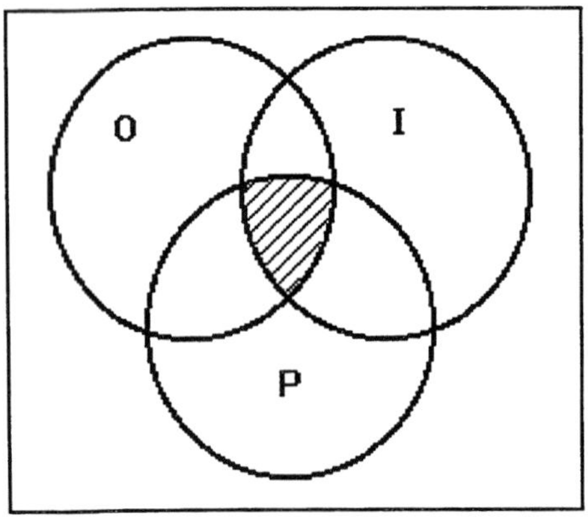

Figure 1. The framework for transformation of engineering curricula.
I = input; P = process; and O = outcomes.

It seems clear that the separation of curriculum and technology will be much more difficult in the future, for after all, it is technology that is bringing about the transformation from an industrial society to an information-based culture. This transformation is being brought about by the confluence of two technologies: communications and computing, previously defined to be telecomputing.

Worldwide communications have been developing over most of this century. Regional telephone companies were interconnected by ocean-crossing cables. The commercialization of radio in the early part of the century followed by television during the latter half of the century has enabled global awareness on an unprecedented scale. Geosynchronous communications satellites permit the interconnection of regional television broadcasting operations in a manner similar to earlier developments in telephony. Today, what happens on one side of the globe may be known by those on the other side of the globe even before the full impact is known at the originating location. But, the real breakthrough has come with the realization that all information, whether audio, video, animation, still images, or simply collections of text and numbers, can be handled in a uniform format, i.e., as digitally encoded information. It is this realization that has paved the way for the merging of communications and computing.

Computing has moved from the ownership and complete control of extremely expensive hardware (so-called mainframe computers) by a very small number of technical professionals to vast numbers of extraordinarily powerful, fully worldwide interconnected machines so inexpensive that they are owned by undergraduate students, in less than thirty years. Today's applications software enables computers to be super-sophisticated typewriters with both spelling and grammar checking; accounting and general purpose business machines; simulation devices addressing "what if?" questions; entertainment systems; automatic controllers; image generators, manipulators and enhancers; programmable learning machines; publishing systems; etc. Yet, the true power of the computer was not realized until computers began communicating with other computers over worldwide networks. The computer enabled banks to transact global business in seconds, documents to be shared among authors distributed across continents and even intercontinentally, experts to be located and questions resolved in minutes because computers are able to communicate at speeds in excess of five orders of magnitude (100,000 times) faster than humans.

The development and application of telecomputing has been an add-on to the engineering curriculum rather than integrated into its design. This is a prime example of the expansion of technological information not adequately handled in current curriculum offerings.

In a recent program description of undergraduate curriculum development, the National Science Foundation (NSF)(1989) described how the base of engineering fundamentals is both expanding and evolving at a rapid rate along with the overall information explosion, much of which is fueled by advances in telecomputing and its impact on traditional sciences and technology. It points out that engineers need to develop better skills in problem definition and in communication, and that there is an increasing demand for engineers to be more informed about risk and uncertainty. Following the results of many other studies, the NSF report points out that there is a consensus that all aspects of engineering education should be re-examined, including the learning and teaching processes, the learning environment, the effectiveness of engineering education, the preparation of young people for lifelong learning, and the use of technology in education.

The proposed NSF undergraduate program emphasizes that such an educational process should produce students with analysis capability, or the critical thinking that underlies problem definition and derives from serious understanding of the physical, life, mathematical and computational sciences, the humanities, and social sciences; innovation and synthesis capability, or the creation and elegant implementation of useful systems and products, including their design and manufacture; integrative capability, where analysis and synthesis are supported with sensitivity to societal need and environmental fragility; and contextual understanding capability, namely the appreciation of the social, economic, industrial, and international environment in which engineering is practiced, and the competence to accept the responsibility for societal leadership effectively. These programmatic outcomes have also been enunciated by Ernst (1989) and Bordogna (1991).

Thus, even within a broad range of educational and professional concerns, a larger social dimension emerges that focuses on engineering and the engineer in society (Knepler, 1977; National Academy of Engineering, 1985; National Research Council, 1985; Vlachos, 1991). Further, the sociology of the profession depends in a complex way on supply versus demand for engineers as an occupational group; the evolvement of the engineer in public policy matters; shifts in government spending (e.g., defense to civilian or commercial programs); changes in the work force brought about by demographics; employment practices and policies; educational support policies; and other factors. Table 2, prepared by the U.S. Office of Technology Assessment (1988), summarizes the factors for and the supply of scientists and engineers.

- **Factors that increase demand:**
    - * Increase in basic or mission research.
    - * Economic growth.
    - * Increasing technology sophistication of U.S. manufacturing and services due to scientific progress, international competition, and demands for a higher standard of living.
    - * Increase in science and engineering higher education enrollments, causing an increased demand for faculty.
- **Factors that decrease demand:**
    - * Sending R&D and engineering offshore.
    - * Decrease in basic or mission research.
    - * Economic recession.
- **Factors that shift demand between disciplines:**
    - * Technological change and scientific advance of all kinds, which render disciplines obsolete while creating new ones.
    - * Using technicians for some tasks now undertaken by scientists and engineers.
    - * Automation of engineering function by means of computer-aided design and manufacturing and other communication and information technologies.
- **Factors influencing supply:**
    - * The size and rate of increase or decrease of demand for scientists and engineers modulated by the salary advantage for scientists and engineers and the national level of R&D expenditure.
    - * The number of births and their racial and ethnic composition.
    - * Education at elementary, secondary and higher levels.
    - * Permanent and temporary immigration of foreign scientists and engineers.
    - * Federal and state initiatives to encourage different types of institutions to award more science and engineering degrees at a higher level.
    - * Legislation and other actions that affect the opportunity to attend and afford college or graduate education.

Table 2. Factors influencing demand for, and supply of, scientists and engineers. (Office of Technology Assessment, 1988).

On the other hand, Armstrong (1990) has contrasted these conventional and popular factors with actual measures of the supply of engineering graduates. He suggests that the supply of baccalaureate degree engineers varies within a surprisingly well-defined range as a percent of the high school graduating class, with the potential for employment being the most important determinant of the choice of engineering as a field of study in college.[3]

From a statistical viewpoint, the number of engineers employed in the nation is a very small part of the total work force over the past generation—on the order of one to two percent. Table 3 presents the number employed by primary discipline as estimated by the Bureau of Labor Statistics and the National Science Foundation (1988). Differences in the data reflect differences in definition and data collection methods.

	BLS 1985	NSF 1988
Electrical	544,000	639,000
Mechanical	272,000	649,000
Civil	221,000	338,000
Aeronautical	95,000	119,000
Chemical	64,000	150,000

Table 3. Engineers employed in the United States.
BLS = Bureau of Labor Statistics; NSF = National Science Foundation (1988).

The labor market varies considerably among engineering specialties. New specialties such as materials, bioprocessing, software engineering, or computer engineering often emerge as offshoots of a traditional specialty. The professionals in these fields are often drawn from the more traditional areas of engineering and sometimes bridge two or more fields, e.g., chemical engineers with advanced degrees in the chemical or life sciences who work as biochemical engineers. The trend toward multidisciplinary career specialties and changes in an engineer's technical focus over a career span appears to be the norm, and it will make characterization of the labor market and engineering work force difficult in the years ahead.

However, all data point out that engineering has been one of the fastest growing occupations in the 1980s, with over 90 percent employed in science and engineering positions. Women comprise slightly more than three percent of the engineering and science work force, and only about three percent of employed engineers hold doctorates. About 80 percent of employed scientists and engineers work in business and industry. More recent estimates by the National Science Foundation (1988) increase the number of engineers to 2.85 million, with the highest proportion in mechanical and electrical/electronic engineering. Such numbers coincide with the fact that the majority of the military budget goes to three industries: aerospace, communications, and electronics.[4]

While the composition of the engineering work force in the year 2000 will look very much the same as in 1986—mostly white males and females—gains will be made by the addition of African American, Asian, and Hispanic Americans. Speculating further, some recent studies on the social characteristics and roles of the engineer find that in the coming years there will be more engineers who are female, Asian, foreign-born, and from dual-career families. Future generations of engineers will predominantly work in service industries, manage and communicate with people, change jobs often, undergo retraining on a sustained basis, and be more visible to the public (Coates, 1987a,b). Table 4 summarizes the social and technical role changes for engineers, and underscores the emerging importance of the profession to the larger society.

CURRENT AND PAST:	EMERGING:
Primarily works in manufacturing.	Continues in manufacturing but also works in service industries.
Male Caucasian.	May be a woman, an Asian, or foreign born.
Sole earner and head of family.	Member of a dual-career, dual-income family with higher and greater financial independence.
A technical expert who solves a problem, writes a report.	Manages people or a process, communicates results to others orally, graphically, or in writing.
Has B.S. degree or worked way into engineering job from technician.	Has B.S. or M.S. degree, or may have begun with two-year technician's degree, earned B.S. on the job. Could have entered work force with Ph.D., especially if foreign born.
Has had one or two refresher courses since college.	Participated in at least one course retraining program. Takes one or more skill or knowledge acquisitions courses annually.
Worked for same large corporation more than twenty years.	Worked with several employers in small and large companies.
Is an employee.	Is self-employed.
Is invisible to the general public, a technical expert in the back room, to be brought out and blamed when things go wrong.	Well-known and acknowledged as an expert on issues and their societal impacts. Frequently consulted.
Has moved up socially by becoming an engineer.	Has other motives for entering engineering than social position, unless foreign born.
Reaches the top executive spot in a corporation with "hands on" skills as a base.	Is studying management and financial skills to gain an advantage in promotion out of engineering.
Experiences rapid obsolescence and retires early if company offers.	Expects obsolescence and work change, looks to self and engineering society for career planning and retraining opportunities.
Uses computers occasionally.	Uses latest telecomputing equipment.

Table 4. A summary of engineering role changes (Coates, 1987).

III. IMPLICATIONS FOR AN INTEGRATED CURRICULUM

The shift towards scientific professionalism has been necessitated by continuing technological change and the interweaving of engineering with science. At the end of World War II, the rapidity of such transformation was coupled with a focus on the scientific content of the engineering curriculum. Also, the theory and practice of engineering has faced new challenges with the more recent environmental movement and the emphasis on social responsibility and participation in policy making. Three strands, then, come together to describe the community involvement of engineering: through professional societies and participation in national affairs; through increasing emphasis on science and telecomputing; and, finally, through increasing involvement in public policy.

Throughout our century, the nature of engineering work has been changing steadily. The corporate engineer has now come to predominate, with work characterized by large project teams, macro-engineering emphasis, relative individual anonymity, and dedication to discrete bits of technology advancement in highly specialized fields (National Academy of

Engineering, 1985). At the same time, the global sphere of corporate operations, the engineering community, and new technologies added a cultural element to the diversification of the profession and the creation of several engineering disciplines. There has been an important, discernible shift in the engineering work force from one characterized according to the employees' functions rather than by the employees' training. The way that employers use engineers can help us understand how easily engineers can move from one discipline to another, the recruitment of young people to engineering and the motivation of those already employed, the extent of the continuing education and training of engineers, the effect of various initiatives on the supply and quality of engineers, and how international attitudes on engineering can be compared (O'Neill, 1990).

Furthermore, traditional engineering practice is becoming more blurred as technologists and technicians today often perform work formerly required of engineers. Figure 2 briefly compares the future roles of engineers and technicians within the framework suggested by Landis (1987). The relative numbers of each population and the degree to which engineers and technicians perform the same tasks are open to question and depend on a great number of economic and educational factors. However, what appears true today is that educational requirements for technicians will increase, and expectations of them will broaden.

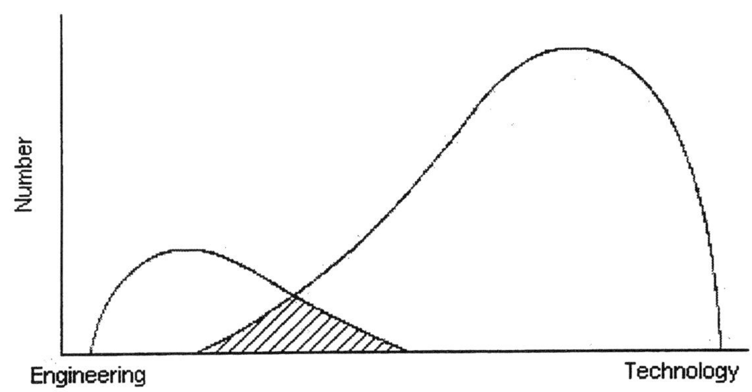

Engineering	**Technology**
Education: B.S., M.S., Ph.D., D.Sc.	Education: B.E.T., M.E.T., B.S.E.
• 4-10 years formal study	• 2-4 years formal study (B.E.T.)
• Analytical orientation	• Minimal basic science preparation
• Broad science and technology base	• State-of-the-art-oriented
• Creates design tools and manufacturing processes	• Uses design tools and manufacturing technologies
• Responsible for product development	• Assists in product development
• Skills applicable to management and planning	• Skills applicable to supervision and planning

Figure 2. Future Distribution Of Engineers And Technologists

For engineers, it may be speculated that their numbers may not increase, while their formal and lifelong educational requirements, as well as expectations of them, will continue to expand. Also, engineers are doing work formerly required of other professionals, so that they become planners and managers, with sensitivity to legal and environmental issues, social participation, need for policy analysis, political feasibility and economic viability. In essence, they provide leadership for complex, interdisciplinary teams.

It is at this juncture that we approach the reformulation of the curriculum and the issue of its downsizing. Such a curriculum implies both an applied science and technology-oriented emphasis and a significant reconstruction of the undergraduate experience. This is a minimalist model that may not retain the baccalaureate as the first professional degree. Education for the baccalaureate degree is viewed as a complementary activity towards both further specialization obtained from post-baccalaureate study and a lifelong educational commitment.

This is the spirit also of a recent report of the American Society of Civil Engineers (1987) that addresses the issue of engineering in the twenty-first century. The report emphasizes that the four-year bachelor of science program should be viewed as the most important step in a system of professional lifelong education. And it continues, "The degree must provide a balance between those skills that enable the civil engineer to handle the complex, scientific, and technological aspects of the profession and those skills that enable the civil engineer to set policy, act as the entrepreneur, and be the organizational leader of the engineering process. Among the strategies proposed for such an approach is instituting the master's degree as the entry-level professional practice degree."

IV. TOWARD A REDUCED BACCALAUREATE PROGRAM

Turning now to a new structure for undergraduate education, we begin with accepting the overarching outcomes desired from the educational process enunciated by Ernst (1989). We assume that entrants to college study are predisposed to engineering and do not require remedial study in mathematics and English composition. We assume that the instructional modality will remain the same as at present, i.e., the passive lecture supplemented by design projects and laboratory experience. However, this latter assumption may be our weakest one because the current instructional modality will come under increasingly greater pressure as numbers of students and educational costs continue to escalate. The economics of engineering education are such that self-paced learning in both engineering science and design must be done using data bases, high-speed computer graphics and its extension, virtual reality, and newly developed languages for expressing design. In this context, telecomputing will play an even larger role than that described herein. In addition, we assume that accreditation requirements will remain essentially unchanged. There might be a change in the professional registration process, and we will elaborate on this later.

For objectives and a new curricular framework, we adopt the following:

a) Reduce total degree requirements to 120 – 123 semester credits.
b) Increase the quality and quantity of laboratory experience.
c) Integrate and effectively utilize telecomputing knowledge throughout the curriculum.
d) Assure active participation of students in design across the curriculum.
e) Increase the component of the curriculum allocated to the humanities and social sciences.
f) Provide a curricular basis for articulation to graduate programs, as well as the professional registration process.

Qualitatively, this new curricular framework can be compared to the existing one as shown in Figure 3. In this illustration, we highlight several features of the proposed system of instruction. Some of them break with current and traditional practices of engineering education, e.g., the licensing process. While our primary aim is to expose the entire structure and not labor over too many details, some time should now be spent on implementation and broad speculation on curricular content.

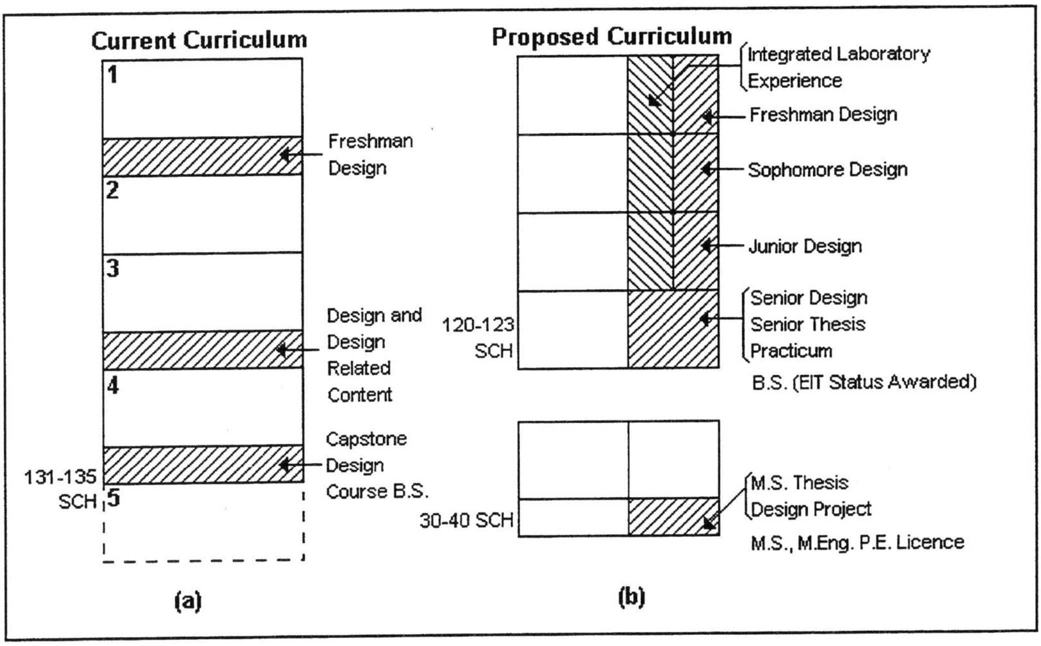

Figure 3. Engineering curricula. (a) The current curriculum. (b) Proposed structure and its articulation with post-baccalaureate study

The reduction of total course requirements implies first that basic science, mathematics, and engineering science courses become better articulated so as to eliminate duplication of material, especially in the first three years of study. Use of computing and applications software can reduce the amount of time devoted to basic science, mathematics, and engineering science courses by providing individualized and more fully integrated instruction. Multimedia applications should enable students to progress through educational materials in more depth and with more individual attention than is currently provided by textbooks. Rather than spend time in class reviewing previously learned materials, multimedia applications can provide detailed reviewing opportunities. Also, the number of technical elective courses needs to be dramatically reduced, or perhaps eliminated in favor of concentrations in more broadly based engineering science areas, e.g., transport phenomena for chemical engineering, solid mechanics and civil or mechanical engineering, groundwater engineering in civil engineering, optical/solid state sciences in electrical engineering, etc. Within this context, engineering knowledge needs to be incorporated into both the content and goals of the teaching process.[5]

An additional benefit of a reduced total credit hour requirement is that the curriculum becomes a bit more manageable within the traditional four years of undergraduate study. If we consider the typical course to be three (semester) credits, the proposed total credit hour requirement reduces the number of courses a student will take from approximately 44 to 41, a reduction of seven percent. With such a downsized curriculum, engineering students might view themselves as having approximately the same formal degree requirement as their counterparts in the arts and sciences. Recruitment of students into engineering as a field of choice might also benefit by this change, and retention might improve as well. Finally, faculties of engineering will necessarily practice a degree of economy in selecting course requirements and specialty areas within their baccalaureate programs.

To increase the quality and quantity of laboratory experience represents, perhaps, the most ambitious objective. While the time spent on laboratory instruction in a typical undergraduate program has certainly fallen since the 1950s, it has been argued that quality has also fallen (American Society for Engineering Education, 1986). A variety of factors have been suggested as causes, including the lack of continuous investment in equipment, technical support staff, and facilities, as well as a low value placed on laboratory instruction by research-oriented faculties and reward systems. Nevertheless, we propose that there needs to be at least one engineering laboratory experience per year that integrates basic science and engineering science courses to emphasize the nature of engineering measurements, data, uncertainty, and analysis of data appropriate to the level of the curriculum. Some of these laboratory courses may well be subject-specific, e.g., the fluid mechanics or hydraulics laboratory course of the third year, but we suggest a more open-ended approach to emphasize the relation of engineering science to design. With this requirement in the curriculum, we estimate that engineering laboratory experience may well consume 20 percent or more of the curriculum, in addition to the traditional laboratory courses in physics and chemistry. Within this framework, instruction in the engineering sciences could be given within the laboratory experience when appropriate, so that some economy in presenting the total curriculum could be achieved.

Here again, computing plays a major role in expanding the quality and quantity of the laboratory experience. Computer interfacing leads to data gathering, data analysis, and ultimately, process control. Laboratory documentation and report generation are also enhanced through scientific and technological text and graphics processing. Reports may be worked on collaboratively over computer networks with both on-campus and distance students. Knowledge domain experts external to the university environment may also participate. The role of computer simulation is only now beginning to be addressed in engineering curricula. But, it would seem that simulation may also help enhance the laboratory experience. In the longer term, virtual reality will enable the student to be integrated into a simulated work/factory design/production environment.

Colleges of engineering will continue to grapple with the dichotomy between teaching skills and basic engineering sciences and design practices. In the short term, computing will remain a topic of debate, but the debate will soften as computing is seen as having a duality that exists in no other element of the educational process. This duality of computing affects the entire curriculum. The results obtained from computing in all aspects of engineering are one of the prime causes for mandating massive curricular reform; yet computing provides the only vehicle for enabling the necessary reform.

Consider as a concrete example, classical automatic controls. For most of the latter half of the twentieth century, courses in automatic control were little more than courses in applied differential equations using transform techniques such as Laplace or Z transforms. Techniques such as root locus, Nyquist and Bode diagrams made up a significant amount of the materials studied. With computers taking on the role of embedded process controllers, the course content needs to shift to the design and development of time critical software, real time programming techniques, and fuzzy and adaptive algorithm design. This is a complete shift from differential-equation-based modeling to modelless applications.

Design experiences in the proposed curriculum would formally occupy a portion of instruction each semester. Generally, first- and fourth-year students receive instruction in design in all engineering curricula.[6] Formal design courses in the second and third years would, along with the parallel laboratory experience, further integrate basic and engineering sciences, as well as accelerate the student's use of judgment through practice.

Tools to support the engineering student in design activities will be the key element in allowing design across the curriculum. These tools will include access to extensive data bases on such things as fluid and material properties; chemical reactions and processes; design correlations; and elementary design elements such as gears, circuit elements, etc. High powered design and manufacturing languages with full support of three-dimensional, animatable graphic entities will enable rapid system development in a simulated business environment. Follow-on design courses will critique student design methods and permit the student to "learn from the project," rather than just receive a grade on a report, as is common today.

Whether "design credit" would continue to be offered in the engineering science courses remains an open question. Our proposal here is based on the belief that the necessary combination of talents for the baccalaureate engineer can be obtained only through formal requirements that focus on the use of knowledge in the setting and context provided by open-ended design.

With respect to the accreditation requirement for design content, this could be met easily by establishing a formal course each year. Under this circumstance, it would be entirely possible to formally meet design requirements under current accreditation guidelines, with instructional contact hours varying each year to meet educational objectives appropriate to a particular point in the curriculum. The implementation of such an approach, however, would imply a paradigm shift in the teaching modality, and this would require broadly based faculty support. This probably would be reached only after extended debate and the building of departmental consensus by advocates and the faculty leadership.

Instruction in the arts and humanities would receive increased emphasis and focus in the restructured curriculum. With but few exceptions today, engineering students meet liberal studies requirements within a framework established by institutional curriculum committees and the limits imposed by the expanded technical content of engineering. Accreditation requirements also play a role, and faculties of engineering have generally been reluctant to exceed minimal standards. What results is breadth without depth and a lack of coherence in the liberal studies programs of most engineers.

Several authors have analyzed this situation and have suggested alternative approaches to liberal education (Cheney, 1989; Johnston *et al.*, 1988), and several engineering programs have been successfully built within either a modified traditional framework or the context of the arts and sciences (Grogan et al., 1988; Swarthmore College, 1991; Harvey Mudd College, 1990). However, little has generally changed with respect to the humanities and social sciences within the vast majority of engineering colleges over the last twenty-five years. Therefore, within the proposed curriculum, we designate 30 percent of formal course requirements for the humanities and social sciences. This will translate to approximately twelve to thirteen courses over the four years of study, much less than the core suggested by Cheney (1989), but yet sufficient, we believe, for engineering students.

More importantly, both breadth and depth can be provided given the number of courses available. However, the necessity of reaching an agreement on coherence and focus within this segment of the curriculum will depend on the involvement of the faculties of the humanities and social sciences. Here we suggest that a workable curriculum can be designed if a limited number of focal study areas are selected as concentrations. Such concentrations might include economics, literature, history, history of science and technology, literature and language of another culture, political science, cultural anthropology, psychology, and sociology.

Although implementation of this aspect of our proposal will be challenging and involve literally every aspect of the proposed structure, we suggest the following as the conceptual framework: (1) Focal study areas would require a four-year, integrated plan of study that would be developed by the second semester of the engineering program. (2) Courses within each study plan would necessarily come from a limited number of candidate courses across the institution and would be selected to provide significant writing experiences in each year of study. (3) Focal study areas would require more than introductory courses, thereby leading the student to material generally reserved for the third and fourth years of study for majors in the humanities and social sciences. (4) Fourth-year courses would be selected so that students would experience considerable verbal communication in the discussion of major texts and authors.

Lastly, we consider the articulation of the proposed undergraduate program to graduate study and the professional licensing process. For graduate study, a change in educational outcomes will be needed. For professional licensing, the policies of the state boards of registration, the National Society of Professional Engineers, and the National Council of Examiners of Engineers and Surveyors would be altered to necessarily reflect the realities of the future environment for the practice of engineering.

One of the aims of the proposed restructured curriculum is to produce a graduate who can practice engineering in a larger technical and social context than he or she can at present. The implication is that a newly graduated engineer will not necessarily have the degree of depth in a specialty needed to meet specific job-related tasks. Post-baccalaureate education would be needed, and the master's degree would be regarded as necessary for the first professional degree.

For faculties of engineering, graduate programs would take on many of the aspects of technical elective programs currently offered within undergraduate curricula. A broadly based but limited core curriculum comprising traditional advanced engineering science and mathematics courses would be required, and the technical specialties could be offered at a higher level of sophistication than at present. Within a typical 18- to 24-month master's degree program, there would be enough flexibility for a good deal of specialization in most of the fields of engineering. Both thesis and non-thesis master of science (or master of engineering) programs would be viable, as well as non-thesis master of engineering programs that might include an advanced design project with industrial sponsorship.

The registration and licensing of engineers can, we believe, be efficiently accommodated within the proposed restructured curriculum. First, we propose that the status of Engineer-In-Training be granted to graduates of the undergraduate program on the basis of a reformulated program of study that brings into better balance information, judgment, and practice through the four years of study. Graduates who enter industry with the baccalaureate degree would not be disadvantaged in their professional sense, and they would enter practice with a motivation to continue their education and become fully licensed. At the conclusion of a master's degree program and an appropriate examination process coupled with an educational outcomes assessment, the professional engineer license could be awarded.

The benefits of adding this feature to the formal education of engineers would be many, including the involvement of the National Society of Professional Engineers, the National Council of Examiners of Engineers and Surveyors, and the professional societies in the educational process in a more meaningful manner than at present. One can speculate that a national licensing commission, similar in scope and mission to the Engineering Accreditation Commission, would be created to provide coordination and review of educational programs and their outcomes. The eventual relations of engineering schools to state governments, industry, and perhaps pension funds and insurance providers could possibly be much stronger and beneficial to both the individual and to the educational institutions.

V. CONCLUSION

To summarize this diversity of data, social perspectives, and professional factors, we conclude that the end of the twentieth century presents engineering education with extraordinary challenges and opportunities. What we have on our hands are: new areas of concern and policy emphasis for a work force with one-third of all engineers involved in military projects and with 80 percent of the federal R&D in defense-related activities; a transition to a new consumer-products-based economy characterized by global competitiveness and increased emphasis on productivity; and, the search for new values and skills in order to make the transition to a post-industrial, information-based society. In short, a context for engineering, engineers, and engineering education has developed with a new framework for technology development and applied science (National Academy of Engineering, 1985; Reich, 1991; Zinberg, 1990).

Thus, we propose that four simultaneous transformations are impacting engineering and engineering education and that they will continue in the future. The first is a conceptual transformation having to do with new concepts, new theories, and new paradigms characterizing the field. The second is a methodological one, particularly the presence and implication of increasing computational prowess as well as multi-objective, multi-purpose analysis. A third is an organizational transformation, with an increasing movement from a defense to non-defense orientation and new management schemes and administrative procedures. Finally, a fourth major transformation has to do with substantive concerns (a new problematique) and the technological breakthroughs and opportunities appearing at the end of the century (Vlachos, 1991).

The conclusion of the above is that an important epistemological shift has been taking place in recent decades. The engineer is being transformed from an answer-giver to a problem architect. Such a combination will require a return to the old tripartite synthesis of knowledge and information (intelligence), judgment (evaluation), and praxis (implementation). Thus, the ultimate coexistence of structured reasoning and disciplined imagination is the reasonable response to the forces of complexification and uncertainty characterizing today's world. The real challenge for engineering educators is the coupling of reasoning with imaginative thinking in the context of increased awareness as to the far reaching social implications of a practicing professional.

At the end, one can talk about the transition to the future from the old practicing and entrepreneurial engineer to the new model of a manager and systems engineer (Semans and Hansen, 1981). In practical terms, the response of the profession is a renewed recognition of the logos in technology and a deeper understanding of the long-range impacts and consequences of the engineering profession. As Semans and Hansen (1981) have pointed out, “Some of our present problems and new goals are reactive: problems created by one technology must be addressed or ameliorated with another technology. Other situations call for creative engineering, particularly for continuing innovation, for most economists agree that our national economic health is strongly dependent upon the invention of products and processes to compensate for our deficiencies in natural resources. There is a growing need for truly innovative technical approaches that satisfy economic and social issues to an unprecedented extent. The future of engineering, and thus the directions for engineering education, will be strongly influenced by these forces.”

There is no doubt that complexity, turbulence, and uncertainty, as well as attendant societal concerns, will continue to expand. In an interdependent global environment, future engineers have the unenviable task of combining specialized knowledge, innovative approaches and technologies to problem definitions and solutions. Expanding the curriculum ad infinitum with more courses and topical material is not the answer. A sensitive core curriculum and integrated praxis in a lifelong educational process of renewal seems a more appropriate combination.[7]

Our thesis is that downsizing and restructuring the engineering curriculum will bring many benefits to the student, educational institutions, and the larger sphere of professional practice. The issue for educators is the manner in which educational inputs, the educational process or curriculum, and outcomes are to be constructed and related in the future. The answer depends largely on answering several overarching questions: What kind of engineers do we wish to have in the future? What is the appropriate balance among information, judgment, and practical experience in educating engineers? What is the appropriate and needed relation of future engineers to engineering technicians? And surrounding these questions are the challenges and opportunities of a quickly transforming global society in which access to information in a variety of forms will be readily available via telecomputing and related technologies.

Our intent here has been to lay out a framework for future debate and to provide perspectives on the curriculum that might be applicable across a wide spectrum of engineering schools. Perhaps the future for engineering education will be characterized by far more variety in curricular structures than at present, with the curriculum proposed here being but one of several that has gained acceptance in both the educational and the professional realm. We think this would be healthy for the profession, for industry, and for engineering education.

ENDNOTES

1. Jones (1989) discusses this in the papers published in connection with the ASEE Frontiers in Education Conference of 1989. An earlier book by Pletta (1984) forms the basis for Jones' summary analysis. From our perspective, engineering programs have evolved to be professional programs but are administered yet within a framework developed for the arts and sciences and general education in most colleges and universities. They are thus constrained by the four-year curricular model developed for the liberal arts. This leads to curricular overload, despite the reduction of laboratory contact hours over the past generation. The natural and practical response of the student is to extend the period of study.

2. Landis (1987) suggests a distribution of engineering tasks by the year 2000 as comprising:

Design, using design automation	55%
Service and maintenance	25%
Programming and design automation	10%
Applied science	5%
Other	5%

   He notes that this list was developed initially by J. Haddad of IBM, who perhaps neglected important engineering tasks in related fields, such as construction, quality control, etc. The point made by such a grouping of tasks is that they are quite different from those for which engineering graduates of the mid-twentieth century curriculum now graduate.

3. Using data for supply versus demand for baccalaureates in chemical engineering, Armstrong (1990) suggests that demographics, recruitment and retention actions by engineering colleges, and curricular innovations are not important in determining the supply of engineers to the work force. In addition to students' perceptions of employment opportunities, he notes that the historical production of engineering graduates varies from 1.3% to 2.6% of high school graduates over the twenty-year period of 1968 to 1988. Other factors include the high degree of complexity in the national and international economy, the complexity in the choice of undergraduate majors and in persistence in undergraduate studies, and the predisposition of engineering freshmen, i.e., students majoring in engineering, to preselect themselves to a great extent.

4. The military budget of the United States emphasized the need for these engineering fields to a great extent throughout the post-World War II era (1945-1990). Whether this will continue in the twenty-first century makes for interesting speculation. In the context of the probable change in military spending, particularly in the development of advanced weapons systems, one might expect decreased economic incentives for the production of large numbers of engineers in these specialties (Vlachos, 1991; Vitale, 1991; Gagné, 1991; Hughes, 1991; Matzkin, 1991).

5. Vincenti (1990) points out the important differences between "engineering" knowledge and scientific knowledge within the context of aeronautical engineering. His thesis divides engineering knowledge into six categories: fundamental design concepts, criteria and specifications, theoretical tools, qualitative data, practical considerations, and design instrumentalities (Ibid., P. 208).

6. The first-year design experience varies greatly across all of the engineering schools today, but it is sufficient to note that an introduction to the "engineering process" and design is usually discussed within the professionalization offered in such courses. Design experiences at the senior level vary from the team approach to individual projects. Formal instruction within the capstone course varies greatly as well, and there is no consensus on what should be offered. One approach that seems to have had good success is the "case study" approach (Zimmerman et al., 1986). Many schools have promoted active involvement with industry where the proximity of industries permit, e.g., Harvey Mudd College.

7. The converse of this is that all other educated people must understand the challenge of engineering as part of a more balanced awareness of benefits, risks, and costs associated with technological change. Courses in "technology and society" and the like are a step in the right direction and typically have been developed in the liberal arts colleges. A broad ranging rethinking of the liberal arts for the technological age seems in order and should be on the national agenda in higher education.

REFERENCES

1. **Atkinson, R. C.**, Supply and Demand for Scientists and Engineers: A National Crisis in the Making. *Science* Vol. 248, pp. 425-432, 1990
2. **Armstrong, R. K.**, *Another Perspective on the Workforce 2000: Supply/Demand*, E. I. DuPont de Nemours and Co., Inc., Wilmington (personal communication), 1990.
3. **Bordogna, J.**, Integrating Education, *Engineering Times*, p. 4, May, 1991.
4. **Cheney, L. V.**, 50 Hours, A Core Curriculum for College Students, National Endowment for the Humanities, Washington, 1989.
5. **Coates, J. F.**, *Forces Shaping the Future of the Engineer and the Engineering Profession*, American Society of Mechanical Engineers, New York, 1987a.
6. **Coates, J. F.**, Forces and Factors Influencing the Recruitment and Retention of Scientists, Engineers and Other Professionals, 1986-2000, J. F. Coates, Inc. (unpublished report prepared for the National Science Foundation), 1987b.
7. **Cunningham, R.**, Unpublished remarks at the Conference on the Future of Mechanical Engineering sponsored by the American Society of Mechanical Engineers, Chicago, 1987.
8. **Davis, J. M.**, *Making America Work Again*, Crown Publishers, New York, 1983.
9. **Dertouzos, M.**, et al., *Made in America; Regaining the Productive Edge*, MIT Press, Cambridge, 1989.
10. **Drucker, P. F.**, *Managing in Turbulent Times*, Harper & Row, New York, 1980.
11. **Drucker, P. F.**, *The New Realities*, Harper & Row, New York, 1989.
12. **Ernst, E. W.**, Innovation in Engineering Education, Position Papers, Emerging Issues in Engineering Education, 1989 Frontiers in Engineering Education Conference, American Society of Engineering Education, Washington, 1989.
13. **Gagné, E.E.**, Engineers' Perspectives on Defense Work and Economic Conversion, Proceedings, Workshop on Engineers and Economic Conversion: Roles, Values, and Attitudes, M.W. Gilliand, Ed., University of Arizona, Tucson, 1991.
14. **Greenberg, D.**, Dangerous Drumbeat on Engineer Shortage?, *Engineering Times*, November, p. 12, 1991.
15. **Grinter, L. E.**, Report on the Evaluation of Engineering Education (1952-1955), *Journal of Engineering Education*, Vol. 46, pp. 25-63, 1955.
16. **Grogan, W. R.**, Schachterle, L. E., and Lutz, F. C., Liberal Learning in Engineering Education: The WPI Experience, Knowing and Doing: *Learning Through Experience, New Direction for Teaching and Learning*, No. 35, Jossey-Bass, San Francisco, 1988.
17. **Hill, C. T.**, New Manufacturing Paradigms, 1991, *The Bridge*, Vol. 21, No. 2., pp. 15-27, 1991.
18. **Hughes, K. H.**, National Technology Priorities and Economic Conversion, Proceedings, Workshop on Engineers and Economic Conversion: Roles, Values and Attitudes, M. W. Gilliand, Ed., University of Arizona, Tucson, 1991.
19. **Hutchinson, C. E., and C. B. Muler**, Educating Engineers: In Praise of Diversity, *Issues in Science and Technology*, pp. 71-74, Summer, 1988.
20. **Johnston, J. S., Jr., Shaman, S., and Zemsky, R.**, *Unfinished Design. The Humanities and Social Science in Undergraduate Engineering Education*, Association of American Colleges, Washington, 1988.
21. **Jones, R. C.**, Alternative Formats for Engineering Education, *Emerging Issues in Engineering Education*, Frontiers in Education Conference, American Society for Engineering Education, Washington, 1989.
22. **Knepler, H.**, 1977, The New Engineers, *Change*, Vol. 9, No. 6, pp. 30-35, 1977.
23. **Landis, F.**, 1987, Reforming the Curriculum to Meet the Challenges of the 21st Century—A View from Academe, *Issues in Engineering Education*, 1987-1, College of Engineering, Colorado State University, 1987.
24. **Layton, E. T., Jr.**, *The Revolt of the Engineers: Social Responsibility and the American Engineering Profession*, The Press of Case Western Reserve University, Cleveland, 1971.
25. **Mark, H., and L. Carver**, The Fourth Revolution: Educating Engineers for Leadership, *Engineering Education*, pp. 104-108, November 1988.
26. **Matzkin, K.**, Defense Economic Adjustment Program. Changes in Defense Spending and Their Impact, *Proceedings, Workshop on Engineers and Economic Conversion: Roles, Values and Attitudes*, M. W. Gilliand, Ed., University of Arizona, Tucson, 1991.
27. **O'Neill, W.**, Who Wants to Be an Engineer? *New Scientist*, Vol. 126, pp. 42-47, 1990.
28. **Perrucci, R., and J. E. Gerstl**, *Profession with Community: Engineers in American Society*, Random House, New York, 1969a.
29. **Perrucci, R., and J. E. Gerstl**, *The Engineer and the Social System*, John Wiley, New York, 1969b.
30. **Pletta, D. H.**, *The Engineering Profession: Its Heritage and Its Emerging Public Purpose*, University Press of America, Lanham, Maryland, 1984.
31. **Reich, R. B.**, *The Work of Nations: Preparing Ourselves for 21st Century Capitalism*, Alfred A. Knopf, New York, 1991.

32. **Rudoff, A., and Lucken, D.**, The Engineer and His Work: A Sociological Perspective, *Science*, Vol. 172, pp. 1103-1108, 1971.
33. **Semans, R. C., Jr., and K. F. Hansen**, Engineering Education for the Future, *Technology Review*, pp. 22-23, Feb./Mar, 1981.
34. **Van Valkenburg, M. E.**, Preparing the Engineers of the 21st Century, *Engineering Education*, p. 103, November 1988.
35. **Vincenti, W. G.**, *What Engineers Know and How They Know It. Analytical Studies from Aeronautical History*, The Johns Hopkins University Press, Baltimore, 1990.
36. **Vitale, E. J.**, Issues Associated with Conversation from an Engineering Perspective, *Proceedings, Workshop on Engineering and Economic Conversion: Roles, Values and Attitudes*, M. W. Gilliand, Ed., University of Arizona, Tucson, 1991.
37. **Vlachos, E.**, The Sociology of the Engineering Profession: Engineering and Economic Conversion, *Proceedings, Workshop on Engineers and Economic Conversion: Roles, Values and Attitudes*, M. W. Gilliand, Ed., University of Arizona, Tucson, 1991.
38. **Walker, E. A., J. M. Petit, and G. A. Hawkins**, *Goals of Engineering Education*, American Society for Engineering Education, Washington, 1968.
39. **Wenk, E., Jr.**, 1988, Social, Economic and Political Change: Portents for Reform in Engineering Curricula, *Engineering Education*, pp. 99-102, November 1988.
40. **Wickenden, W. E.**, Report of the Investigation of Engineering Education, 1923-1929, Society for the Promotion of Engineering Education, Vols. I and II, 1930/1934.
41. **Womack, J. P., D. T. Jones, and D. Ross**, 1990, *The Machine That Changed the World*, Harper Perennial, 1990.
42. **Zimmerman, J. R., J. G. S. Billingsley, and F. A. Kulacki**, The Case Study Approach in the Capstone Design Course, Proceedings, American Society of Engineering Education Annual Conference, Beidenback, J. M., et al., Eds., p. 638-644, 1986.
43. **Zinberg, D. S.**, The Next Generation of Engineers: A New Breed? *Aspen Quarterly*, pp. 50-57, Spring, 1990.
44. Demographic Trends and the Scientific and Engineering Work Force--A Technical Memorandum, Office of Technology Assessment, U.S. Government Printing Office, Washington.
45. Engineering Education and Practice in the United States, Committee on the Education and Utilization of the Engineer, National Academy of Engineering, National Academy Press, Washington.
46. Engineering in Society, Panel on Engineering Interactions with Society, National Academy of Engineering, National Academy Press, Washington.
47. Engineering Education and Practice in the United States: Foundations of our Techno-Economic Future, National Academy of Engineering, National Academy Press, Washington.
48. Engineering Education and Practice in the United States: Engineering in Society, National Academy of Engineering, National Academy Press, Washington.
49. Engineering Undergraduate Education, National Research Council on Undergraduate Engineering Education, Committee on the Education and Utilization of the Engineer, Commission on Engineering and Technical Systems, National Academy Press, Washington.
50. Engineering Education and Practice in the United States: Engineering Undergraduate Education, National Academy of Engineering Academy Press, Washington.
51. Quality of Engineering Education, Final Report of the Quality of Engineering Education Project, American Society for Engineering Education, Washington.
52. Civil Engineering in the 21st Century, American Society of Civil Engineers, New York.
53. Educating Scientists and Engineers: Grade School to Grad School, Office of Technology Assessment, OTA-SET-377, U.S. Government Printing Office, Washington.
54. Foreign Engineers on the Rise, *Science*, January 29, p. 455.
55. Undergraduate Curriculum in Development in Engineering, National Science Foundation, Washington.
56. Scientists and Engineers: 1988 Estimates, Detailed Statistical Tables, NSF 88-322, Washington.
57. Harvey Mudd College Catalogue, Harvey Mudd College, Claremont.
58. Swarthmore College Bulletin, Swarthmore College, Swarthmore.

Chapter 3

ARTIFICIAL REALITIES, VIRTUAL COMMUNITIES, AND INTELLIGENT ARTIFACTS: IMPLICATIONS FOR ENGINEERING EDUCATION

C. Dede
George Mason University

I. ABSTRACT

The sophisticated capabilities of emerging information technologies enable re-engineering current models of instruction to achieve greater effectiveness. Rather than teaching-by-telling, engineering educators can use visualization techniques and immersion interfaces to create artificial realities that scaffold learning by doing. Video links and machine-based intelligent agents can enrich distributed synthetic environments, amplifying learners' motivation and accomplishments by applying new learning models from cognitive science. As a national information infrastructure evolves, educators can facilitate the development of virtual communities that enhance students' knowledge by bridging engineering research and practice. Eventually, virtual learning environments can be intermingled with the real world via artifacts with embedded intelligence. Since reconceptualizing the profession of engineering is vital to America's success in the global marketplace, implementing these educational innovations is essential to developing the human resources our nation needs for economic prosperity.

A. FORCES DRIVING NEW PARADIGMS FOR ENGINEERING EDUCATION

Three sets of forces are driving the re-engineering of engineering education: a reconceptualization of the engineering profession; a growing divergence of work place and school educational models; and new ideas from cognitive science about the nature of learning. Emerging national economic development policies are creating an opportunity for universities and other instructional providers to implement alternative engineering education paradigms that respond to these incentives for change.

1. Reconceptualizing the Nature of Engineering

An informal consensus has emerged in the engineering profession about major goals for engineering education in the 1990s. Few would argue with the need for increased instructional effectiveness in attaining the following objectives:

- Fostering the creative design of entire systems from a commercial as well as a research perspective
- Developing improved abilities for information access, communications, and teamwork

- Enabling just-in-time evolution of practicing engineers' skills

This study argues that the best method to reach these outcomes is not by automating conventional engineering education approaches to improve their efficiency, but instead by applying sophisticated information technologies to scaffold alternative, more effective models of teaching/learning.

These educational objectives are priorities because of shifts in the nature of engineering as a role within society. Over the past several decades, three major changes have occurred in the practice of engineering. First, actional competence has significantly increased in importance relative to informational knowledge. Civilization is evolving too fast to wait for a detailed model before acting. Instead, to a greater extent than ever before, engineers must engage in decision making, design, and assessment under conditions of incoherent objectives, incomplete data, and imperfect understanding [Denning, 1992].

Second, the range of roles engineers must play and the scope of knowledge they need have widened dramatically. As corporations struggle to compete in the volatile global marketplace and government agencies wrestle with rapidly shifting societal priorities, the ability of organizations to quickly reconfigure internal processes via top-down, bottom-up, and middle-out innovation is crucial to their survival. On a deep level, engineers must understand how "quality" is defined from various perspectives, both by stake-holders in the design process and by those affected by the artifacts this process produces. They must also comprehend how complex, market-driven forces in the public and private sectors shape the evolution of civilization's technologies [Dertouzos et al, 1989].

Third, to translate this knowledge into improved effectiveness, engineers need more sophisticated communication work skills so that they can serve as expert partners with their clients. True innovations are not simply improved tools or machines, but instead are shifts in standard organizational practices empowered by reconfiguring institutions around the capabilities of those new devices [Kash, & Willenbrock, 1992]. Promoting learning at the institutional level, as well as the individual level, demands types of communications expertise and psycho-social knowledge not historically stressed in engineering education.

a. The Traditional Definition of "Engineering"

One way of summarizing what is involved in "re-engineering engineering" is to contrast the emerging role of engineers with classic descriptions of the profession as typified by Vincenti's *What Engineers Know and How They Know It* [1990]. Vincenti modifies G.F.C. Rogers' definition of engineering into "the practice of organizing the design, construction, and operation of any artifice which transforms the physical world around us to meet some recognized need." Within this definition, he depicts design as the key role and indicates that *normal* design (incremental, evolutionary changes to well-understood devices) is what most engineers do for a living.

Contextual factors—such as economic, military, social, personal, and environmental needs/constraints—create an ambiance that shapes project definition and high-level design, but Vincenti presents the tensions between this ambiance and technical issues as peripheral to standard engineering practice. Engineers' problem solving may focus on individual devices (e.g., an airplane) or technological systems (an airline), but the emphasis is on creating an artifact rather than its implementation as a solution in a larger context. From an engineering perspective, whether or not something "works" is defined at the device level rather than at the contextual level.

Vincenti describes a generalized form of the engineering design process as:

1) Familiarization with the problem;
2) Identification of analytic criteria;
3) Development of experimental instruments and techniques;
4) Refinement of evaluative measures;
5) Defining a research scheme;
6) Measurement of performance characteristics; and
7) Assessment of results.

Communities of technological practitioners form the social context within which this design process occurs; their traditions of practice are the primary source of judgment about the quality of the artifact produced.

The major contrast in roles explored by Vincenti is between the engineer and the scientist. Engineers are seen as producing artifacts; scientists, as striving to understand the workings of nature. Engineers must worry more than scientists do about cost issues and the consequences of errors. Engineers develop theories to aid practical calculations; scientists, for explaining known phenomena or predicting new ones. Engineers use realistic and detailed models far more frequently than scientists, who focus on approximate modeling of simplified or partial phenomena to aid theoretical understanding. Optimization is more central to engineering than to science.

b. Emerging, Alternative Conceptions of "Engineering"

The industrial economy developed mechanistic models of organizations based on hierarchical authority, bureaucratic information flows, and the cascading of specialized jobs to achieve the institutional mission. Within this context, traditional conceptions of engineering function quite effectively. In the knowledge-based economy, however, mechanistic organizational approaches are gradually being displaced by organismic models instantiated through teams with flexible roles, ad-hocratic information flows, and rapidly shifting authority based on complementary knowledge. One outcome of this shift is that, in itself, the introduction of an improved artifact is likely to cause little improvement in institutional functioning—thus driving changes in the traditional role of the engineer.

As a result of this transformation in the nature of work, the concept of "design" is shifting from a focus primarily at the artifact level to the creation of contextual innovations. In organismic institutions, design entails the need-driven evolution of devices in the context of reconceptualizing the organization's mission, information flows, and psycho-social structures to take advantage of emerging artifacts' potential capabilities. Simultaneously, the process of design is altering from a waterfall model in which specialists at each stage make incremental enhancements to an organization-wide collaborative endeavor that involves synthesizing a variety of perspectives on quality through teamwork and iterative improvement.

These shifts in institutional functioning and the nature of design are driving an alternative conception of engineering. In contrast to Vincenti's traditional formulation, engineers increasingly must participate on design teams in which an artifact's success is defined in contextual terms—a successful shift in organizational practices—rather than simply whether the new device meets technical performance specifications. Moreover, as the pace of scientific and economic change increases, normal design (incremental, evolutionary changes to well-understood devices) is becoming a smaller part of engineers' practice. Instead, organizations must continually re-engineer their mission, products, and operations under conditions of incoherent objectives, incomplete data, and imperfect understanding.

An emerging, alternative definition of engineering is "the practice of organizing the design, construction, and operation of any artifice-based systemic innovation that transforms both our physical and psycho-social environment to meet some recognized need." A revised, generalized form of the engineering design process can be depicted as:

1. Team-based exploration of a contextual need amenable to an artifice-based solution;
2. Collaborative identification of analytic criteria based on multiple perspectives about quality;
3. Development of experimental instruments and techniques for both physical capabilities and psycho-social impacts;
4. Refinement of evaluative measures based on performance specifications and on organizational re-engineering;
5. Defining a research scheme and an implementation strategy;
6. Measurement of performance characteristics when contextualized in pilot environments; and
7. Assessment of results via iterative, evolutionary implementation.

Such an alternative conception of design demands an expanded set of knowledge and skills on the part of engineers. In particular, analogies between engineers and scientists become less useful, as the focus of engineering shifts from an emphasis purely on physical reality to a role bridging the functional capabilities of devices to their consequences for organizational effectiveness in meeting societal needs. Instead, analogies to medicine are more appropriate: Physicians must understand both the biological reality of diseases/therapies and the psychosomatic context of personality and lifestyle. This expansion in the role of engineers opens up exciting possibilities of capturing a greater number of talented young people selecting a career path.

The evolution of sophisticated information technology has helped to create the need for this shift in engineering by driving the formation of global markets, knowledge-based economies, and societal change too rapid to allow precise modeling. Simultaneously, these technologies are empowering the transformation of the engineer's role by enabling both sophisticated design tools and the complex information flows essential to organismic institutional structures.

Thus, one set of pressures for changing engineering education stems from shifting societal needs and advanced information technologies driving a reconceptualization of the profession. Along a different dimension, the emergence of organismic institutions is creating a growing divergence between work place and school models of learning, building pressures for change in ways universities and other instructional providers educate engineering students.

2. Diverging Work place and School Models of Learning

While school-based models of instruction have been relatively static for more than a century, the evolution of the global market and our country's shift from an industrial to a knowledge-based economy are transforming work place learning [Dede, 1992]. In organismic models of institutional functioning, just-in-time learning-while-doing targeted to authentic, novel problems is displacing the classroom-based, discipline-focused, learning-by-listening approaches characteristic of schooling and of work place training in mechanistic organizations. The widening differences between work place and school educational situations can be highlighted by contrasting contemporary examples of each.

Illustrative Work place Teaching/Learning

A paradigmatic example of idealized work place learning is an architect starting a new assignment designing an office building for a corporate client who wants a physical environment tailored to maximize total quality management (TQM) principles. The architect has no prior background in TQM, so she is placed in a learning-while-doing mode situated in the work place context. Since her employer wants to keep costs down and minimize time-off-task, the architect's goal is to immediately master only as much knowledge about this philosophy of work as is needed to enable her to design the building.

She is an experienced person motivated to learn by desires to continue her employment and to advance her career. As a member of a team involved in designing the building, her understanding must be communicable to the rest of the group, if it is to be useful. The immediate challenge is how to locate content both on the spatial implications of TQM and on how architectural design can reinforce organizational culture. Unfortunately, university-based learning experiences are unlikely to provide much leverage on this constellation of needs and constraints.

Illustrative School-based Teaching/Learning

A paradigmatic example of typical school-based learning is an eighth-grader reading in his textbook about Eli Whitney's design of the cotton gin. The student is studying this topic because the public believes that literacy in a broad range of discipline-based content will prepare him to be a better worker, citizen, and family member. This learner has little life experience and is uninterested in the material, as he sees no need to acquire this knowledge. The teacher must struggle to build some type of intrinsic motivation, as the only extrinsic incentive is his parents' reaction to periodic reports from the school on his individual progress.

Because of financial constraints caused by the necessity of holding total schooling costs relatively constant, the student is learning by assimilating rather than by doing. He listens to large-group lectures, reads texts, and fills out short-answer worksheets for practice and assessment. Once in a while, he goes to the library or on a field trip, but almost all his learning is done in custodial classroom settings that are decontextualized from real-world situations. Occasionally, he has a brief writing assignment, engages in collaborative problem solving with his peers, or takes a standardized test that measures low-level skills and knowledge across a wide spectrum of material. Regardless of how well or poorly he does on these tests, the learner must study roughly the same content as the rest of the group.

Unfortunately, despite the fact that everyone in this system of teaching/learning is trying hard, this student is much more interested in the physiological and psychological aspects of puberty than in the cotton gin's design. This would likely be true even if his writing assignments, collaborative problem solving experiences, or tests were more frequent.

a. Highlighting Contrasts Between Work Place and School Models

The contrast between these two learning situations must be qualified by indicating that the school-based model codifies the regressive end of typical current practice—which is not what many educators would wish—while the work place model represents a progressive, idealized version of what many employers want as the knowledge-based economy evolves. Thus, the comparison is neither between two idealized approaches nor between two models of typical practice; rather, the contrast is between the direction of the leading edge work places and where the least effective schools are. This is a useful differentiation from which to explore ways that engineering education can best serve students moving from school to work place settings, as well as methods of aiding professional engineers in updating their skills and knowledge.

The table on the next page provides a summary of contrasts highlighted in the paradigmatic examples. In each situation, the *Givens* are factors outside the control of educators and the *Consequences* are outcomes that flow from the nature of the *Givens*. For example, in the work place model the given goal of learning is mastery of some role-related problem, which creates the consequence that the educational content should be authentic, problem-based, and transdisciplinary.

b. Implications for Engineering Education

Imagine a continuum with these two different models of teaching/learning at either end:

Work Place <--> School

In this contrast, higher education occupies an intermediate position between the school and the work place, because the *Givens* for university-based education are a mixture of the extrinsic factors that shape these two different instructional models. For example, college-level learners attend voluntarily—unlike public school students—but many are inexperienced and lack intrinsic motivation beyond getting a diploma.

Industrial training environments in mechanistic organizations, which prepare for narrow job skills and cascading production processes, can function in a similar manner to school-based instructional settings. As a result, higher education, until recently, had little difficulty in bridging school to work place educational needs. However, with the advent of the knowledge-based economy and organismic organizational models, work place learning situations are markedly diverging from the traditional model of schooling. As a result, institutions of higher education are experiencing serious difficulties in developing an intermediate teaching/learning approach that straddles the contrasting models exemplified in Table 1 above.

The reconceptualization of the engineering profession discussed earlier is reinforcing this shift toward teaching/learning models at the work place end of the continuum. For example, engineering is increasingly done by teams rather than individuals, and organismic organizations necessitate just-in-time learning. However, typical university practices in engineering education are located substantially on the school side of the spectrum. Current models of teaching/learning are often misaligned with the evolving needs of engineering students and the profession, creating pressure on universities and other instructional providers to alter their educational approaches.

Beyond a redefinition of the nature of engineering and a growing divergence between school and work place educational situations, new learning models from cognitive science are also driving the re-engineering of engineering education.

Work Place		
Givens	Cost-benefit	employer determines
	Goal	problem-solving mastery (criterion-referenced)
	Learner	experienced need-driven team member
Consequences	Content	authentic, problem-based, transdisciplinary issue-focused rather than wide-ranging
	Pedagogy	learning-while-doing (any place; just-in-time) extrinsic motivation assumed
	Assessment	improvement in role performance minimum cost (minimum time-off-task)
School		
Givens	Cost-benefit	public determines
	Goals	custodial (fixed place and time) socialization (cover all material just-in-case) screening (norm-referenced)
	Learner	inexperienced low-perceived need for knowledge individual in competition with other students
Consequences	Content	discipline-based wide-ranging more than deep or issue-focused
	Pedagogy	learning by assimilation decontextualized rather than situated large group more than individual or small team intrinsic motivation must be fostered
	Assessment	standardized tests literacy-oriented rather than mastery-centered no increases in current public funding

Table 1. Contrasting Work Place and School Models of Teaching/Learning

3. Learning Models from Cognitive Science

New ideas from cognitive science about the nature of learning are inducing educators at every level to restructure instructional environments. Some basic principles that underlie these models of learning are listed below [Dede, 1990]:

- Learning involves an evolutionary series of progressively more sophisticated conceptions of reality. Learners interpret every instructional experience through an existing mental model; they are not empty vessels to be filled.
- New concepts and skills are best remembered and integrated with existing knowledge if learning is active and constructive rather than passive and assimilative.
- Learning is motivated by multiple factors, including intrinsic curiosity, social interaction, extrinsic rewards, and the joy of accomplishment. Because of this mixture of motivations, cooperative learning in groups is sometimes the optimal strategy; in other situations, individual learning is superior. Often, the ideal environment for learning is very diverse, mixing different ages, developmental levels, and cultural backgrounds.
- Learning is dependent on an individual's beliefs and attitudes. For example, understanding a student's cultural perspective is vital to communicating knowledge and to measuring learning. Also, each individual has a unique style of learning, based on cognitive, sensory, psycho-motor, and social factors. Tailoring instruction to this style is vital for educational effectiveness.
- The activities through which learning takes place and the context within which learning occurs are integral parts of the knowledge gained. Educational effectiveness is enhanced by situating learning in an environment similar to that in which the knowledge will be used.
- Once learners have mastered the fundamentals of a subject, using an interdisciplinary approach to subject matter is more effective than breaking material into the traditional disciplines, because real-world problem solving involves comprehending richly interconnected systems.
- Each step in learning requires time for reflective ideation. Compressing the time necessary for an individual to master an educational experience results in little or no learning. Learning is continuous and unbounded; people who treat every situation as an opportunity for growth learn more than those who limit their education to classroom settings.
- Reality is richer and more complex than people can fully comprehend. The goal of learning is to participate meaningfully in a cultural context rather than to attain absolute truth. Teachers are learners who are adept at facilitating growth.

A detailed discussion of the theoretical underpinnings of these alternative approaches is beyond the scope of this study.

These principles are centered on learning rather than teaching. For three decades, intensive efforts to improve teaching based on traditional models of learning have had little effect on student outcomes. Increasing instructional effectiveness by focusing on cognitive science's new theories about learning is an important shift in strategy.

a. Technology-based Instantiations of these Principles

Set in the late 1990s, the three vignettes below illustrate educational environments based on these learning principles from cognitive science. Each vignette exemplifies how sophisticated information technologies can scaffold alternative, more effective models of teaching/learning in design-based technical fields.

The first vignette is set in a college classroom:

Vignette I: Multimedia as a Vehicle for Knowledge Webs

> As a semester-long project, three students in Professor Roentgen's freshman course on technology and society have chosen to create a collaborative presentation on the development of the atomic bomb. To accomplish this task, they will use an inquiry-based multimedia system that presents alternative historical interpretations (from Rhodes' *The Making of the Atomic Bomb* to Renais' film *Hiroshima, Mon Amour)* in the context of original source documents. Different teams within the class are utilizing similar multimedia systems to learn about other technology/society topics.
>
> Through Professor Roentgen's lectures and multimedia lab sessions, the students will explore exemplary historical perspectives and discuss their underlying epistemology and rhetoric. Building on this background, this team of learners will use the system's multimedia production console to create their own interpretation of how the atomic bomb project illustrates the broad socio-historical theories the instructor is presenting. Professor Roentgen's purpose in using this teaching approach is to help students understand the concepts and principles underlying history as an inquiry-based, interpretive activity.
>
> Today, the team is exploring a guided tour through multimedia database materials on how engineers utilized competing purification technologies to purify fissionable isotopes as rapidly as possible. Based on ideas presented by Professor Roentgen in his lectures, they have formed a hypothesis about the ways that the relative political leverage of Lawrence, Seaborg, and Oppenheimer interacted with economic considerations in determining the mix of technical approaches applied. Using the production console, the learners are collecting clips of video documentaries, original source documents, and historical analyses that support or call into question their hypothesis. A context-sensitive, machine-based coach is available to help with interface problems, navigation challenges, and inquiry strategies.
>
> Over the course of the semester, about half the class time is spent in exploring this web of historical interpretations and original source materials, with Professor Roentgen and his teaching assistants circulating about the lab to help individual teams. The other half of the course content is devoted to large-group lectures in which the instructor describes overarching sociological theories and illustrates how historians utilize inquiry approaches to find patterns in incomplete and contradictory data.

With minor modifications, such multimedia applications for exploring webs of knowledge can present a variety of topics across the full spectrum of inquiry-based disciplines [Fontana, Dede, White, & Cates, 1993]. The richness of this expository medium helps to build learner motivation, an important issue for required subjects like science in which many students may initially be uninterested or phobic. With the guidance of the instructor, who remains at the center of the educational experience, learners progress from assimilating exemplary mental models to constructing their own interpretations of the concepts and principles underlying the material. Working in collaborative teams allows students to combine complementary backgrounds and learning styles and to develop psycho-social and communication skills.

The idealized model of work place learning presented earlier about an architect with a TQM assignment forms the basis of the second vignette:

Vignette II: Learning-While-Doing with ARCHIE

> Yvette hums quietly to herself as she and ARCHIE skim through a sea of stories, harvesting metaphors and analogies. Some reminiscences discuss removable room partitions incorporated into 1960s British Open Schools to facilitate collaborative learning-by-doing. Unfortunately, gains in flexibility were often undercut by problems with noise and lack of permanent storage space. Other tales describe 1990s experiences in reconfiguring hospitals based on virtual reality approaches to surgery. In-patient operations and extended hospital stays have been dramatically reduced by these techniques, but some of the savings from smaller patient wards have been offset by elaborate information infrastructures in the treatment areas.
>
> At this moment, her case-based architectural design tool, ARCHIE, is describing how Deming (the originator of the Total Quality Management philosophy) configures his seminar room before conducting a workshop. On the monitor, a room diagram is shown using standard graphical conventions for architecture. Several paragraphs of text are displayed at the bottom of the screen, ignored by Yvette. Since her learning style is predominantly visual and auditory rather than symbolic, ARCHIE is vocalizing this textual material while she watches a graphical pointer maneuver about the blueprint. Three figurines are gesturing near the top of the display, indicating that they know related stories. On the right-hand side of the monitor, an interest-based browser shows index entries grouped by issue, spatial configuration, functional system, stake holder, and building life-cycle.
>
> Yvette is in the early stages of a new assignment: designing an office building for a corporate client who wants a physical environment tailored to maximize Total Quality Management (TQM) principles. In five hours, her findings are needed as grist for a design team meeting in the firm's decision support complex. Through computer-guided brainstorming and model-building, the group will forge a collective plan of action by synthesizing Yvette's insights with complementary knowledge contributed by other team members.
>
> ARCHIE is an architectural design tool based on the concept that new knowledge is most easily mastered and applied when presented in the context of its use. This system converses with a person engaged in problem solving, responding to problem descriptions with recollections about similar situations encountered by other practitioners. These concrete examples are pedagogically valuable in suggesting possible approaches, warning of potential pitfalls, and fleshing out abstract rules of thumb. Beyond a repertoire of war-stories, ARCHIE also provides access to documentation and evaluations of existing buildings, linked to more abstract guidelines and codes describing good building practice. [At Georgia Tech, Eric Domeshek and Janet Kolodner are currently developing such an architectural tool. Their work and that of similar projects served as the inspiration for this description of ARCHIE, although I have modified and enhanced their system in this vignette.]

Of the team members assigned to this project, Yvette's ESP (Educational Strengths Profile) has the best case-based and analogical reasoning skills, so she has been assigned the role of using ARCHIE. To get contracts, her firm often submit bids with perilously low profit margins, so an employee's ability to learn while doing (minimizing time-off-task) is an important asset for promotion. Today, Yvette is finding issue-focused learning somewhat frustrating, as time constraints prevent her from exploring some fascinating, but peripheral ideas about how architectural design can reinforce organizational culture. Perhaps when this job is completed she can find the time to take a university course on this topic....

The vignette leaves open whether ARCHIE would be a stand-alone application, purchased outright and updated occasionally, or would exist as an on-line service updated continuously with metered access. A university could deliver these types of case-based tools as an alternative form of instruction for work place learning, linking their use to professional development and certification. Of course, consulting firms such as Arthur Anderson could also develop this type of educational service, undercutting institutions of higher education.

The third vignette takes place in a student's home, about 10:30 in the evening:

Vignette III: Navigating Through Cyberspace

Roger was unobtrusively sidling across the Bridge of the Starship Enterprise when the Captain spotted him out of the corner of his eye. "Take the helm, Ensign Pulver," growled Captain Jean-Luc Picard, "and pilot a course through the corona of that star at lightspeed 0.999. We have astrophysical samples to collect. You'll have to guard against strange relativistic effects at that speed, but our shields cannot stand the radiation flux we would experience traveling at a slower speed."

As Picard glared at him from the screen of his home ITV *(interactive-television set)*, Roger drummed on his Cyberspace Console with his fingers and cursed quietly to himself. He had hoped to avoid this type of pop quiz in his senior physics class. The two margaritas he had drunk at dinner were still having some effect on his cerebral processes. One thing for sure, Einstein had never made a major breakthrough in general relativity while loaded.

[This vignette is based on distributed simulation technology such as the U.S. Department of Defense SimNet, a current training system that allows users at geographically remote sites to interact together on a virtual battlefield. This "dial-a-war" approach uses commercial telephone lines to interconnect a variety of computer platforms, creating a virtual world that includes some artificial-intelligence-based opponents. In this vignette, I have modified and extended SimNet's basic capabilities, which will likely soon be commercially available via the Defense Conversion funding the Clinton administration is providing.]

Roger had intended to sneak onto the Ecology Deck of the Starship and put in a little work on his biology class project in controlling closed-system pollution levels—but no such luck. Worse yet, he suspected that the Vulcan communications officer watching him while she translated a message in French was in fact the "avatar" *(computer-graphics representation of a person)* of a woman he admired who sat three rows behind him in his languages class. Of course, he could be wrong; she might be someone teleporting into this simulation from who knows where or she could even be a "knowbot" *(a machine-based simulated personality used to simplify the job of the faculty Mage directing an instructional simulation).*

Buying a little time by summoning up the flight log, Roger glanced curiously around the bridge to see what new artifacts his fellow students had added since yesterday to this MUD *(Multi-User-Dungeon, a current type of adventure game in which participants mutually evolve an elaborate, shared synthetic environment by continuously modifying its contents).* In one corner, an intriguing creature was sitting in a transparent box, breathing a bluish-green atmosphere—maybe this was the long-awaited alien the anthropology and biology majors were creating as a mutual project. The 3-D goggles from his Nintendo++ set intensified the illusion that the lizard-like countenance was staring right at him.

"Impulse Engines to full speed, Mister!" barked Captain Picard. This Mage seemed rather grumpy for a regular faculty member, thought Roger. Maybe he's a visiting fireman from the new Net-the-Experts program. On his console, Roger rapidly selected equations that he hoped would yield the appropriate relativistic corrections for successfully navigating through the star's corona.

Automatically, a cognitive audit trail of his actions began streaming to his factual-knowledge assessment file for physics. Each time he requested help from the computer-based coach, the performance score displayed on his console dropped. Why, Roger asked himself, couldn't I have lived in the days when students got to take multiple-choice tests?

This third vignette is based on the emerging information infrastructure championed by Vice President Gore. As will be discussed in more detail later, avatars, knowbots, and distributed synthetic environments, within which virtual learning communities form, are attractive potential outcomes of scaling today's Internet to the high-bandwidth, wide-area capabilities of the proposed NREN (National Research and Education Network).

A related development not presented as a separate vignette is "classrooms with virtual walls." An example would be a flexible manufacturing systems class visiting a factory to observe computer-integrated manufacturing strategies, with students and professor linked in a virtual learning environment via notebook computers with wireless modems.

b. Themes Underlying These Vignettes

These three vignettes are designed to exemplify a spectrum of learning situations. Table II below indicates the learning strategy, pedagogic strategy, knowledge type, and outcome type dominant in each vignette.

	Vignette I	Vignette II	Vignette III
Learning Strategy	assimilative	analogical	constructive
Pedagogic Strategy	mimetic & didactic	knowledge as design	transformative & facilitative
Type of Knowledge	explicit	case-based	implicit
Type of Outcome	underlying concepts/principles	skilled performance	factual knowledge

Table 2. Illustrative Types of Learning Situations

Explaining these descriptors in detail is beyond the scope of this study. Perkins [1986] depicts methods of using design-based pedagogic strategies for various subjects. Gardner [1991] contrasts mimetic versus transformative approaches to teaching and presents a detailed categorization of learning outcomes based on skilled performance, underlying concepts and principles, and factual knowledge. (Gardner and Perkins are co-directors of Harvard's Project Zero, an ambitious attempt to develop new precollege curricular and assessment models based on multiple dimensions of human intelligence.)

Of course, optimal learning experiences draw on all these strategies and types without isolating them in subsets. For example, mimesis is a powerful means to teach skilled performance, and constructive learning strategies are valuable in generalizing acquired skills to related situations. Also, the first vignette on multimedia spans all the approaches across Table II, although its primary emphasis is on the strategies and outcomes in its column. Thus, the vignettes are constructed not to indicate rigid groupings, but to illustrate a repertoire of ways that technology can scaffold the learning principles from cognitive science presented earlier.

In summary, three sets of forces are driving the re-engineering of engineering education: 1) a reconceptualization of the engineering profession, 2) a growing divergence of work place and school educational models, and 3) new ideas from cognitive science about the nature of learning. Emerging national economic policies for human resource development recognize the importance of preparing engineers who can bridge the functional capabilities of devices to their consequences for organizational effectiveness in meeting societal needs. These policies are creating opportunities for external support to aid universities and other instructional providers in implementing alternative, technology-intensive paradigms for engineering education that respond to these incentives for change.

B. OPPORTUNITIES FOR IMPROVING ENGINEERING EDUCATION

This study examines how an interrelated subset of emerging information technologies and applications—distributed simulation, knowbots and avatars, virtual communities, artificial realitiesvisualization,, intelligent artifacts—can leverage new approaches to teaching/learning. Through improved learning environments scaffolded by these advanced technologies, engineering educators can inculcate the knowledge and skills needed to function in the context of a knowledge-based economy and global marketplace.

As a national information infrastructure evolves, instructors can use distributed simulation to facilitate the development of virtual communities that enhance students' knowledge by bridging engineering research and practice. Video links and machine-based intelligent agents can enrich distributed synthetic environments, amplifying learners' motivation and accomplishments by applying new learning models from cognitive science. Rather than teaching-by-telling, engineering educators can use visualization techniques and immersion interfaces to create artificial realities that scaffold learning by doing. Eventually, virtual learning environments can be interspersed throughout real world settings via artifacts with embedded intelligence.

1. Technology-mediated Communication

Enabling people to communicate effectively across barriers of distance and time by forming virtual communities is an emerging goal for information technology. During the last decade, the field of computer-supported cooperative work has established that developing productive technology-mediated human interaction involves complex psycho-social issues beyond simply transmitting data along a channel from sender to receiver [Sproull, & Kiesler, 1992]. Communication depends on affective as well as cognitive interchange, and researchers studying collaborative educational environments are just beginning to map the complicated emotional dynamics of peer teaching and learning. Training learners to use technology-mediated communication effectively—whether over e-mail or in artificial realities—involves more than simply mapping skills from face-to-face encounters into the virtual environment [Dunlop, & Kling, 1991].

The act of learning is always constrained by the characteristics of the communications channel between the student and the content to be mastered [Dede, 1991a]. For example, the wider the bandwidth of a communications medium, the more immediate and rich the learning experience can be: Seeing a video tape on how to ride a bicycle conveys more information than reading an article about that subject. The greater the interactivity of a medium, the more feedback can be communicated to motivate and individualize learning: Having a friend teach you bike-riding is more effective than is watching a video tape on the topic.

Historically, instruction over a distance has been hampered because its communications channel is intrinsically narrow compared to forms of education that involve immediate contact with the learning situation. As an illustration, in correspondence courses, textual materials are used to inform students about a subject, and interaction with the instructor occurs through the physical transportation of written signals. This method of learning is severely limited in its immediacy, richness, motivation, and individualization.

Because advances in information technology increase the bandwidth and interactivity of communications channels, they strongly influence distance learning. As the telephone, the radio, and finally the television were developed, new distance learning approaches emerged. Much of current research in distance learning centers on how the limitations of these media can be minimized to involve students deeply with subject matter [Goldman, & Newman, 1992; Scardamalia et al., 1992].

An important issue in instruction at a distance is how learning can be optimized by tailoring content to the communications characteristics of the medium. For example, in situations where "a picture is worth a thousand words," radio will be less effective than television. On a deeper level, regardless of its content, the medium itself sends a message that influences learning. The imagination and involvement of a learner are engaged in different ways when reading a fairy tale in a book than when watching a movie depicting that story. The interaction between medium and message is a complex phenomenon important to distance learning, but still largely unexplored.

These ideas about technology-mediated communication are important background material for the sections that follow on new types of learning environments for engineering education. Design principles and constraints from the concepts expressed above are vital to the successful evolution of virtual communities, artificial realities, and intelligent artifacts that promote the learning of complex concepts and skills.

2. Distributed Synthetic Environments

The 1990s will be a time of multi-billion dollar initiatives to build national information infrastructures (e.g., the National Research and Education Network; the competition among telephone, cable, and computer companies for the information services business). A new type of educational message made possible by high-end media is distributed simulation. For example, SimNet [Orlansky, & Thorp, 1991] is a training application that creates a virtual battlefield on which learners at remote sites can simultaneously develop both individual and collective military skills.

Complex data-objects that indicate changes in the state of each piece of equipment are exchanged via a telephone network interconnecting the training workstations ("dial-a-war").

The appearance and capabilities of graphically represented military equipment alter second-by-second as the virtual battle evolves. Through this tele-presence approach, a widely distributed group of personnel can engage in simulated real-time warfare without the necessity of gathering the participants at a single site to conduct combat. Such a learning environment has many possible applications in engineering education (e.g., virtual factories, construction sites), as the Star Trek context in Vignette III earlier illustrates.

Distributed simulation enhances students' ability to apply abstract knowledge by situating education in virtual contexts similar to the work place environments in which learners' skills will be used. When the material involved has psycho-social as well as intellectual dimensions, the design of authentic experiences to embed in synthetic environments for education becomes more complex. By using sophisticated telecommunications, learners can interact in rich machine-mediated psycho-social environments populated both by video-links to other people and by simulated beings. These simulated beings may be "knowbots" (machine-based agents) or "avatars" (computer graphics representations of people); each adds an important dimension to education in synthetic environments.

Knowbots and Avatars

One example of a training application that incorporates machine-based agents involves software engineering education [Stevens, 1989]. Using hypermedia, Digital Video Interactive (DVI), and rule-based expert systems, the Advanced Learning Technologies Project at Carnegie Mellon University has created a virtual environment similar to a typical corporate setting. The trainee interacts with this artificial reality in the role of a just-hired software engineer still learning the profession. Through direct instruction and simulated experience, students are trained in a technical process, code inspection, which is one stage of a formal methodology for software development.

The learner can access various rooms in the virtual software company, including an auditorium, library, office, training center, and conference facility. Machine-based agents (knowbots) that simulate people, such as a trainer and a librarian, facilitate the use of resources to learn about the code inspection process. Via specialized tools in the office, the student can prepare for a simulated code inspection, in which he or she can choose to play any of three roles out of the four roles possible in this formal software review process. For each inspection, a rule-based expert system utilizes DVI technology to construct knowbots that simulate the three roles not chosen by the learner. This knowledge-based system controls the topic of conversation; determines who should speak next; and models the personalities of the knowbots in the inspection meeting, altering their cognitive and affective perspective depending on what is happening.

The learner uses a menu-based natural language interface to interact with these simulated beings, who model behaviors typical in code inspection situations. The student can choose from a wide range of options of what to say, as well as determine when to make remarks. Emotional inflection of his or her utterances, from a calm passive tone to an angry snarl can be selected. By mimicking the reactions likely from human participants in a real simulation, the knowbots provide the learner with a sense of the strengths and weaknesses of different intellectual/psycho-social strategies for that role in a code inspection.

Without using artificial realities and knowbots, this type of authentic experience is very difficult to simulate in classroom settings. Not only is the training environment dissimilar from the institutional context in which software development skills will be used, but also students do not know how to role play exemplary, typical, and problematic participants in code inspections. Through knowbots, the instructional designer can provide paradigmatic illustrations of how to handle a variety of situations, without the expense of having teams of human actors perform for each individual learner. This training approach could generalize to many other forms of engineering education.

Distributing among simulated beings the intelligence incorporated into artificial realities raises challenging instructional design issues. The quality of what students learn is determined by the accuracy with which a knowledge-based system can simulate not only human reasoning, but also people's personalities and emotions. This is a very difficult task, necessitating knowledge acquisition about psycho-social dynamics as well as intellectual issues.

Another design issue posed by knowbots is whether students will profit more by interacting with a variety of partial machine-based intelligences (a collaborative learning/peer teaching strategy) or with a single, omniscient source of knowledge (an intelligent tutoring systems approach). Students may empathize more readily with flawed simulated beings (e.g., an "adept, but still learning" moderator for code inspections) than with a perfect intelligent coach who has mastered every aspect of the educational domain.

Programming several machine-based agents with highly focused skills is also easier than constructing a single, broad-based knowbot; knowledge-based systems work best for narrow, well-specified domains. Early research into educational interfaces based on conversation with human-like entities indicates that this strategy leverages learning along a variety of dimensions [Laurel et al., 1990]. In Vignette II, the figurines gesturing near the top of the display to indicate that they know relevant architectural cases illustrate such an educational usage of knowbots.

The simulated beings in artificial realities for engineering education need not all be machine-based agents. Military and aerospace research projects have created tele-robotic technologies that allow operators to control devices across distance [Fisher, 1990]. These enable the creation of educational environments that provide a virtual link between remote learners and real world settings, For example, a student can manipulate industrial process technology without physically traveling to the factory [Uttal, 1989]. An extension of tele-operations is tele-presence, which allows people to interact across distance via avatars, as opposed to direct video-links.

Some of the representations in the SIMNET artificial reality discussed earlier are avatars controlled by human beings; others are knowbots whose actions are generated by a knowledge-based system under general human guidance. Similarly, the virtual world for code inspections described above could be redesigned to incorporate both avatars controlled by human agents and machine-based knowbots. For artificial realities in engineering education, balancing the different types of beings interacting in the virtual environment (avatars, knowbots, video-links) is an interesting issue for instructional design research. The Star Trek learning environment in Vignette III illustrates some psycho-social dimensions of virtual communities that include knowbots and avatars.

3. Virtual Communities

This study's Appendix presents a forecast of when various aspects of distributed synthetic environments, artificial reality interfaces, and intelligent artifacts will be technically and economically feasible. As these technologies evolve, enabling learners to interact in virtual worlds, intriguing interpersonal dynamics may emerge that are quite different from typical real-world encounters. People participating in virtual worlds often feel as if knowbots are real human beings, an illustration of the general principle that users tend to anthropomorphize any type of machine-based agent. Joseph Weitzenbaum's ELIZA program, which simulates a Rogerian therapist, is an example of this tendency. Some users type in responses to ELIZA's prompts for hours, attributing human understandings to a pattern-matching natural language program that has no conception of people's emotions or behaviors.

As a complement to responding to knowbots as if they were human, participants in a virtual world interacting via avatars tend to treat each other as imaginary beings. An intriguing example of this phenomenon is documented in research on Lucasfilm's *Habitat* [Morningstar, & Farmer, 1991]. *Habitat* was initially designed to be an on-line entertainment medium in which people could meet in a virtual environment to play adventure games. Users, however, extended the system into a full-fledged virtual community with a unique culture; rather than playing pre-scripted fantasy games, they focused on creating new lifestyles and utopian societies.

As an entertainment-oriented cyberspace, *Habitat* provided participants the opportunity to get married or divorced (without real-world repercussions), start businesses (without risking money), found religions (without real-world persecution), murder other's avatars (without moral qualms), and tailor the appearance of one's own avatar to assume a range of personal identities (e.g., movie star, dragon). Just as SIMNET enables virtual battles, Habitat and its successors empower users to create artificial societies. What people want from these virtual societies that the real world cannot offer is magic, such as the gender-alteration machine (Change-o-matic) that was one of the most popular devices in the *Habitat* world.

Users learned more about their innermost needs and desires by participating in Habitat than they would have by spending an equivalent amount of time listening to psychology lectures. Similarly, social scientists are discovering more about utopias by studying Habitat's successors than they did by researching communes, which were too restricted by real-world considerations to meaningfully mirror people's visions of ideal communities. Giving users magical powers opens up learning in ways that educators are just beginning to understand. As with any emerging medium, first traditional types of content are ported to the new channel: Then alternative, unique forms of expression—like *Habitat*—are created to take advantage of expanded capabilities for communication and education.

Unlike passive, linear media such as television, synthetic learning environments center on interaction and collaboration. Engineering education in the future may balance learning experiences in three types of environments: contemplation and introspection via books and symbolic media; shared discovery in virtual communities populated by knowbots and avatars; and assimilation in conventional instructional settings.

4. Immersion Interfaces and Artificial Realities

Trainers have long used technology-based simulations to mimic simplified real-world environments. However, virtual realities are qualitatively different than small-screen models because entering and exploring a "world" is a more intense mode of human-machine communication than "conversational" interfaces in which the user commands the computer to do something, then waits for a reply [Walker, 1990]. The key capability that artificial realities for learning add to current simulations is immersion: the subjective sensory impression that the user is "inside" a synthetic environment comprehensive and realistic enough to induce the willing suspension of disbelief.

Much research is needed to clarify all the variables that affect this sense of immersion [Dede, 1993]. At present, most work by investigators studying artificial realities centers on achieving sensory (predominantly visual) fidelity to shift the user's frame of reference and sense of personal space. However, this focus on objective realism neglects other, more subjective dimensions (consciousness, meaning, mood) that underlie immersion. For example, dreams are immersive even though their sensory and epistemological fidelity may be low. As another illustration, a well-written horror novel read alone at night can induce strong suspension of disbelief even though no visual images are utilized.

A weak analog to immersion interfaces that many readers will have experienced is the IMAX theater, in which a two-story by three-story screen and high resolution images can generate in the observer strong sensations of motion. Adding three-dimensionality, highly directional and realistic sound, an even wider visual field, and the ability to interact with the virtual world through natural physical actions produces a profound sensation of "being there," as opposed to watching. Because common sense responses to physical stimuli work in artificial realities, the learner quickly develops feelings of mastery, rather than the perception of helplessness and frustration typical when first attempting to use an unfamiliar computer interface or operating system.

Also, these types of educational environments are strongly motivating, since designers of virtual worlds can readily incorporate four powerful levers for emotional involvement: fantasy, challenge, curiosity, and collaboration [Malone, & Lepper, 1985]. Artificial realities can be tailored to the affective needs of individual learners, incorporating fantasy characters matched to the student's personality in challenging games or complex adventure-worlds. An example is the Star Trek learning environment in Vignette III earlier. Of course, these fantasy settings must be carefully designed to suppress the learning of behaviors harmless in a virtual environment, but hazardous in the real world (e.g., walking through a buzz-saw).

Full-fledged immersion in a virtual world is certainly not essential for many aspects of engineering education. Two-dimensional, small-screen views and indirect manipulation via mice or similar input devices are adequate in most situations—and much less expensive. Being "inside" a learning environment is crucial primarily in developing 3-D manipulation and navigation skills (e.g., surgery, military maneuvers); in recognizing complex patterns and configurations (e.g., visualization of fluid flows, computer-aided design); and in remediating misconceptions induced by real-world experiences (as discussed in the next section).

Virtual environments that simply mimic reality have their uses, such as allowing learners to experience activities that are dangerous or expensive in the real world. However, adding the ability to magically act in ways impossible in the real world opens up new dimensions for engineering education. One powerful type of representational strategy that transcends reality is visualization, through which learners can manipulate typically intangible entities such as molecules and mental models.

5. Visualization

The fundamental idea underlying visualization is displacing cognitive complexity into the human visual system. Human beings have very powerful pattern recognition capabilities for images. Spatial data management systems and scientific visualization have established that; when symbolic meanings are mapped into visible attributes such as shape, texture, size, color, and motion—increased insight into underlying structural patterns of information is attained [Tufte, 1990].

Artificial realities can support two emerging types of visualization: "sensory transducers" that allow users' eyes, ears, and hands to access previously imperceptible phenomena (such as a molecule) and "cognitive transducers" that perform a similar function for intellectual entities. Sensory transducers provide a means of grasping reality through illusion [Brooks, 1988]. Using computers to expand human perceptions (e.g., allowing a medical student—like Superman—to see the human body through X-ray vision) is a powerful method for deepening learners' intuitions about physical phenomena.

Realistic, directional sound is attainable today to enable auditory transducers, and visual transducers already exist through computer graphics. However, tactile forcefeedback is a difficult and expensive hardware challenge. For the early 1990s, cost will likely restrict haptic transducers to applications, such as exploring molecular docking sites, that necessitate people's very delicate and important sense of touch [Minsky et al, 1990]. Together, all these types of sensory visualization modalities can combine to intensify immersion in a magical learning environment of previously intangible physical entities.

A second form of visualization, cognitive transducers, make intellectual entities such as knowledge structures visible and manipulable. They are a logical extension of sensory transducers in enhancing the power of artificial realities for engineering education. Transforming the symbolic into the geometric via data visualization is useful in situations where the amount of data is large and interacting with the data to shape its presentation can aid in interpretation.

A leading-edge illustration of making cognitive entities tangible is the Information Visualizer, an experimental interface that uses color and three-dimensional, interactive animation to create information objects [Robertson, Card, & Mackinlay, 1991]. If incorporated into the multimedia application depicted earlier in Vignette I, this cognitive transduction strategy would enable users to see and manipulate the webs of knowledge woven through the multimedia database. Such an approach aids learners in mastering patterns of interrelationships among chunks of information, a vital skill in the design of complex artifacts [Dede, 1991b].

The author has recently begun collaborative research on a Virtual Physics Laboratory that illustrates the use of sensory and cognitive transducers in an artificial reality for technical education [Loftin & Dede, 1993]. We are designing a prototypical immersion environment to address well-documented misconceptions held by physics and engineering students, even those who receive excellent grades in multiple technical courses [Halloun, & Hestenes, 1985a]. These include misconceptions about the nature of mass, acceleration, momentum, charge, energy, potential difference, and torque (*key concepts*), as well as Newton's laws, conservation laws, the atomic model, and electron flow models for circuits (*fundamental principles and models*).

Conceptual primitives form mental constructs, the understanding of which is prerequisite for higher-order concepts. However, many science-oriented students have difficulty understanding these concepts even at the qualitative level, let alone the problems that occur with quantitative formulation. Among common misconceptions are the "position-speed confusion" (i.e., ahead = faster) and the "motion implies contemporaneous force" notion [Halloun & Hestenes, 1985b; Reif & Larkin, 1991].

These mistaken ideas stem from misinterpreting common personal experiences with complex real-world phenomena in which many forces are simultaneously acting. For example, although students with technical backgrounds can parrot Newton's First Law (only the application of a force can alter an object's motion), they frequently misapply this principle in novel problem-solving situations. Such errors are due to the deceptively universal presence of friction in real-world phenomena; objects in motion seem to slow and stop "on their own."

Not only are these misconceptions strongly held by students entering physics courses, but they are very difficult to remediate with conventional approaches to instruction. In group teaching, mistaken ideas at the qualitative level can easily go undetected, since a student's superficial knowledge of formula manipulation techniques can mask a misunderstanding of underlying concepts. We are studying whether an immersive artificial reality that allows learners to alter the laws of physics, using visualization to underscore the consequences of their actions, can rapidly correct these pervasive misconceptions.

In incipient form, the Virtual Physics Laboratory provides support for changing within its walls the magnitude and direction of gravitational acceleration, as well as the magnitudes of atmospheric drag, frictional coefficients between surfaces, coefficients of restitution of objects, and the length of a simple pendulum. In addition, observations and measurements are made possible by controls that "freeze" the passage of time while objects are positioned and/or given an initial velocity. A *trace* feature displays the trajectories of objects in motion, and a digital display can provide both time and displacement data.

As an illustration of how such an immersion environment can aid technical education, a BallWorld has been created that enables learning experiences not possible in real-world classrooms. Users access BallWorld via computerized clothing (head-mounted display, gesture glove) and experience a strong subjective sense of being inside a synthetic environment. Recoil phenomena can be experienced by throwing a small (relative) mass ball, with an almost negligible impact on the motion of the thrower, then contrasting this with throwing a large (relative) mass ball. The shift in the high-mass ball's motion with respect to the laboratory will be slight, while the thrower will experience rapid motion and collisions with the lab's walls, floor, and ceiling.

To extend this example further, a learner can launch two balls so that they collide, observing their collision in the laboratory frame of reference. The experiment can be replayed with the observer's frame of reference attached first to one ball and then to the other. Finally, the learner can observe the collision from the center-of-mass frame of reference; this is the most useful for carrying out conservation of momentum observations, but is quite difficult to visualize with purely internal imagery.

Over the next two years, we will add a variety of sophisticated sensory and cognitive transducers to the Virtual Physics Lab. Controlled trials will be conducted to assess the leverage that immersion and visualization provide in remediating misconceptions induced through real-world experiences. Experiences in this artificial reality will be compared to student outcomes using more traditional approaches (e.g., small-screen, two-dimensional simulations) to clarify the relative utility of high-end virtual environments and to suggest design principles for optimizing the educational value of artificial realities.

The generic term applied to rich virtual worlds for visualizing information is "cyberspace," from a science fiction novel by William Gibson [1984]. The vision of our civilization a couple decades hence that Gibson presents is both intriguing and plausible; the nervous system of global business is based on workers manipulating huge virtual structures of data in a shared artificial reality. Teleoperation (performing activities over distance) and telepresence (mimicking face-to-face contact through video/graphics representations) dominate human activity. Some "cybernauts" are interested in having life imitate art and are building computational tools that would enable Gibson's imagined future.

Visualization as a means to enhance engineering education is not confined to artificial realities accessed through immersion interfaces, but is an important feature in many types of synthetic environments. Artificial realities themselves are likely only a transitional medium for incorporating virtual environments into every aspect of our physical world. During the next decade, increases in the power of computers coupled with decreases in size and cost will enable personal environments replete with intelligent objects having embedded microprocessors.

6. Artifacts with Embedded Intelligence

Just as motors shrunk in size and effectively "disappeared" during the Industrial Revolution, so computers are vanishing into our everyday context, supplementing physical reality with a virtual overlay of data. Intelligent objects for learning may first appear as smart manipulatives in pre-college classrooms, providing individualized coaching for students. As one illustration, imagine a child stacking blocks in order of size from biggest to smallest to form a tower. When he picks up a block whose size is out of sequence, it could say, "not me," while the correct block lights up and says, "my turn." Designing embedded virtuality (speech, imagery, intelligence) into all the components of engineering laboratories to promote learning will be feasible no farther in the future than the advent of the personal computer in the past.

To illustrate some potential implications of artifacts with embedded intelligence for engineering education, a fourth vignette depicting the daily routine of a faculty member fifteen years from now is presented below. The ideas and situations in this fourth vignette draw heavily on a scenario from Weiser [1991]. Brief descriptions *in italics* explain neologisms that may be confusing to readers. This scenario's purpose is not to predict how engineering education will evolve. Instead, the vignette is a baseline future that extrapolates from trends in how we use telecommunications today to illustrate the types of smart devices that will permeate society in the future and some instructional capabilities that they enable.

Vignette IV: Engineering Education via Intelligent Objects

Vesper is driving to work through heavy rush hour traffic. She is a faculty member in computational engineering at a university located far from her home in the suburbs. Despite the long drive, the position was irresistible because the campus serves as a test site for advanced networking technologies.

She glances in the "foreview" mirror to check the traffic. *{Commuters' automobiles are hooked into a large network that uses data sent by cars and highway sensors to monitor and coordinate the flow of traffic. The foreview mirror presents a graphic display of what is happening up to five miles in front of her car on Vesper's planned route to work.}* Noticing a traffic slowdown ahead, Vesper taps a button on the steering column to check for alternate routes that might be faster. A moment later, she cancels the request for rerouting as the foreview mirror reveals the green icon of a food shop on a side street near the next exit from the freeway. The foreview mirror helps her to find a parking space quickly, and she orders a cup of coffee while waiting for the traffic jam to clear.

While drinking her coffee, Vesper calls up some work on the screen of her notepad computer. *{This machine has the approximate processing power of today's supercomputers and is about the size of a thick pad of paper. It is linked via wireless networking and fiberoptic cable to a large web of computers, including those at Vesper's campus.}* The university's diagnostic expert system for debugging prototype Ultra Large Scale Integrated (ULSI) circuit designs can handle the routine misconceptions typical of most sophomore engineering students, but occasionally is stumped by an unusual faulty procedure that some learner has misgeneralized. *{A computer program trained to mimic human experts can handle many routine aspects of evaluating student performance, but complex assessments still require human involvement.}*

Vesper has an uncanny ability to recognize exotic error patterns by quickly scanning a complex schematic. She diagnoses three sets of student misgeneralizations before resuming her trip to school. Her new "bug collection" will automatically be sent to the national database on design misconceptions to be entered into its statistical records. Her notepad also forwards her diagnoses to the university's expert system on ULSI design, which incorporates the new bugs into its knowledge base and begins preparing tutorials to correct those particular errors. Later in the day, this instructional material will be forwarded to the appropriate learners' notepads to provide individualized remediation.

As Vesper walks into the engineering complex on campus, her personalized identity tab registers her presence on the university's Net of security sensors. *{In a clip-on badge displaying her picture and name, a small device is embedded that broadcasts information about Vesper's movements. Such an identity screening procedure is part of the university's security system. In this future world, these elaborate precautions have unfortunately become necessary.}* A moment later, the machines in her office initiate a log-in cycle in preparation for her arrival. She realizes that she has left her car unlocked, but does not bother to retrace her steps; from her office, she can access the network to lock her car via a remote command.

As Vesper gets to her desk, the telltale by her door begins blinking, indicating that the department's espresso machine has finished brewing her cafe au lait. *{A telltale is a remote signaling device that can be triggered to blink or make a sound, advising people in its vicinity of some event happening elsewhere.}* Vesper drinks a cup of cafe au lait every morning upon arriving. She heads down the hall to get the coffee; the espresso maker's brew will be much better than the vile stuff she had consumed at the food shop. On returning to her office, she instructs her desktop workstation to remind her not to stop there again. A copy of her evaluation is automatically forwarded to the food shop's manager and to the local consumer ratings magazine.

In the hour before class, as her sophomore students arrive, they wander around the halls visiting friends and faculty, gradually congregating in the engineering lab to work on projects for their exhibition portfolios. Vesper will join them in about half an hour to begin face-to-face instruction. She takes a break from viewing her videomail to fast-reverse through the electronic trails their movements have left on the security system. Valerie is still dallying too long before getting down to work; Vesper will have to speak with her. Richard has not arrived at the engineering complex, but no message has come in to indicate why he is later than usual. Vesper decides to wait another fifteen minutes before taking action about his unexpected tardiness.

Her desktop workstation conducts a brief dialogue with the "intelligent" equipment in the lab, then reports its findings. *{Just as with the identity badge and the telltale, classroom devices can have microprocessors and wireless networking embedded. This gives each set of objects limited abilities to sense what is happening and respond, guiding some forms of student learning.}* Ronald is redoing activities he has already mastered rather than moving on to the new work she had assigned; Vesper notes that she needs to talk to him about this. Everyone else seems to be on task and involved enough to be occupied for another fifteen minutes before she arrives at the classroom.

Skimming an engineering education journal, she notices a case study that resembles a problem student in one of her colleague's classes. His student appears to have a rare type of learning disability that interferes with developing a spatial sense of geometric relationships, an important skill in his branch of engineering. Vesper sends an excerpt from the article to her colleague's machine with voice-mail appended explaining its significance.

She tends to avoid video-mail, even though its greater bandwidth empowers more subtle shades of meaning. It is too much trouble to assume a professional demeanor just to send a simple message. The knowbot in her journal-reading application notes that she found the article useful and reinforces the pattern recognizers that triggered its selection.

A chime sounds in Vesper's office; the Holographic Animation Lithographer, an expensive piece of lab equipment, has predicted the likelihood of a near-future malfunction and shut itself off as a precaution. With a sigh, Vesper takes her Virtuality Goggles out of desk drawer and slips the strap over her head. Once at the lab, she can use the goggles to "see" the interior of the device through different electromagnetic modalities (infrared, X-ray) in an attempt to verify the problem.

A small light on the edge of Vesper's glasses begins blinking. A phone call is coming in; must be from someone not on the network. "Activate," says Vesper (the only word her glasses can recognize). A voice begins speaking in her ear; Richard's girlfriend, informing her that he is sick again. With a sigh, Vesper makes a note to prepare hardcopy homework that will be sent off by snailmail—what a hassle! *{Many network users refer to traditional delivery services for paper copies as "snailmail," preferring the rapid transmission of electronic mail via telecommunications.}* She will be glad when the government finally recognizes that access to basic network services is a fundamental right of all citizens, even if that does mean subsidizing subscriptions for the poor. Her notepad automatically informs the attendance monitor in the Registrar's office of the reasons for Richard's absence.

Before walking down to the lab to join her students, Vesper decides to have a conversation with Dimitri. They had agreed to do virtual team-teaching for a couple weeks, as each has skills the other would like to learn through apprenticeship. In particular, Dimitri has some good ideas for interdisciplinary teaching combining traditional mechanical engineering with nanotechnology.

She and Dimitri have given each other temporary access to their location detectors and to the ongoing contents of the other's desktop screen, as well as to the large liveboards and video-cameras in both their labs. *{Liveboards are large flat-panel displays that can present computer graphics or video in a big-screen format readily viewed by everyone in the room. They automatically archive what is presented and can display multiple windows of information simultaneously.}*

Dimitri appears on her screen, but explains that he has to rush off for a meeting. Vesper deliberately had not bothered to check his schedule; she finds that keeping track of hers is enough hassle. However, she keeps forgetting that he is seven time zones distant, so her accuracy in predicting when Dimitri might be free has not been good. Their knowbots automatically block off the first mutually available time when they could talk at length.

Before signing off, Dimitri promises to send her a video of his lecture on electron tunneling as a challenge in ULSI packaging. Using an annotation tool, he will append comments to different segments of the video giving his assessment of the strengths and weaknesses of that particular rendition. Dimitri also suggests that Vesper get in touch with Hiroshi Sakashito, the representative from a vendor sponsoring an experimental approach to minimizing this problem.

Vesper calls up biographical information on Hiroshi to get a sense of his background and interests. *{People who wish to do so can make available a limited amount of personal information on the network to facilitate their interactions with others.}* She wants to start off on the right foot, as she would like to request an unrelated favor: an extra notepad for her husband. Since her immersion in the world of embedded virtuality, Vesper has felt a little remote from people not on the network. Feeling that way about her spouse is making her uncomfortable. Reminding her notebook to place a call to Hiroshi when she has time, Vesper sets off to class. Another humdrum day...

Activities with multimedia, analogical databases, and distributed simulation are deliberately not included in this fourth vignette—although they would be part of Vesper's daily activities—because these have already been depicted in prior vignettes.

As discussed earlier, the purpose of this vignette is not to suggest that Vesper's world is the only possible future for engineering education, but instead to illustrate the types of smart devices that will permeate society in the future and to eliminate the human capabilities that they can enable. Artifacts with embedded intelligence will be affordable and widely available in the work place within two decades. The barriers to such an infrastructure for engineering education are more psycho-social than technical and financial.

Many readers may find the above vignette unattractive from a affective and ethical perspective. This particular scenario images the ways an engineering educator immersed in sophisticated technology might extend her productivity. Vesper's environment may seem implausible—why would a person choose to live in such a machine-centered environment with so little direct human contact—but how would today's world of cellular phones, facsimiles, electronic mail, and answering machines have seemed two decades ago?

The vignette also deliberately incorporates a high level of surveillance; instruction is individualized by monitoring students' activities and intrusively intervening if these do not match some predetermined pattern. From my perspective as an educator, such a "Big Brother" world would be profoundly unattractive. However, I deliberately incorporated some dystopian aspects into this vignette—including concerns about equitable access—to underscore that the design of powerful technologies must be carefully considered to avoid unfortunate side-effects. Of course, artifacts with embedded intelligence could be incorporated into engineering education without the types of behavioristic manipulation with which I have crafted this future depiction. However, since the other three vignettes are optimistic portrayals, interjecting some pessimism into this vision seemed a good way to balance the scenarios.

7. Immediate Priorities for Engineering Education

Suppose that, over the next few years, the engineering education community were to extensively experiment with the emerging technologies and applications described in this study, including conducting pilot implementations. How should a systematic assessment of these technologies' potential value for teaching/learning be initiated? In the early stages of their evolution, what types of advances in engineering education might be achieved?

Since the institutional contexts and resources of this study's readers vary widely, this section briefly sketches generic priorities for immediate exploration by engineering educators:

- *Build Alliances with Vendors.* The war is underway for which providers (cable, telecommunications, computer) will dominate the information services market. Each coalition is willing to provide substantial resources to strategic partners who can influence regulators, legislators, and judges by demonstrating educational benefits of information infrastructures.
- *Develop a Critical Mass of Human Resources.* By intermingling the learning experiences of potential and practicing engineers, educators can draw on the substantial knowledge the profession already has about emerging technologies. In addition, partnerships that provide access to complementary strengths from other fields (e.g., cognitive psychology, pedagogy) are essential to developing a requisite spectrum of ideas and skills.
- *Study Initiatives in Domains Beyond Education.* The Multi-User Dungeon environments discussed earlier are a fertile source of design knowledge for virtual communities, as are related efforts in entertainment, business, and scientific visualization.
- *Create Partnerships with Business and Government.* Both the public and private sectors will gain multiple benefits from a reconceptualization of engineering and from new models for work place training. These provide a strong rationale for business and government participation in experimental efforts to assess the value of emerging technologies for teaching/learning.
- *Acquire Funding for Research and Assessment.* Under the new federal administration, the Department of Commerce, the National Science Foundation, and the Defense Conversion Initiatives are all targets of opportunity for funding. In addition, private foundations are intrigued with the possibilities of the information infrastructure and interested in exploring its potential for nontraditional forms of education.
- *Distribute Risks with Other Innovators.* Emerging technologies are intrinsically high-risk, high-gain endeavors. By coordinating small-scale experiments, conducted at different sites across the country, engineering educators as a community can gain considerable knowledge without any single group having to take extraordinary risks on unproven approaches.
- *Alter Incentives to Encourage Innovation.* Institutionalizing innovations necessitates shifting an organization's reward structures. For assessing the myriad outcomes of a radical shift in teaching/learning, alternative student credentialization via portfolios and design exhibitions is important. For faculty, organizational incentives that reward gaining generalizable knowledge from pilot implementations are essential. Too often, judgments about an experimental approach are made on the basis of whether it worked well the first time rather than on what was learned about designing evolutionary models.

The construction of more detailed strategic plans for exploring the potential of these technologies depends on circumstances within a specific institutional situation.

If these instructional innovations prove to be valuable in their early stages of evolution, the following types of advances in engineering education seem likely:

Impacts on Attributes

- *faster mastery of material* (through individualization, higher motivation, and contextualization in authentic situations)
- *higher student/faculty ratios* (leveraging faculty resources via enhancing learner access and automating rote performances)

- *wider variability of class sizes* (using savings from the instruction of large, distributed groups to fund individualized coaching)
- *just-in-time, on-location education* (tailored to work place situations rather than the traditional school model)
- *broader range of clients served* (reaching potential and practicing engineers now underserved due to barriers of distance, time, and cost)

Impacts on Effectiveness

- graduates have mastered leading-edge knowledge and skills for the new roles engineers are assuming, increasing their marketability and satisfaction
- stakeholders in engineering perceive major benefits from increased investments in education
- faculty experience greater fulfillment from their profession; recruitment of high quality personnel is enhanced
- a wider range of promising students are attracted to become engineers.

Outcomes such as these are core criteria for assessing the potential value of further investments beyond the stage of pilot implementation.

C. CONCLUSION

Technology evolves in waves of innovation and consolidation. The advent of "motion pictures" about a century ago ushered in civilization's fourth medium, another dimension to communication beyond spoken language, written language, and still images. Later, new technologies appeared to embellish the capabilities of moving images: broadcast and narrowcast television, video tapes, videodiscs, multimedia, hypermedia. Now all of these are merging into a synthesis so far beyond its individual components that it constitutes a new meta-medium.

In Schrage [1990], Alan Kay analyzes the implications of new media through the question, "What does a medium ask you to become in order to use it?" Print requires an imagineer, constructing elaborate internal models from a scanty set of symbols. Television implicitly inculcates a passive observer; the telephone, a conversationalist. Some educational implications of the meta-medium sketched in this study center around its channel, which is rich and powerful enough to mimic the real world. Other instructional consequences come from the new types of content that can be communicated via this channel, such as virtual communities. Together, channel and content form the message of this immersive meta-medium, which is still too indistinct to fully comprehend, but potentially is both fascinating and frightening.

Any powerful information technology is a double-edged sword, a source of either propaganda or education. Through advances in computers and telecommunications, synthetic environments can be created that seem intensely real to participants, and artifacts with embedded intelligence can intersperse the virtual and real worlds. Artificial realities, virtual communities, and knowbots are emerging technologies that have enormous potential to improve engineering education. However, faculty and learners must recognize that these instructional vehicles carry intrinsic attributes that can empower or subvert the goals of an educational experience. Careful research is needed to understand how to optimize the design and utilization of the emerging meta-medium.

Given uncertainties about the availability and effects of these emerging technologies, engineering educators may conclude that intensive experimentation and pilot implementations should be deferred until these devices are further developed and better understood. In the next few years, the best approach may appear to be improving teaching/learning primarily by fine-tuning and automating traditional models of instruction. This strategy is far from optimal for several reasons.

First, reconceptualizing the engineering profession necessitates educational approaches that promote the transfer of knowledge from academic to commercial settings, build sophisticated skills in institution-wide communications, and bridge the functional capabilities of devices to their consequences for organizational effectiveness in meeting societal needs. Distributed synthetic environments, virtual communities, and intelligent artifacts enable a seamless intermingling of university and work place settings to enhance the exchange of ideas and to situate learning in authentic problem solving contexts. At present, multinational corporations are creating an infrastructure for the virtual work place; engineering educators will benefit greatly from early inclusion in this process.

Second, both potential and practicing engineers are seeking instructional situations closer to the work place learning end of the continuum than the traditional schooling model now practiced by universities. If engineering educators do not implement technology-based teaching strategies that empower just-in-time learning-while-doing, other competitive providers will step in to meet the need, using powerful ideas from cognitive science to produce instructional outcomes significantly better than those from conventional pedagogy. Moreover, both the federal government and the major information technology vendors are currently anxious to fund demonstrations of information infrastructures enhancing teaching/learning, but this window of opportunity will not last for long.

Finally, over the next generation, some of the most important artifacts for the evolution of civilization will be those discussed in this study. Engineering students need intensive experience with these technologies so that they can better design and utilize them in work place settings. Emerging instructional media are becoming a large part of the message prospective and practicing engineers must master. In implementing these emerging technologies to foster a new paradigm for engineering, we as educators must model for our students decision making, design, and assessment under conditions of incoherent objectives, incomplete data, and imperfect understanding, so that they can value and master these skills.

REFERENCES

1. **Brooks, Jr.**, F.P. Grasping reality through illusion: Interactive graphics serving science in *CHI '88 Proceedings*, 1-11, 1988.
2. **Dede, C.**, Evolving from multimedia to virtual reality. H. Maurer, Ed., *Educational Multimedia and Hypermedia Annual,* Charlottesville, VA, 1993.
3. **Dede, C.**, Education in the 21st century in *Annals of the American Academy for Political and Social Science* 522, 104-115, July, 1992.
4. **Dede, C.**, Emerging technologies: Impacts on distance learning. *Annals of the American Academy for Political and Social Science* 514, 146-158, 146-158, 1991a.
5. **Dede, C.**, Designing a tool for imaging mental models underlying training. *Proceedings of the International Conference on the Learning Sciences,* Charlottesville, VA, 1991b.
6. **Dede, C.**, Imaging Technology's Role in Restructuring for Learning. K. Sheingold & M.S. Tucker (Eds.), *Restructuring for Learning with Technology.* New York, 1990.
7. **Denning, P.J.**, Educating a new engineer. *Communications of the ACM,* 35(12), 83-97, 1992.
8. **Dertouzos, M.L.**, Lester, R.K., & Solow, R.M. *Made in America: Regaining the Productive Edge.* Cambridge, MA: MIT Press, 1989.
9. **Dunlop, C., & Kling, R.**, Social relationships in electronic communities in C. Dunlop & R. Kling (Eds.), *Computerization and Controversy: Value Conflicts and Social Choices,* Academic Press, New York, 322-330, 1991.
10. **Fisher, S.**, Virtual environments, personal simulation, & telepresence in *Multimedia Review: The Journal of Multimedia Computing* 1(2), 34-44, 1990.

11. **Fontana, L., Dede, C., White, C., & Cates, W.**, Multimedia: Gateway to higher-order thinking skills in *Proceedings of the 1993 Association for Educational Communications and Technology National Conference*, Arlington, VA, 1993.
12. **Gardner, H.**, *The Unschooled Mind,* Basic, New York, 1991.
13. **Gibson, W.**, *Neuromancer,* Ace Books, New York, 1984.
14. **Goldman, Shelley V. and Newman, Dennis**, Electronic interactions: How students and teachers organize schooling over the wires in *Interactive Learning Environments* 2(1), 31-44, 1992.
15. **Halloun, I.A., & Hestenes, D.**, The initial knowledge state of college students in *American Journal of Physics*, 53, 1043-1055, 1985a.
16. **Halloun, I.A., & Hestenes, D.**, Common sense concepts about motion in *American Journal of Physics,* 53, 1056-1065, 1985b.
17. **Kash, D.E., & Willenbrock, F.K.**, Engineering education for the workers of the future in *Issues in Science and Engineering*, 30-34, 1992.
18. **Laurel, B., Oren, T., & Don, A.**, Issues in multimedia interface design: Media integration and interface agents in *Proceedings of the ACM Computer-Human Interface Conference*, New York, 133-139, 1990.
19. **Loftin, B., & Dede, C.**, The virtual physics laboratory. Unpublished proposal to the National Science Foundation, 1993.
20. **Malone, T.W., & Lepper, M.R.**, Making learning fun: A taxonomy of intrinsic motivations for learning. In R.E. Snow & M.J. Farr (Eds.), *Aptitude, Learning, and Instruction: III. Cognitive and Affective Process Analysis,* 176-189, Lawrence Erlbaum, Hillsdale, NJ, 1985.
21. **Minsky, M., Ouh-Young, M., Steele, O., & Brooks, F.P. Jr.**, Feeling and seeing: Issues in force display in *Computer Graphics: Proceedings of the 1990 Symposium on Interactive 3-D Graphics,* 24(2), 235-244, 1990.
22. **Morningstar, C., & Farmer, F.R.**, The lessons of Lucasfilm's Habitat in M. Benedikt (Ed.), *Cyberspace: First Steps,* 273-302, MIT Press, Cambridge, MA, 1991.
23. **Orlansky, J., & Thorp, J. SIMNET**, An engagement training system for tactical warfare in *Journal of Defense Research*, 20(2), 774-783, 1991.
24. **Perkins, D.**, *Knowledge as Design,* Lawrence Erlbaum Associates, Hillsdale, NJ, 1986.
25. **Reif, F., & Larkin, J.**, Cognition in scientific and everyday domains: Comparison and learning implications in *Journal of Research in Science Teaching*, 28, 743-760, 1991.
26. **Robertson, G., Card, S., & Mackinlay, J.**, The information visualizer in *Proceedings of CHI 91,* 181-188, New York, 1991.
27. **Scardamalia, M., Bereiter, C., McLean, R.S., Swallow, J., & Woodruff, E.**, Educational applications of a networked communal database in *Interactive Learning Environments* 2(1), 45-71, 1992.
28. **Schrage, M.**,*Shared Minds: The New Technologies of Collaboration,* Random, New York, 1990.
29. **Smith, R.B.**, Experiences with the alternate reality kit: An example of the tension between literalism and magic in *Proceedings of CHI+GI 1987,* 324-333, New York, 1987.
30. **Sproull, L. and Kiesler, S.**, *Connections: New Ways of Working in the Networked Organization,* MIT Press, Cambridge, MA, 1992.
31. **Stevens, S.**, Intelligent interactive video simulation of a code inspection in *Communications of the ACM* 32(7), 832-843, 1989.
32. **Tufte, E.R.**, *Envisioning Information,* Graphics Press, Chesire, CN, 1990.
33. **Uttal, W.R.**, Teleoperators in *Scientific American,* 261(6), 124-129, 1989.
34. **Vincenti, W.G.**, *What Engineers Know and How They Know It,* Johns Hopkins University Press, Baltimore, MD, 1990.
35. **Walker, J.**, Through the looking glass in B. Laurel (Ed.), *The Art of Computer-human Interface Design,* 213-245, Addison-Wesley, 1990.
36. **Weiser, M.**, The computer for the 21st century in *Scientific American*, 265(3), 94-104, 1991.

APPENDIX: THE EVOLUTION OF EDUCATIONAL TECHNOLOGY

For at least another fifteen years, the information technologies will continue to increase in power while decreasing in cost. The table below presents a forecast of when some technological capabilities useful for learning will be affordable on high-end personal computers and in telecommunications infrastructures.

Functionality	Uses	Time Frame
Hypermedia (nonlinear traversal of multimedia information)	Interlinking of diverse subject matter; easier conceptual exploration, training, collaboration	Now
High quality voice synthesis	Auditory natural language output	Now
Cognitive audit trails (automatic recording of user actions)	Support for finding patterns of suboptimal performance	Now
Advanced manipulatory input devices (e.g. gesture gloves with tactile feedback)	Mimetic learning that builds on realworld experience	Now
Optical-disc systems with multiple read/ capabilities	Support of large data- and knowledge-bases; very cheap secondary storage; facilitation of artificial realities	Now
High-bandwidth fiber-optic networks	Massive real-time data exchange	Now
Synthesis of computers, telecommunications	Easy interconnection; realistic simulation	mid to late 1990s
Standardization of computer and telecommunications protocols	Easy connectivity, compatibility; lower costs	mid to late 1990s
Sophisticated User Interface Management Systems	Easier development of applications; reduced time for novices to master a program	mid to late 1990s
Intelligent computational agents embedded in applications	Support for user-defined independent actions	mid to late 1990s
Computer-supported cooperative work (collaborative design, collective problem solving, group decision support)	Mastery of team task performance	mid to late 1990s
User-specific, limited-vocabulary voice recognition	Restricted natural language input	mid to late 1990s
Information utilities (synthesis of media, databases, and communications)	Access to integrated sources of data and tools for assimilation	mid to late 1990s

High-resolution color monitors with 3-D graphics	Vivid simulation of reality; easy reading of text	mid to late 1990s
Microworlds (limited, alternate realities with user control over rules)	Experience in applying theoretical information in practical situations	Late 1990s
Consciousness sensors (input of user biofeedback into computer)	Monitoring of mood, state of mind	Late 1990s
Microcomputer performance equivalent to current supercomputers	Sufficient power for simultaneous advanced functionalities	Late 1990s
Knowledge processing (contextually-linked data storage with embedded inference)	Goal-oriented, context-specific mastery of concepts and skills	Late 1990s
Intelligent tutors and coaches for restricted domains	Models of embedded expertise for greater individualization	Year 2000+
Artificial Realities (computer-maintained virtual worlds)	Intensely motivating simulation and experience	Year 2000+
Artifacts with embedded intelligence	Intermixing of virtual environments and the real world	Year 2000+

Chapter 4

AN ANCHORED INSTRUCTION PERSPECTIVE ON EDUCATIONAL PRACTICE WITH A FOCUS ON ENGINEERING EDUCATION

Charles K. Kinzer, Vanderbilt University

This chapter is written from an educational perspective, advocating a construct that has become known as "anchored instruction." Education in a general sense, including undergraduate engineering education, is failing too often in a critically important area—problem solving. Teaching and learning in this area can be enhanced through an anchored instruction approach—specifically, the application of learned knowledge in real-world settings. Our traditional educational approaches (usually thought of as lecture with some field experience or laboratory work) can and do effectively educate if education is defined as learning a set of facts or procedures. But if education is defined as learning to apply facts and/or procedures, especially in contexts that differ from the context in which learning occurred, or in contexts that require an extension or repurposing of learned facts, then education is failing badly. In advancing this argument, we first suggest a view of what education attempts to achieve, followed by a discussion of problems that are common in classrooms and that anchored instruction attempts to mitigate. We then more specifically define anchored instruction and provide examples of anchors that have been used in various domains. We conclude by summarizing ways that anchored instruction might be used in engineering education.

I. A VIEW OF EDUCATION AND EDUCATIONAL GOALS

Generally, engineering education attempts to achieve the same things as any other field: the transmittal of a body of knowledge which, on reaching a critical mass, enables the student to function as do other professionals in the respective field. Education does not aspire only to transmitting factual knowledge. Rather, its goal is to facilitate application of learning within appropriate settings. It does little good to know that a hammer can drive nails unless one remembers this information and uses it when encountering a situation where a nail must be driven.

Consider the analogy of learning a new language, where the components of instruction (the new language's syntax, vocabulary, pronunciation, and so on) are analogous to the content that must be learned in any domain. Before one can "work" with other professionals (who, by definition as professionals, are already fluent in the domain, e.g., are already-fluent speakers of the language), the content within the domain must be mastered. Each field has a core of knowledge that can, in some sense, be viewed as a language that must be learned if "speakers" of that language are to communicate with each other, i.e., perform tasks that are in the purview of that language group. Just as speakers of French have difficulty communicating with speakers of English (in effect, one group has not learned the requisite knowledge allowing them to function as "professionals" within the other group) so, too, learners of the content domain comprising "engineering" are learning concepts that will allow them to perform tasks that other professionals in the field already do fluently. Teaching the fundamental body of knowledge within a domain, together with its application, is the goal of education and creates a functioning body of expert professionals who can communicate with each other and work within the given area.

I use the language analogy because if we consider a body of knowledge as analogous to a language, then other relationships between learning in the real world as opposed to learning in school can be readily grasped. Consider that language is best learned in the real world rather than in school. Partly, this is because learning in school is often decontextualized and lecture-based. Learning in the real world, however, is often contextualized and social. Learning a language in the real world involves its use in social situations for authentic purposes; words, syntax, and concepts are used when appropriate to a context. Learning a language in school, however, is often contrived, lecture-based, and based on analysis of syntax and vocabulary in the abstract. Just as in engineering laboratories, language learning laboratories attempt to inject a real world context and setting, often with group learning activities to simulate social aspects of learning, but these settings are generally artifacts and additions to compartmentalized, lecture-based approaches and are still far removed from learning in the real world. This difference from learning in the real world, largely from a social, constructivist perspective, exacerbates several problems that are found in classes at all grade levels from primary school through college, and that are discussed below.

II. A BRIEF EXAMINATION OF CLASSROOM BASED PROBLEMS AND ANCHORED INSTRUCTION'S RESPONSE TO THEM

Assuming that students come to class with the necessary prerequisites, several educational theorists have noted that there are two major areas where instruction can break down, whether teaching facts or problem-solving skills. The first area deals with the diversity of background knowledge represented in a classroom of students, and the effect that background knowledge differences have on comprehension. Much of this research comes from the areas of psychology and reading education, where it has been demonstrated for some time that comprehension and interpretation of a text is greatly influenced by a reader's prior knowledge. For example, schema theorists have shown the vital importance of background knowledge in one's ability to answer higher-level as opposed to only factual comprehension questions. In passages such as the following, it is easy to answer the lower-level, factual questions (numbers 1 and 2 after the passage), but is difficult if not impossible to answer the higher-level (and, most would argue, the more important) questions: numbers 3 through 5.

> The procedure is actually quite simple. First you arrange things into different, related groups. Of course, one might be sufficient depending on how much there is to do. If you have to go somewhere else, due to lack of facilities, that is the next step—otherwise, you are pretty well set. It is important not to overdo things. That is, it is better to do too few things than too many. In the short run, this may not seem important, but complications can easily arise. A mistake can be expensive, as well.
>
> At first the whole procedure will seem complicated. Soon, however, it will be just another facet of life. It is difficult to foresee any end to the necessity for this task in the immediate future, but then, one can never tell. After the procedure is completed ,one arranges into different groups again and puts things in appropriate

places. Eventually they will be used once again and the whole thing will have to be repeated. However, this is a part of life!

1. Is the procedure simple or complex?
2. Is it better to do too few things, or too many?
3. Why is it difficult to foresee an end to this task?
4. Why is a mistake expensive?
5. Provide a comprehensive title that indicates specifically what this passage is about.

(Modified slightly from Bransford and Johnson, 1972).

People who have difficulty answering the last three questions generally are able to do so after the passage title is provided: Washing Clothes at a Laundromat. Providing the title allows one's background knowledge to facilitate higher-order applications and to answer inferential questions. Note, however, that if students do not have appropriate background knowledge (and there are always some students who do not), that they will still have difficulty, even with the provision of the title—not with reading the words, but with understanding the concepts.

Similar research has shown that background knowledge is important not only for understanding and interpreting a passage, but that one's interpretation depends specifically on the nature of one's background knowledge and on one's interests. For example, the following passage was given to college music and physical education majors, who were asked that they read through it once only, and then state what the passage was about.

> Every Saturday night four good friends got together.
>
> When Gerry, Mike and Pat arrived, Karen was sitting in her room writing some notes. She quickly gathered the cards and stood up to greet her friends at the door. They followed her into the living room but, as usual, they couldn't agree exactly on what to play. Gerry eventually took a stand and things were set up. Finally, they began to play.
>
> Karen's recorder filled the room with soft and pleasant music. Early in the evening, Mike noticed Pat's hand and the many diamonds. As the night progressed, the tempo of play increased. Finally, a lull in the activities occurred. Taking advantage of this, Gerry pondered the arrangement in front of him. Mike interrupted Gerry's reverie and said, "Let's hear the score!"
>
> They listened carefully and then commented on their performance. When the comments were all heard, tired but happy, Karen's friends went home.
>
> (Anderson, Reynolds, Shallert, & Goetz, 1977).

The majority of music majors felt that the passage was about friends getting together to play music; the majority of physical education students felt that the passage was about friends getting together to play cards. Clearly, these students' interests and backgrounds influenced their interpretations.

Consider the relevance of such findings to education in general and to engineering education specifically. Engineering education students are increasingly coming from backgrounds that do not reflect their instructors'. Many are foreign students. But even if the students are natives, backgrounds and interpretations differ. When course instructors use an example to make a point in class, the appropriateness of the chosen example depends on the background knowledge of each student. The differences in backgrounds and expertise between the college student and the course instructor mitigate against the example being appropriate in facilitating understanding of the concept under discussion.

This issue of background knowledge and its match to an instructors' examples is related to issues of shared, or common knowledge and its importance in learning. Largely unconsidered outside of education, the degree of shared knowledge between instructor and student directly influences both the manner of teaching and the success with which concepts are learned. Some have called this the relationship between "given" and "new" knowledge in discussing learning from, and comprehension of, texts. Others have simply called this "shared" knowledge. In any case, it has been well documented that the degree of shared knowledge between teacher and learner is very important.

Providing an anchor that is experienced by the whole class, including the instructor, mitigates background knowledge differences. The anchor serves as a common background and makes the students and instructors less dependent on their own, individual background knowledge sources by serving as a commonly experienced item of prior knowledge. This allows the instructor to draw examples from it, and allows students to make reference to it when questions are asked. In short, the anchor provides a common reference point between instructor and student, and allows shared knowledge to facilitate both dialogue and comprehension.

The second problem mitigated by anchored instruction is what Whitehead (1927) has called the inert knowledge issue. Inert knowledge can best be described as knowledge that can be demonstrated as having been learned, but that is not applied when appropriate in new situations. This is often seen in school situations, where knowledge learned (and demonstrated on tests) in mathematics is not applied in science, or knowledge learned in science is not used in social studies. Even in English, students often demonstrate their knowledge of vocabulary definitions for learned words, but do not access these words or use them spontaneously when appropriate in speech or writing.

Glick and Holyoak (1980) demonstrated the presence and effect of inert knowledge through the use of two passages presented to college students. Here is the first.

> A general wishes to capture a fortress located in the center of a country. There are many roads radiating outward from the fortress. All have been mined so that while small groups of men can pass over the roads safely, a large force will detonate the mines. A full-scale direct attack is therefore impossible. The general's solution is to divide his army into small groups, send each group to the head of a different road, and have the groups converge simultaneously on the fortress. (p. 309)

Students memorized the information in the passage. They were then asked to use it to solve the problem illustrated in the next passage.

> You are a doctor faced with a patient who has a malignant tumor in his stomach. It impossible to operate on the patient, but unless the tumor is destroyed, the patient will die. There is a kind of ray that may be used to destroy the tumor. If the rays reach the tumor all at once and with sufficiently high intensity, the tumor will be destroyed. At lower intensities the rays are harmless to healthy tissue, but they will not affect the tumor, either. What type of procedure might be used to destroy the tumor with the rays, and at the same time avoid destroying the healthy tissue? (pp. 307-308).

Only a few of the college students were able to solve the problem initially, without prompts. However, when explicitly asked to use the information in the fortress passage, over 90% were able to solve the tumor problem, realizing the similarity between dividing the troops into small units that converge and using a number of small dosage rays to converge on the tumor. Of interest is that the students memorized the information in the fortress passage.

But, until explicitly told to use this information, only 20% spontaneously accessed this information to help solve the ray problem—the knowledge they had acquired was inert.

Partly, inert knowledge can occur because of the decontextualized nature of learning in school. That is, since learning in school is often presented as an end in itself, rather than as a means to an end, the learned information is "filed" as being accessed specifically when called for. In a test that pertains specifically to to-be-learned knowledge, for example, the student expects to use the learned knowledge on a test, and accesses it to do so. Outside this domain, however, expectation for using the knowledge is not as great, and access is impaired.

Anchored instruction, with its emphasis on learning in highly contextualized situations and on explicitly discovering reasons for actions that help solve problems posed in the anchor, mitigates the inert knowledge problem. Because anchored instruction emphasizes finding information and making explicit why needed information is useful to solving the target problem, reasons for later access can be more closely linked to application at the time of learning. Anchored instruction also uses analog problems (problems that require applying to-be-learned knowledge in situations similar to the anchor) to extend the problem and to reinforce that the knowledge is useful beyond the specific (though realistic) context within which it is learned.

III. A DEFINITION OF AN ANCHOR AND CHARACTERISTICS OF ANCHORED INSTRUCTION

As implied above, an anchor may be thought of as a shared experience to which instructors and students can refer during instruction. However, the anchor should be of sufficient depth as to present a reason for learning—it should provide a challenge so that students must learn smaller pieces of information and master them in order to solve the larger problem, or challenge, posed by the anchor. In this way, instruction is grounded, or anchored, in a real-world problem that requires learning in order to arrive at a solution. In some contexts, this anchor is thought of as a common, hands-on project or activity that is central to the course content.

Consistent with this view, Daly (this volume) points out that anchored instruction begins with a problem, or challenge, that involves all class participants. As students realize that certain knowledge is required in order to solve the problem, this knowledge is learned and applied. This close tie between learning and application, and the fact that the problem is closely related to real-life contexts, helps students to achieve the link between acquisition and use of knowledge. Typically, engineering education has used anchored instruction in a hands-on situation as Daly describes. That is, a project is assigned to the class, and the requisite knowledge (or course content) is learned to solve the problem or to complete the project. In some ways, this is similar to just in time learning, where students learn information because it is needed to solve a particular problem. This differs substantially from more traditional course curricula, where students are in essence told that they must master chunks of apparently unrelated content (both within and across courses), which only becomes relevant much later, perhaps as undergraduate students become seniors and need to solve more problems that require integrating knowledge from across these classes.

In addition to hands-on anchors, however, videodisc-based anchors have been successfully used in several domains, and appear to have several benefits to a purely hands-on, project-based anchored approach. The basic definition of an anchor remains the same: a shared experience that poses a central question that requires learning. However, the definition and design of the anchor when used with videodiscs, especially when under computer control, must be considered in light of design principles that move beyond the hands-on, or project-based view of anchored instruction. The following section discusses the use of videodisc-based anchors.

IV. GENERAL PRINCIPLES OF VIDEODISC-BASED, ANCHORED INSTRUCTION

McLarty et al. (1990) have pointed out several important aspects of using videodisc anchors, if the inert and background knowledge issues noted above are to be mitigated. Among these is the necessity for students to become "experts" about the content of anchor. By this, McLarty et al. mean that students must view the anchor and work with it to the extent that its various scenes can become a reference point for everyone in a class. That is, an anchor is not an anchor if one student refers to "the library scene" in a question to the teacher or the class, and some students do not know what happened in that scene. To those students the importance of the library scene to the point under discussion would be lost, because they are not yet experts with regard to the content of the anchor. Only if the entire class has enough facility, or expertise, with the anchor so that a reference to the library scene is meaningful to all—so that all know that the main character in the anchor called *The Occoee River Adventure* looked at a particular Table within a particular book in the library and took notes from that Table—will there be appropriate shared knowledge for learning to occur, and for background knowledge to come from the common anchor rather than the disparate backgrounds of individual students.

To develop expertise in the anchor, students typically must watch the anchor in its entirety once. They then go through the anchor again, stopping where the class, by consensus, agrees that logical breaks, or scenes, occur in the video. These scenes are then named, with all members in a class noting the scene names and the corresponding frame numbers for each scene. From then on, the teacher and/or student(s) can refer to scenes by name, and if one or more students is unsure about what occurred in that scene, the frame numbers allow quick search and a replaying. In fact, replaying and revisiting the anchor to gather information and to refresh the communal memory with regard to the anchor should be a natural occurrence in an anchored instruction classroom.

In addition, a videodisc-based anchor has the advantage of being non-linear in nature. It allows students to revisit and re-explore situations where they might need additional practice, thus acquiring fluency through repetition. And it allows the posing of problems in domains that may be dangerous or impossible for novices to work on. A videodisc-based anchor includes the following features that are more difficult to include in a hands-on anchor: a narrative format, embedded data design, analog problem situations, comparisons of solutions to experts (CTGV, 1993). Each of these is discussed below; each is different from a hands-on, or project-based anchor.

A. NARRATIVE FORMAT

Videodisc-based anchors are created in narrative, that is, story-based formats. The narrative format, as opposed to an explanatory format, allows the creation of realistic "plot lines" or problem situations. The story-like format also allows the insertion of a more well-developed situation, and can include features of characterization and setting (much like is done in novels or other story forms) so that the problem becomes important from a multidimensional perspective. For example, when following a problem to its conclusion, the personality of the solver, and the impact of the problem's solution on others in the relevant community become important. The narrative format enhances students' motivation, interest in, and understanding of the issues surrounding the problem. The problem to be solved is placed in a realistic and interesting story that facilitates students' involvement.

In contrast are videos that present a problem and ask the student to watch while the steps taken to solve the problem are explained. While descriptions of the relevance of the problem are provided (usually in the form of a voice-over while the solution is being performed), there is no need for active engagement by students because the solution can be watched rather than participated in. The reasoning behind the steps taken toward the solution is also usually provided in such "non-anchor" videos, but students are simply told why actions occur—they need not decide for themselves why certain actions might be relevant or what their impact might be. Thus, anchored instruction involves viewing a narrative that will require (as noted in the embedded data section that follows) active participation and reasoning. Other video, usually in non-narrative format, describes the solution to a problem while outlining the reasons for steps leading toward it's solution. The difference might be thought of as the difference between a story that intends to engage and a documentary that intends to inform.

B. EMBEDDED DATA DESIGN

One aspect that is critical in the view of anchored instruction as described by the Learning Technology Center investigators at Vanderbilt University is the principle that anchors incorporate embedded data as opposed to explicit data. Consider the difference between having to find data and explain its relevance to solving a particular problem, as opposed to being told that a piece of information is important. The educational literature is replete with examples of enhanced learning and retention when students must find and use information as opposed to simply using information that is provided to them. A related example is the axiom that one really learns something only when asked to teach it to someone else. This deep comprehension, or deep understanding is fostered by active search for information (which must at times, be guided by the instructor), rather than the passive intake of information. Thus, video anchors include data that are important to solving the problem or challenge, but that are embedded in the narrative, real-world context that is part of videodisc-based anchors.

One example from *The Occoee River Adventure* centers around the main character going into a library, choosing a book off the shelf (the title is visible), looking up a Table of flow rates (the Table is readable on the screen), making some notes from the Table (the notes are not visible), then leaving the library and continuing on to a power plant. The student must decide whether or not the book is important to the solution of the problem posed by the anchor. This is decision that the student must make consciously, because anchors attempt to emulate everyday situations and, as in real-life problem situations, students confront false paths and extraneous or even false information that may be present in the anchor.

Once the student decides that the library book and Table contain relevant information to the problem's solution, the Table can be examined and a decision must be made about how to use the information in the Table to facilitate the problem's solution. Note that the student knows only that the main character accessed the Table. The student does not know why, or how the information from the Table was used—these are things that must be discovered and implemented as part of the problem solution by the student. Recall that the student in anchored instruction does not simply watch the problem being solved, but must consider the actions of the characters in the anchor and use those actions as guides in their own problem solving exercise. Remember also that the notes made by the main character from the Table are not known to the student, so the student, who is attempting to solve the problem together with the main character in the anchor, must choose information from the Table and use it. This is quite different from having the student simply watch the video that is <u>not</u> an anchor—a video where one simply watches the main character solve the problem; where information is provided as relevant, and its use is simply watched. Such a video must be considered either entertainment (because passive viewing and not decision making is required) or is simply a modified lecture on tape. It requires little of the active decision making and generative learning that is part and parcel of videodisc-based anchored instruction, which includes embedded rather than specified data.

C. ANALOG PROBLEM SITUATIONS

A third feature of videodisc-based anchors is the provision of analog problems within the same context as the originally-presented anchor. One way to think about this feature is that it is similar to a problem in the real world, where a change in one factor can occur (for example, a change in the weather might occur during construction) which requires the reconsideration (or, at very least, the recalculation) of the original problem solution. Similarly, one might consider that experts within a domain often operate from a set pattern of responses, which are modified and varied according to the demands of specific situations. These patterns of responses are internalized through repeated exposure to similar sets of problems. Expertise is developed partly from such repetitions.

Videodisc-based anchors allow students to solve an initial problem. Then, the problem is repeated, but with one aspect of the problem changed, and the student recalculates a new solution. The problem is then changed again, and the solve-change one or more variables-solve pattern can be repeated as often as desired. Such analogue problems allow students to practice appropriate problem-solving routines within similar contexts so that the routines become more automatized. They also enhance the student's understanding of the many and varied factors that must be considered in any real-world problem solution.

D. COMPARISON OF SOLUTIONS

The fourth major aspect of videodisc-based cases is the ability to provide solutions to the problem on the disc, to which the students can compare their own work. This occurs in two areas. First, the solution to the problem is provided by the main character in the anchor, within the specific context in which the anchor occurred. Second, the disc includes comments and solutions by experts in the field, to which the student can function as a type of apprentice, comparing his/her answer to those of the experts, and listening to the experts' reasoning processes. But this only occurs <u>after</u> the student has provided a solution. The comparison to experts allows differences and similarities between the experts and the novice to become explicit, and points of difference facilitate the learning process.

Comparison of solutions is also desirable from an interdisciplinary perspective, where students consider the solutions and reasons of experts from a variety of areas in order to understand that different perspectives can result in different, but perhaps equally valid, solutions to similar problems. This insight mitigates tunnel vision toward problems—something that is important for novices to avoid—and facilitates recognition that many problems require a team approach, and that experts from various domains and fields of engineering and science must work together and consider alternative solutions to a given problem.

V. SUMMARY OF FEATURES OF VIDEODISC-BASED ANCHORS

The previous section has noted that there is a difference between videodisc-based anchored instruction and anchors that are hands-on, or project based. While hands-on anchors, where actual projects are completed and problems are solved and concepts are learned as a result of project completion are very powerful, videodisc anchors have the ability to provide repeated experiences within a similar context and can model social situations in ways that are not possible with hands-on projects. The use of embedded data, the ability to visit and revisit the problem through the use of analogs, the ability to compare one's solution to those of multiple experts, and the use of an engaging narrative all facilitate learning and mitigate against the lack of background knowledge discussed previously. All of these features converge to facilitate learning that is generative rather than passive,

Thus, while hands-on projects are extremely beneficial and provide concrete, grounded examples, the cost effectiveness and additional benefits of videodisc anchors make them a viable addition to any class, including classes that use hands-on, project-based anchors. The next section presents the use of videodisc anchors in various domains, providing additional support for their use.

VI. VIDEODISC-BASED ANCHORS IN VARIOUS DOMAINS

Over the past several years, videodisc-based anchored instruction has been tested and found to be effective in various domains. For example, anchored instruction has been used to embed vocabulary instruction, and literacy concepts such as plot structure, characterization, and cause and effect relationships (see Bransford, Sharp, Goldman, Kinzer, & Soraci, 1992; Bransford, Vye, Kinzer, & Risko, 1990). It has also been used to teach social studies concepts related to the Victorian Era and world history. In each of these cases, students' knowledge of facts increased, but more importantly, their <u>use</u> of learned knowledge spontaneously in appropriate problem situations also increased markedly. For example, it is well documented that vocabulary taught traditionally is usually "learned" at the level of students being able to provide definitions of words on vocabulary tests, yet the words are rarely used in conversations or writing outside the test situation. This is one example of the inert knowledge issue, discussed earlier.

A control-group study with two fifth-grade classes provides an interesting finding that addresses the inert knowledge issue (Kinzer, McLarty, & Martin, 1989; Risko & Kinzer, 1989). Two classes, one taught traditionally with vocabulary embedded in sentences, the other taught the same vocabulary through the use of an anchor, showed no statistically significant difference on paper and pencil tests of vocabulary definitions. Students in both classes were able to provide definitions and use the words in sentences equally well. However, ethnographic observations of these classes (two observers recorded field notes and analyzed video and audio tapes of the students while they were in class), found that the anchored instruction class spontaneously used the learned words 13 times more often than did the control group. In other words, for the anchored instruction group the knowledge was accessible; it was not inert. Similar findings are available for social studies and the literacy content noted above.

Perhaps more related to the engineering domain, however, are studies performed through the Diabetes Research and Training Center (DRTC) of the Vanderbilt University Medical Center (Pichert, Snyder, Kinzer, & Boswell, 1994; Pichert, Meek, Schlundt, Flannery, Kline, Hodge, & Kinzer, 1994; Pichert, Smeltzer, Snyder, Gregory, Smeltzer, & Kinzer, 1994). Over the past three years, several videodisc-based anchors have been developed and tested through the DRTC. These have ranged from teaching problem-solving and decision making relating to insulin management, to learning sick-day rules. Studies have most often occurred with adolescents ranging from 12 through 18 years of age. The issue addressed is that diabetics must recognize potential problem situations and deal with them, and that this population must also anticipate problems before they come up, perhaps through recognizing patterns of blood sugar readings in a logbook or through application of knowledge about the effects of overeating or unplanned exercise.

The DRTC studies have been performed in control-group designs and have resulted in the overall finding that learning of facts is not significantly different for groups taught with and without the video-based anchor (see, for example, Pichert, Snyder, Kinzer, & Boswell, 1994; Pichert, Meek, Schlundt, Flannery, Kline, Hodge, & Kinzer, 1994; Pichert, Smeltzer,

Snyder, Gregory, Smeltzer, & Kinzer, 1994). However, one study in particular (Pichert, Meek, Schlundt, Flannery, Kline, Hodge, & Kinzer, 1994) indicates that students who were taught in the anchored instruction condition were significantly better able to link reasons for actions to actions. That is, these students were able, even after a lengthy delayed follow-up, to both state an action and an appropriate reason, while students in the more traditional group essentially memorized actions but did not articulate reasons behind their actions. Being able to link reasons to actions facilitates the use of what is learned, and is critical in addressing application and problem-solving abilities.

Two years ago the Center for Intelligent Systems at Vanderbilt University created a video anchor *(The Occoee River Adventure)* that was used in the introductory section of an fluid dynamics class. In an 18-minute video, one follows a consulting engineer who receives a telephone call from a power company executive who wants to know how much to increase the flow rate over the Occoee River dam in order to meet increased demands for electricity at certain times of the day and certain months of the year. As the engineer solves the problem, he inspects documents, interviews various members of the community, visits the library for references sources, visits the dam's control room and interviews the technician on duty, and reads a file of correspondence pertaining to the problem. The video-anchor ends with the engineer looking into the camera and restating the problem/challenge, while reminding the student that all of the information that is required to solve the problem in the video.

Although on its face the problem seems relatively simple, the engineer in the video is confronted with various issues that must be considered in the problem's solution, including the need to address the requirements of different community groups who would be affected by the problem's solution, just as engineers in the real world must customize their solutions to fit into the real-world community. This embedding of social data facilitates undergraduate student's learning of something that experts in a domain already know: that the best technical solution may not be the best practical solution. This must be learned in addition to simply acquiring the formulas and factual knowledge that comprise so much of the curriculum, and is a natural outgrowth of the narrative style that comprises videodisc-based anchors. For example, in *The Occoee River Adventure*, through reading correspondence and interviewing various people in the community en route to solving the problem, the engineer discovers that there is a fish farm above the dam, and that the owner has lodged complaints in the past about the adverse effects of low water levels on his business. In a recent letter, the owner voices a concern about rumors of possible increases in the flow rate over the dam and points out that his fish will die and his fish farm will go out of business (perhaps resulting in a lawsuit) if the water levels become too low.

In counterpoint, however, is correspondence from several whitewater rafting outfitters. These business owners, downstream from the dam, have often voiced complaints that not enough water is let over the dam. They point out that the whole community benefits greatly from tourism and tax dollars that are generated by their businesses, and not only applaud the possibility of increasing the flow rate over the dam, but advocate that the increase be significant. The task of the engineer in *The Occoee River Adventure* is to find the optimal flow rate that will allow the power company to meet customer demands for power, while considering the concerns and the social impacts that the solution might have.

This particular anchor was used for a two-week period. Students in the course watched the video, became experts about its content, and solved the problem. In order to do so they had to identify relevant and irrelevant information, do research about flow rates, learn formulas and theories that pertain to the problem, and so on. Data gathered as part of the course and course evaluations indicated that students spent more time on this multistep problem than they typically did in more traditional word-based, non-anchored problems, that they referred to the video multiple times, and that the activity was well received.

Although videodisc-based anchored instruction has also been successfully used, and evaluated in other domains, such as mathematics instruction at various grade levels in public schools (in contrast to typical word problems at the end of chapters, in a manner similar to what is found in college classes), these will not be reviewed here. Interested readers are referred to CTGV, (1993a, b, c, d, e, f) for a discussion of studies and their implications. The three examples provided above, in the areas of literacy and social studies, medicine/patient care, and engineering are intended to provide evidence that the videodisc-based, anchored instruction approach has been successfully used in a wide variety of domains. Its potential for use as a supplement in undergraduate engineering curriculum, perhaps together with anchored, hands-on approaches seem to be effective and can be supported.

In summary, Table 1 provides general guidelines and expectations, on the part of the learner as well as the instructor, for teaching and learning within videodisc-based anchored instruction contexts.

Notice that learners must change their mind-set from passive receivers of knowledge to active participants in a problem-solving process. To some students, this is initially disconcerting. They are much more used to "being told" and taking notes in lectures rather than learning for a specific, highlighted, ongoing goal that closely links information to its purpose and use. Instructors must also modify their courses from what is traditionally found to allow much more interaction by students with information. Implementing anchored instruction also requires instructors to modify traditional teaching approaches to implement cooperative learning approaches.

Steps six through nine on Table 1 relate to practice and analog problems. In these steps, alternative strategies are considered, the influence of varying factors in the problem are presented and discussed, and clear relationships to similar problems are noted, with similar problems also being solved. Steps 10 and 11 note that homework must be done between class sessions, with the instructor ensuring that homework relates to the problem in the anchor, and that sufficient practice is taking place to solidify the learning required to solve the anchor's challenge. Notice, however, that there must be repeated opportunity to revisit the anchor—not to review the <u>entire</u> video, but to watch specific segments as needed for the instructor to clarify points and examples, and for the students to check embedded data. Table 1 reflects areas where the instructor must be prepared to show segments on an ongoing basis. One of the major difficulties that people initially implementing video-based anchored instruction seem to have is that many opportunities to revisit the video are missed. Instructors need to be consciously aware that the technology supports rapid and accurate access, and that the anchor should be revisited whenever possible.

Notice also that videodisc-based anchors are intended to be viewed once in their entirety by the whole class, after which the anchor is segmented and expertise (shared knowledge and reference points) become available to all classroom participants. Problems and sub-problems must then be identified, and the knowledge required for solution (and the embedded data in the anchor) must be captured. Here the instructor becomes a facilitator who guides students to embedded data and becomes a resource for students' learning required content.

Learner Goals and Procedures	Instructor Goals and Procedures
1. Actively participate in a problem-solving group	Prior understanding of problem-based learning and cooperative, group-learning approaches
2. View a problem-based anchor once in its entirety via videodisc	Allow learners to view the anchor without interruption or comment. After initial viewing, pose a central problem that students will solve through analysis of data embedded in the video
3. View the anchor again, segmenting it into logical parts, labeling each part, and becoming facile with the story	Facilitate consensus on segments and their descriptive labels
4. Identify everyday situations in the anchor that are obstacles to solving the challenge or central problem presented	Play appropriate segments and help learners to identify obstacles not previously noticed
5. Define the nature of each obstacle using information in the story or using supplemental information (either on videodisc and/or through other sources)	Facilitate analysis of obstacles; provide supplemental information as necessary (make sure reference sources are available; play appropriate segments)
6. Generate alternatives for handling identified obstacles	Encourage brainstorming of alternatives; be non-judgmental
7. Discuss criteria for evaluating each alternative's merits and feasibility	Provide information; refer learners to available references and video to evaluate alternatives
8. Relate obstacles and alternatives to other, similar problems	Challenge learners to identify in their own lives obstacles and problem situations similar to those in the video
9. Develop concrete plans for applying what has been learned to other situations	Facilitate development of personally meaningful plans of action for self care
10. Do "homework" activities between class sessions (based on new learning or to prepare for future discussion)	Prepare and suggest homework activities to help learners safely practice new knowledge and skills; exhort learners to practice applying new knowledge and personalized action plans in daily life
11. Discuss homework as completed, in terms of outcomes and insights	Facilitate relating homework outcomes to obstacles identified in the video and to learner's lives

Table 1. Outline of the Anchored Instruction Problem-based Learning Method as used with a Videodisc Anchor (Modified from Pichert, Meek, Schlundt, Flannery, Kline, Hodge, & Kinzer, p. 1036).

VII. A VIEW TO THE FUTURE: POSSIBILITIES AND ENHANCEMENTS THROUGH SIMULATIONS AND MULTIMEDIA

The discussion throughout this paper has focused carefully on videodisc-based anchors. This was done purposefully, since there is as yet little use (and thus, with the exception of hands-on or project-based anchors, little evidence) that relates to anchors in other contexts. In order to adhere to the principles of anchored instruction, as described above, video tape does not provide the same capabilities as do the random access, clear freeze-frame and exact control of starting and stopping points that are part of videodisc technology. The revisiting of scenes when students have a question is simply not feasible with video tape, which takes too long to rewind and is difficult to control exactly enough when trying to replay a specific scene. In addition, the fact that videodiscs have two, independent audio tracks has allowed the creation of anchors that can use video with audio that accompanies the video in two languages. The second audio track has also been used to provide audio-based interviews with experts or audio-based explanations for portions of video that appear on the disk. This feature of videodiscs also moves the use of anchored instruction beyond the capabilities of video tape (see also Bransford, Vye, Kinzer, & Risko, 1990; Bransford, Sherwood, Hasselbring, Kinzer, & Williams (1990).

There is, however, great potential for using multimedia, rather than simply videodisc, anchors in instruction. Newer multimedia technologies, such as CD-ROM and Quicktime movies, can enhance anchored instruction because they incorporate the features of videodiscs and add the capabilities of viewing the video on individual computers without requiring a videodisc player. This allows instructors to more easily create and require homework assignments using anchored instruction which, to date using videodiscs, has essentially been limited to small group learning situations in a computer lab equipped with individual computers and videodisc players, or to use by an instructor in front of an entire class. CD-ROMs, of course, allow students to take the anchor to their own personal computers and complete assignments outside the class or laboratory setting. Newer multimedia technologies also enhance the embedded data and extension capabilities of the anchor through the use of text and graphics that can surround the video on the screen.

Finally, future uses of multimedia anchored instruction will merge principles of video-based anchors with high-level simulations. Current uses of videodisc-based anchored instruction provide common background knowledge, shared reference points, grounded (and challenging) problems that require information to be learned and used, opportunities for problem identification, and opportunities to solve related analog problems all within a generative rather than a passive learning environment. Instruction, however, occurs mainly away from the anchor. While the data to solve the problem are embedded in the anchor, learning the formula or other tool that use the data, typically takes place outside of the video (in lectures that relate the learning to the anchor or through text materials). Simulations that can teach and allow practice with regard to the anchor can include the playing and replaying of pertinent scenes (to find embedded data, for example). They can model the outcomes of students' proposed solutions and provide individualized instruction based on ongoing, individualized, student modeling as the simulation is completed. Simulations will have a significant effect on both the delivery of instruction and on the extension of anchored instruction beyond its current applications.

Such a simulation was initially pilot tested with the *Occoee River Adventure* discussed throughout this paper. In part, the simulation modeled the flow rates through various diameter pipes, showing effects on turbines and, ultimately the amount of power generated. The simulation taught concepts of fluid dynamics that were accessed by students as needed to solve the problem posed in the anchor. Students searched the anchor for the embedded data to which their learning would apply. Analog problems changed the model so that effects of different options and factors could be easily seen. And, in analyzing students' use of the multimedia simulation-anchor it became apparent that students took many different paths toward reaching the same goal. For example, some students had more sophisticated knowledge than others, and the simulation allowed them to spend less time on certain aspects of the solution than on areas where more learning was needed. It is the coupling of multimedia anchors with simulations that appear to have the greatest potential for improving instruction in the future.

VIII. SUMMARY

Following a discussion of educational goals, we have presented the major principles of anchored instruction and its current uses. We also related videodisc-based anchors to hands-on, project-based anchors and examined some of the difference between these two approaches. The paper pointed out that narrative structure, embedded data, analog problems, and the comparison of solutions to experts' solutions and comments, done within a generative learning environment, are key aspects of video- or multi-media based anchored instruction.

REFERENCES

1. **Anderson, R. C., Reynolds, R. E., Schallert, D. L., & Goetz, E. T.**, Frameworks for comprehending discourse. *American Education Research Journal*, 14, 367-382, 1977.
2. **Bransford, J. D., Sharp, D., Goldman, S., Kinzer, C. K., & Soraci, S.**, Literacy in an age of integrated media. *Elementary school literacy: Critical issues*, M. J. Dreher & W. Slater, Eds., pp. 183-210, Norwood, MA: Christopher Gordon Publishers, 1992.
3. **Bransford, J. D., Vye, N. J., Kinzer, C. K., & Risko, V**, Teaching thinking and content knowledge: toward an integrated approach, in *Dimensions of thinking and cognitive instruction*, B. F. Jones & L. Idol, Eds., pp. 381-413, Hillsdale, NJ: Lawrence Erlbaum Associates, 1990.
4. **Bransford, J. D., Sherwood, R. D., Hasselbring, T. S., Kinzer, C. K., & Williams, S. M.**, Anchored instruction: why we need it and how technology can help, in *Cognition, education and multimedia*, D. Nix & R. Spiro Eds., (pp. 115-141). Hillsdale, NJ: Lawrence Erlbaum Associates, 1990.
5. **Bransford, J. D., & Johnson, M. K.**, Contextual prerequisites for understanding: Some investigations of comprehension and recall. *Journal of Verbal Learning and Verbal Behavior, 11*, 717-726, 1972.
6. Cognition and Technology Group at Vanderbilt (CTGV), The Jasper experiment: An exploration of issues in learning and instructional design. *Educational Technology, Research, and Development, 40(1)*, 65-80, 1993a.
7. Cognition and Technology Group at Vanderbilt (CTGV), An anchored instruction approach to cognitive skills acquisition and intelligent tutoring, in *Cognitive approaches to automated instruction*, J. W. Regian & V. J. Shute Eds., pp. 135-170. Hillsdale, NJ: Lawrence Erlbaum Associates 1993b.
8. Cognition and Technology Group at Vanderbilt (CTGV), Toward integrated curricula: Possibilities from anchored instruction, in *Developing instruction and assessing competence* , M. Rabinowitz (Based.), pp. 33-55, Hillsdale, NJ: Erlbaum 1993c.
9. Cognition and Technology Group at Vanderbilt (CTGV). An anchored instruction approach to cognitive skills acquisition and intelligent tutoring, in *Cognitive approaches to automated instruction*, J. W. Regian & V.. J. Shute Eds., pp. 135-170, Hillsdale, NJ: Lawrence Erlbaum Associates, 1993d.
10. Cognition and Technology Group at Vanderbilt (CTGV), Anchored instruction and science and mathematics: Theoretical basis, Developmental projects, and initial research findings, in *Philosophy of science, cognitive psychology, and educational theory and practice*, R. Duschi & R. Hamilton, Eds., pp. 244-273, New York: SUNY Press, 1993e.
11. Cognition and Technology Group at Vanderbilt (CTGV), Designing learning environments that support thinking: The Jasper series as a case study, in *Designing environments for constructive learning*, T. M. Duffy, J. Lowyck, & D. H. Jonassen Eds., pp. 9-36, New York: Springer-Verlag, 1993f.
12. **Kinzer, C. K., Gabella, M. S., & Rieth, H. J.**, An argument for using multimedia and anchored instruction to facilitate mildly-disabled students' learning of literacy and social studies. *Technology and Disability Quarterly*, 1994.
13. **Kinzer, C. K., McLarty, K., & Martin, G.**, Some effects of macrocontexts on vocabulary learning, paper presented at the annual meeting of the American Educational Research Association, San Francisco, CA, 1989.

14. **McLarty, K., Goodman, J., Risko, V. J., Kinzer, C. K., Vye, N. J., Rowe, D. W., & Carson, J.,** Implementing anchored instruction: Guiding principles for curriculum development, in *Literacy theory and research: Analyses from multiple perspectives,* J. Zutell & S. McCormick Eds., 39th NRC Yearbook, pp. 109-120. Chicago: National Reading Conference, 1990.
15. **Pichert, J., W., Snyder, G. M., Kinzer, C. K., & Boswell, E. J.,** Problem solving anchored instruction about sick days for adolescents with diabetes, *Patient Education and Counseling, 23,* 115-124, 1994.
16. **Pichert, J. W., Meek, J. M., Schlundt, D. G., Flannery, M. E. Kline, S., S., Hodge, M. B., & Kinzer, C. K.,** Impact of anchored instruction on problem-solving strategies of adolescents with diabetes. *Journal of the American Dietetic Association,* 94(9), 1036-1038, 1994.
17. **Pichert, J. W., Smeltzer, C., Snyder, G. M., Gregory, R. P., Smeltzer, R., & Kinzer, C. K.,** Traditional vs. anchored instruction for diabetes-related nutritional knowledge, skills, and behavior. *The Diabetes Educator,* 20(1), 45-48, 1994.
18. **Risko, V. J., & Kinzer, C. K.,** Effects of providing videodisc-based macrocontexts to anchor vocabulary learning, paper presented at the 34th annual meeting of the International Reading Association, New Orleans, LA, 1989.

Chapter 5

ANCHORED INSTRUCTION AND ENGINEERING EDUCATION

James Dally
Mechanical Engineering Department
University of Maryland at College Park

I. INTRODUCTION

Anchored instruction involves a method of teaching that can be applied to most courses in the engineering curriculum. The method requires that the instructor relate the theory that is being presented to some realistic example. The example can be an exercise, an experiment involving measurements or demonstrations, computer simulations or the design of a product. Most instructors employ anchored instruction to some limited degree in many courses. For example, statics textbooks incorporate some exercises that the student can relate to the real world. However, the theory is usually advanced before the problem is introduced and in most instances the problem that is selected to illustrate the theory is more abstract than real. The basic approach in anchored instruction is to present the problem, describe its engineering and social significance, and *then* introduce the theory that is necessary to approach the problem. The concept is to motivate the student by relating the material to be covered to something real and known to the student. The student can relate to the known quantity and is more able to accept and understand abstract knowledge that is introduced together with the familiar material.

In the next section we review the recent literature that describes efforts to employ anchored instruction using several different techniques. The most common method for teaching new concepts using anchored instruction is based on a project, in which the student engages in the design of a product or a process. The length of the project can vary from one course to one year, or across several years of the curriculum. Other techniques stress simulation and introduce hardware to show the connection between theory and practice. Hands-on experiences involving either assembly or dissecting common products are often used to introduce vocabulary and to illustrate engineering principles. Finally, plant trips afford yet another opportunity to relate classroom solutions to real field applications.

A section is presented to assess the engineering curriculum and educational processes as practiced today. We have been critical of our current status because of the high attrition rate of the freshman and sophomore students and because of our lack of achievement in developing B.S. graduates with skills that are of significant value to industry. We have characterized the curriculum as rigid, fragmented, repetitive, and in some instances, obsolete. The delivery system that is most frequently employed, chalk and blackboard, is discussed, and its impact on faculty productivity and escalating educational costs is described.

A section on reversing the process used in engineering education is presented. By reversing the process, we advocate that real elements of the engineering process be advanced before even beginning the scientific and mathematical foundation. An excellent way to reverse the educational sequence is to give the first semester freshman an opportunity to seriously engage in the product realization process. The students design, manufacture, assemble, and test a simple product in one semester. Their response to this approach has been enthusiastic. They are motivated by the need to create their own design and to manufacture their own piece parts.

The spectrum of the product realization process is very broad and there is a tendency to stop the learning experience at the conclusion of the paper design prior to building hardware. We discuss reasons for continuing the product realization process until a prototype of the product is built and tested. The favorable response of most students to the opportunity for a hands-on experience in which they actually build something is remarkable. Through this effort they couple the design process and the manufacturing and assembly processes. We talk about concurrent design, but they are practicing it in developing the prototype. They also learn firsthand to appreciate the quality of a drawing package and the meaning of form, fit, and finish.

In developing a product in a classroom environment, we suggest the use of student teams. The students appreciate that the team approach is a simulation of the industrial environment. The team provides the opportunity for the students to develop personal skills involving both leadership and cooperation. The benefits of increased communication from student to student are described. Finally, the social setting afforded by the formation of a team and its enhancement of the learning process is described.

Demands on the student, the faculty, and the institution are discussed in some detail. The student is faced with the need to work as a team member on a project that is extremely time-consuming. The faculty member must assume a new role—that of a coach instead of an instructor. The faculty member must learn to teach with little structure and respond to unanticipated questions with off-the-cuff mini-lectures. The institution must provide more time for the faculty to spend with the students and change its reward structure to better balance research activity and undergraduate instruction.

A section is included to show that anchored instruction can be employed in mechanics. The emphasis here is significantly different because the course material in mechanics is relatively fixed and has not changed much for several decades. We suggest an integration of mechanics, materials, and design to reduce the effects of fragmentation of the subject on knowledge retention. We also discuss the use of a multipurpose studio where experiments can be conducted and teams can design, build, and test models in conjunction with these integrated courses that combine mechanics, materials, and design.

We conclude with a brief discussion of anchored instruction.

II. LITERATURE REVIEW

There have been many efforts to modify the engineering curriculum during the past five years. These modifications address many educational issues that include: integration of education and research; strategies for curricular innovation; computer-aided instruction; design in engineering education; total quality engineering education; outreach; recruitment; and retention. Since anchored instruction is a method that can be used in teaching most courses, we can find contributions and developments related to this method under all of the topics listed above.

Perhaps the most ambitious example of anchored instruction has been underway since about 1980 at Allborg University in Denmark [1,2]. Allborg is a new university that was established with a project-based educational system. The studies are centered around a project theme for each semester. Lecture courses, study courses, and laboratory work for that semester are to be used in the analysis of the problem selected for the project team. The project themes are selected by Study Boards that are elected annually and consist of equal numbers of students and faculty. The faculties in the engineering departments and in the business school have shared responsibilities for the curriculum in both divisions. The system is popular with the students and enrollment has tripled in the past fifteen years in a country with a constant number of entering students. Other countries are considering employing this approach and currently a system based on the project theme is being installed in Nigeria. The authors indicate that it is difficult to implement this approach in existing university systems, but they encourage an attempt to introduce a program based on project themes in the U.S.

Projects that involve design, manufacture, assembly, and test are common vehicles to introduce engineering fundamentals using anchored instruction. Dally and Zhang [3-5] at the University of Maryland have developed a first semester freshman course, which involves the product realization process. While the development of a real product following a design specification is the central theme, engineering principles covered just-in-time are also presented. Since a drawing package is required, the students are eager to develop entry level skills on a CAD program. Because they are preparing drawings of their own creations, the students are much more interested and attentive than they are in traditional graphics classes where they copy unknown and unrelated parts from a workbook. Simple products are selected so that the students working in teams can actually manufacture or procure the parts that they have designed. This course provides a new experience in coupling design and manufacturing which is essential in the development of an education program that is to prepare our graduates to design for competitive advantage [6].

Another approach employed at the University of Rhode Island, described by H. Viets [7-8], involves a modular approach to engineering design. An ambitious multi-year project is assigned to students in their first or second year which is to be completed during their final year. The project is divided into modules that are included in several of the engineering courses and spread across the curriculum. The experience with the approach is generally favorable although the students are uncomfortable with open-ended problems and problems involving costs of materials or production.

Vagners [9], at the University of Washington, has improved the undergraduate controls system course by using elements of anchored instruction. He has emphasized the design of control systems so that the instruction includes selection of sensors, actuators, and feedback paths in addition to the traditional modeling and analysis. Throughout the course he stresses simulation and hardware-in-the-loop environments to validate and test the models of the control systems.

Another interesting approach employing the concepts of anchored instruction was developed by Sheppard [10]. She developed a new course involving the dissection of several artifacts including a printer, fishing reel, and a ten-speed bicycle. The purpose was to give the students a hands-on experience that would expand their engineering vocabulary and illustrate the engineering principles involved in the design of common products. The students responded favorably and the exposure to real hardware motivated the students in the more traditional problem-solving methods that were also covered in the course.

Plant trips offer still another technique for using anchored instruction methods to facilitate learning. Regan [11] described two different plant trips where design problems were posed prior to the visit to the industrial facility. In both cases the problems were developed jointly by the faculty and the industrial representatives. The students submitted written solutions and small cash prizes were awarded by the host firms to the students with the best papers. The exercises gave the students the opportunity to address vague concepts that are not usually covered in typical class room assignments. Even in "design" classes where open-ended solutions are sought, the opportunity to visit the solution and observe the unique characteristics of a process or item of equipment is missing. The students responded very favorably to the field trip and the opportunity to see their work as it related to the real world in an actual refinery.

Simulation [12] represents a significant movement toward anchored instruction. Processes can be modeled and software written that permit the student to explore solution space. The graphical output of these programs allows characteristics of the many solutions to be visualized. In some of the simpler cases, the optimum solution may be apparent from the graphical display. Since this approach is more real and less abstract than working toward unique solutions of narrow problem sets, it enhances learning, and, more importantly, it brings the design and analysis processes much closer together.

III. CURRENT CURRICULUM AND PROCESSES

The current curriculum in the freshman and sophomore years in most colleges of engineering in the U.S. is remarkably similar. The mechanical engineering curriculum for the first two years at the University of Maryland is shown as a representative example in Table 1.

In examining this course listing, we note that 74 percent of the credit hours are serviced by departments outside of the College of Engineering (51 percent in the mathematics and science departments, and 23 percent in CORE departments). CORE is the acronym for the general university requirements that are common to all programs at the University of Maryland. Only 26 percent of the curriculum in these first two years is actually taught by the faculty in the College of Engineering. It is little wonder that as many as 50 percent of our students either transfer out of the College of Engineering during this period or simply quit and leave the University. We give them very little attention, farm them out to the service departments, expose them to a flood of abstract material, and teach them in a relatively hostile environment often with indifferent faculty or poorly trained TAs.

Let us look more closely at the seventeen credit hours that are taught in the College of Engineering. Most of these credit hours (13) are devoted to engineering science courses, if we classify thermodynamics in that category. Only the engineering analysis and programming (EAP) course and the mechanical engineering project are outside of the engineering science offerings. However, if we carefully examine the description for the EAP course we find that it really involves numerical methods and programming and should therefore be classified as a mathematics and computer science course rather than as an engineering course. The bottom line is clear, that only one credit hour (the mechanical engineering project), or less than 2 percent of the curriculum, can be defined as engineering in the first two years of the engineering curriculum. One may argue that engineering science courses have engineering content. There is some degree of truth to this assertion; however, the mathematics and science content is often so significant and the engineering content so meager that these courses should really be classified as science/mathematics courses taught by engineering faculty.

The courses offered to the lower division students are usually taught in large lecture classes supported with recitations and/or laboratories. Classes of often 100 or even 200 students are in attendance at the lectures. Lectures covering abstract material in such a setting rarely prompt questions, and meaningful discussion almost never occurs. The delivery is not only one-way but the content is often covered in a single pass. The recitations provide a better environment because the class size is reduced and questions with meaningful

Freshman and Sophomore Curriculum Mechanical Engineering, University of Maryland			
FRESHMAN YEAR			
Course	**Credit Hours**		
	Spring	**Fall**	
Chemistry	4	4	
Physics I	-	3	
Calculus I & II	4	4	
Engineering Science I & II	2	2	
CORE	6	3	
Credit hours	16	16	
SOPHOMORE YEAR			
Calculus III & Differential Eqs.	4	3	
Physics II & III	4	4	
Engineering Science III & IV	3	3	
Engineering Analysis and Programming	3	-	
Thermodynamics I	-	3	
Mech. Engr. Project	-	1	
CORE	3	3	
Credit hours	17	17	
Total credit hours			66

Table 1.

discussion can take place. The difficulty, however, usually arises with the teaching assistants who are usually responsible for handling the recitation sections. Some TAs are superb teachers and are more effective than the faculty member that is presenting the lecture. But other TAs are poorly trained and often not even familiar with our language or culture. This latter group communicates poorly and is largely ineffective in dealing with the undergraduate students in the classroom, in the laboratory or in interacting with students during office hours.

The effect of nearly total exposure to science, mathematics, and CORE for 98 percent of the first two years causes disillusionment of a significant fraction of the student body; many students leave the program. Exit interviews with these students clearly define the problem. They have grown tired of the abstract material which is delivered in large lecture halls by indifferent teachers. They reject the recitations and laboratories offered by poorly trained TAs. They have not yet been exposed to engineering as it is practiced. Our engineering students never intended a career in mathematics or in science; otherwise, they would have enrolled directly in those departments. They seek other academic programs that appear more interesting and when they find them they leave the College of Engineering.

Attrition is significant. About 30 percent of the students give up in the first year and another 15 to 20 percent leave during the second year. This attrition is costly for the student, his or her family, the educational system and society in general. If a student's departure is because of academic failure, it is a very traumatic and unpleasant experience. In some cases the student never completely recovers and the damage imparted is permanent. Potential is lost never to be regained. The educational system has lost a valued customer, as well as the opportunity to develop an engineer. A career—or worse—a personality may have been seriously damaged.

For those students who succeed, manage to stay with the program and obtain passing grades, the experience is more satisfying. But many students with grade point averages of 3 or better (based on a perfect 4.0) have not acquired the knowledge or proficiencies that we had intended. They have passed our examinations and courses, but their understanding is limited and they quickly forget the abstract material when it is not employed in subsequent courses in the following semester or even the next year. In future attempts at restructuring the curriculum, we must recognize that classroom knowledge is perishable with a very limited shelf life.

Some of my colleagues believe the first two years is a weeding out period that is valuable in eliminating those students that are not serious and talented. This is an unfortunate attitude, because it helps continue a forty-year-old process that is seriously flawed. More importantly, this process inflicts serious damage on the educational opportunities of many students with a natural talent for engineering.

We have stated that our current process is old (about forty years) and flawed. This judgment is based on three premises:

1. The curriculum lacks flexibility.
2. The program is fragmented and repetitive.
3. The program does not adequately recognize technology.

To show that the curriculum lacks flexibility, consider an analogy between our curriculum and a brick wall. The entire wall represents our undergraduate program and the bricks represent our courses. As we observe the wall, it is evident that all of the bricks are about the same size. Our courses are mostly three or four credit hours with remarkably few exceptions. With programs of 120 to 132 credit hours, we build our educational wall with about forty bricks. Moreover, the bricks are mostly primary colors with, say yellow for mathematics and science, blue for engineering, and red for humanities, arts and social studies. With the exception of our engineering science bricks, which are green, we have no other colors or hues. Finally, our prerequisites imply that the bricks must be laid in a carefully prescribed order with yellow and green on the lower rows and blue on the upper rows. Of course, we must toss in a red brick on every row to create diversity in the color pattern.

For a given program our educational walls are remarkably similar from one college of engineering to the next. We may use different textbooks, but we all offer about the same courses. Some may attribute this similarity in program (the cookie cutter syndrome) to ABET, but the accrediting board provides a free year and a strong design requirement that gives more flexibility in structuring the curriculum than most institutions usually elect to take. The rigidity and the age of the current programs are due to the administrators and the ultra-conservative faculties who ignore the exit interviews of the departing students and make incremental curriculum revisions after months of committee deliberations. The "new" curriculum these committees adopt is remarkably similar to the old one except for the date of the revision.

As we observe our educational wall with its different color-bricks, it is clear that it is divided, fragmented, and repetitive. We have separated mathematics from science and both from engineering. We have taken major areas of classical knowledge and fragmented it into very small topics. Rather than teach mechanics, we divide it into statics, dynamics, fluid mechanics and mechanics of materials. The knowledge imparted to the student by this fragmented approach often appears unrelated and abstract. The divisions between the different courses are often exaggerated by the use of different textbooks by different authors. The material, which is really very simple in concept, is presented in a cloud of elegant mathematics that is appreciated by the instructor. Unfortunately elegant mathematics, particularly in the freshman year, adds another layer of complexity for the student to overcome. Even those students who pass are often confused and do not clearly understand some of the concepts. As we move into the upper division and begin to offer engineering design courses where knowledge from several science and engineering science courses must be used and synthesized, we find the students markedly deficient. They have not retained the basic knowledge and they do not know how to apply it in a design analysis. They have been taught in the abstract using problem sets where they are given clearly described input and asked to find a precisely specified output. It is rare indeed that anyone ever interprets the answer to a problem to indicate its engineering significance.

In design, the problem statement does not exist, and it is necessary for the students to formulate the problem on their own. After formulating the problem, they find that the input data needed for a solution are not given. They often are at a loss to find the information necessary to solve problems without aid. The students must not be criticized for their inability to cope with the real problems that arise in design. They have been taught in the abstract using methods that are similar in many respects to those used in teaching mathematics in high school.

The changes that must be made in the curriculum to resolve these deficiencies are significant. Clearly, we should attempt to integrate and teach much of the material in physics, mathematics, and engineering science as a combined body of knowledge. The need to integrate has been argued conclusively by Bordogna [14] in his definition of the engineering task, "The engineer is to integrate knowledge across ever-changing disciplines and apply it to create shared economic wealth with a respect for the quality of life." A significant effort to integrate the curriculum during the freshman year has been made at Rose-Hulman Institute of Technology [15]. Real progress has been made in this pilot effort, but unfortunately it is a development that no other college is duplicating. The title of a recent paper, "Do We Go It Alone?" by Anderson et al. [15] reflects their concern about being the first and only college of engineering in the U.S. to integrate the first year of the curriculum.

The current engineering curriculum does not reflect the rapid changes taking place in technology. There are two important aspects driven by changes in technology that we have been slow to recognize in structuring the curriculum. The first is in the content of our courses and the second is in the method of delivery of the material. The content of our courses has changed to some degree to reflect new computation capabilities. However, the changes made to date do not adequately reflect either the availability or the speed of the personal computers (PCs), the high memory capacity of the hard disks, and the enormous capacity of the optical disks for data banks. In the profession, we are rapidly moving toward a practice where a very large segment of the normal engineering activity (drawing, analysis, scheduling, reporting, etc.) is performed on a computer with special applications software. Today and in the future, the practicing engineer deals with questions of what and why. The question of "how to" perform a specific analysis task will for the most part be handled by a team of specialists that write the application software. Unfortunately, many of our courses still concentrate on "how to" solve an equation with little emphasis on the formulation of the problem or in the interpretation of the solution. The course content must change to reflect the fact that solving equations is no longer an issue. We need to address methods of modeling complex systems, ways of writing equations that describe the behavior of the model, and interpretation of the response of the system. What we model and why we model it is much more important than exploring methods for solving equations.

The most common technology that we use to deliver our material is still the blackboard and chalk. We have film and video, which has been used to a limited extent, but which is still largely ignored even when it can be used very effectively to supplement live instruction. More recently we have developed computer-aided instruction to better illustrate engineering principles and concepts by using interactive feed-back and animation. Soon we will have data bases organized for rapid access to unbelievably large amounts of information. It is difficult to understand why we have ignored so much technology for so long in developing delivery systems engineering courses.

One of several different reasons for the poor acceptance of technology in education is the budgeting process. Like the curriculum, the budgets of most academic departments are all about the same. Ninety percent of the department budget is fully committed on the first day of the fiscal year to salaries, and these moneys are mostly to support the faculty. The remaining ten percent of the budget is not sufficient to cover everything else (equipment, travel, supplies, services, etc.). Implementing technology today is not expensive because costs of electronic processors have decreased markedly in the past decade. However, unless we change the ratio of salary to equipment in the budgets, technology will continue to be too costly for the typical department. Another reason related to the budget process is the typical ratio of support personnel to faculty. We tend to have about ten faculty members for every support person in most departments. Clearly, this level of support is not adequate for the operation and maintenance of the equipment that we have acquired, which often malfunctions at critical times.

As a consequence of continued heavy reliance on the traditional chalk and blackboard delivery, we have not improved faculty productivity. We continue to support a labor intensive process with little effort being made by either the faculty or the administration to introduce technology that will reduce the cost of delivery and improve the learning process. Much is made in the news about the escalation of health care cost in the past decade. We should recognize that the cost of education has increased at a rate even higher than health care [16].

IV. REVERSE THE PROCESS

We have described an educational system that is functioning poorly in most colleges of engineering in the U.S. We have argued that the system does not properly educate even the successful students, and that it may do serious damage to those students that cannot cope with the process. We have described a system that lacks flexibility in its offerings and a system that has fragmented large blocks of knowledge into narrow courses that are often repetitive. We have also implied that the system is not properly utilizing technology, and its costs have escalated more rapidly than those of our health care system.

The agenda to reform the system is massive, and this paper deals with only one topic of many in that agenda, namely anchored instruction. To begin the discussion of anchored instruction, we need to reverse the educational process. We need to start building the educational wall with blue bricks that represent studies in engineering and as the need arises bring in the yellow, green, and red bricks in a context that is real. In reversing the process, we will provide the first semester freshmen with the opportunity to design. Indeed, they will participate in a product realization process that involves designing, manufacturing, assembling, and testing a product.

To realize a product in a single semester for freshmen is an ambitious challenge, because there is too much work involved in even very simple products for any one student to complete the task in the limited time period. This excessive work load creates an educational opportunity. Teams of five to ten students are formed and each team has the responsibility to develop a prototype of the product. The fact that design teams are almost always employed by industry in product development is well known to the students, and they recognize the importance of learning teamwork skills very early in the design process.

When we discussed the possibility of teaching a freshman design course, many members of the engineering faculty expressed their skepticism. Their concerns are listed and then addressed below:

- *Freshmen know little or nothing.* This statement is a false assessment made by many faculty members. The typical freshman brings a background of eighteen or more years of experience to the classroom. He or she also brings a fresh can-do attitude that is superior in many respects to the jaundiced attitude of some junior and senior students.
- *Freshmen cannot design a real product.* This is also a false assessment. Freshmen are very creative. We have not stifled that creative urge that brought them to the college to study engineering with countless hours of lectures on abstract and often irrelevant material. The difficulty freshmen encounter is in performing a thorough and rigorous analysis of their designs. They realize that their analytical skills are inadequate as they attempt to refine the initial design concepts and as they attempt to defend their ideas in open debate with their team members. They are very receptive to short presentations by the instructor that provide theory that relates to their problem. They learn quickly and appear to retain the knowledge, because it relates to a real problem that affects them directly.
- *Freshmen cannot prepare engineering drawings.* This statement is partly true, since only about one in four or five students had the opportunity to take a graphics course in high school. (Incidentally, the same statement can be made of our seniors, because they have equal difficulty in preparing engineering drawings of adequate quality to use in manufacturing piece parts.) Again, this difficulty presents an opportunity. The students realize they must sketch and sketch well to communicate with their team members. They are interacting with their design team, and they must be able to convey their ideas with sketches. They also recognize the need to prepare detailed drawings that they can use later in the manufacturing phase of the product realization process. They are highly motivated, and they grasp the fundamentals of graphics quickly. They are producing a drawing of their own part, their own creation. They own the design; it is their personal intellectual property. They want to learn the skills necessary to describe their design in an acceptable format so that it can be produced. This experience is much different from the normal graphics assignment where the student prepares a drawing of some object, out of a book which has no significance to them. They will never see the object again after they complete the drawing and they consider it abstract and of little interest.
- *Freshmen do not have manufacturing skills.* This statement is certainly true, but it applies to almost all of our students and not just freshmen. Only those students who have been fortunate enough to have industrial or shop experience acquired off campus have any degree of manufacturing skills. We must recognize this deficiency and keep the products that are to be developed as simple as possible while retaining sufficient complexity to challenge and maintain the interest of the students. Our initial experience in teaching an introductory design course is to consider products that can be constructed primarily from wood and/or plastic. The parts can be fabricated in an inexpensive woodshop using relatively simple tools such as saws, sanders, and drills. It is also possible to consider components fabricated from steel or aluminum, if they can be procured locally at reasonable costs.
- *A student cannot design and build a prototype of a product in a semester course because it is too much work.* True, but again it affords the opportunity to form teams and to have the students work together toward a common goal. Teamwork skills are developed that are as important as some of the analytical methods we pursue with such vigor. There are also social benefits to the

formation of the teams. Bonding takes place that helps many students adjust to the university environment. The teams provide another communication path—student to student—and learning takes place under social conditions that are much more favorable for true understanding than the traditional lecture delivered in a large auditorium.

- *The students cannot work together in teams without conflict.* This statement is mostly false. Our experience with teams is that the students usually work well together. Problems will occur, but the students are usually capable of resolving these problems without outside assistance. Indeed, intervention by the instructor is usually not appreciated and often does more harm than good. When conflicts arise, valuable time is lost in the product development cycle, but the students eventually work out the kinks in their groups and bring every team member to bear on the problem. This experience has significant value in developing people skills which are sometimes missing in some of our current graduates.
- *It is not possible for an instructor to know enough detail about the product development of each team to assign individual grades for each student.* This is often true, but why is giving an individual grade so important? The product realization process is a team effort, and as such it is more appropriate that all team members receive the same reward for their performance. A few students complain about not controlling their grade, but they are the exception. If the instructor has convinced the class that team effort is important and that industry uses teams to develop products, then the students readily accept their roles as team members. If for some reason individual grades are critical, it is possible to assess the contributions of each student by having the team prepare a written description of the role and the relative value of the accomplishments of each member. If the students all sign this assessment, they indicate their concurrence with the relative ranking of their contributions.

A freshman design course offers opportunities to present material in six different areas. The first and probably the most important is the product realization process. Some of our most successful companies have developed a well-defined product realization process [6]. This is a systematic approach to developing a product that begins with the customer and leads initially to a performance specification. We modify this initial phase of the product realization process by identifying the product for the class and then by developing the performance specification through classroom discussion. The concept here is to have each team in the class assume ownership of the product by having them write the product specification. We make the point that new inventions cannot be a part of the process because new inventions bring risk and uncertainty to the product realization process. This point is much more difficult for faculty to accept than for students, because many faculty members spend most of their productive time performing high risk research that will probably never be employed directly in a product or a process. We also stress the importance of developing many design concepts and then selecting the most suitable concept based on trade-off studies.

Graphics, always an important element in the freshman program, retains its importance when design is introduced in the first course. One cannot design without illustrating a concept or a piece part. This fact becomes apparent to the students very early in the design process, and they ask for instruction. How do you sketch? How do you prepare a drawing that is suitable for the shop? How do you dimension the drawing? We respond to these questions with short lecture-like presentations that cover only the most simple examples. We then introduce the students to a computer-aided drawing program (CAD) that is as user-friendly as possible [17]. The students are formally instructed on the use of the CAD package in a single two-hour, hands-on presentation. Following the formal presentation, they study and learn to use the CAD program efficiently on their own time. (A skills exam conducted about a month after this single instruction period showed excellent results.)

A goal of the introductory design course is to cover design for manufacture. This goal cannot be completely achieved because the typical student does not have sufficient knowledge about manufacturing tools or processes to adequately perform concurrent design. However, the opportunity is presented to at least introduce the concept. In reviewing the drawing package with the team early in the design process, the instructor can begin to ask questions about the methods used in fabricating the part and to question the wisdom of designing parts with complex shapes. When the student actually begins to manufacture these parts within a few weeks, these questions become paramount, and the student realizes that design and manufacture are closely coupled. Drawing changes are often made at this stage of the process to simplify the fabrication and assembly processes. While we have not been able to effectively teach concurrent design, we have been able to instill the awareness of the concept as the students' design and then build their first prototype.

In addition to the CAD program, there are a number of important application software programs that most practicing engineers employ, including word processing, spreadsheets, math solvers, plotting programs, etc. We provide computer facilities for the students to use, and require that all assignments be prepared using a word processor. Instruction in word processing was not required because we found that nearly all of the students had acquired this skill in high school. We also provide them with a single two-hour period of hands-on instruction on the use of a spreadsheet. We show the simple computational capabilities and the graphical output that can be generated with the spreadsheet. A few homework problems are assigned, related to the analysis of the design of the product, that requires the use of the spreadsheet. By the end of the semester, the students are all familiar with the PC laboratory, and have achieved at least entry-level skills in the use of a CAD package, a spreadsheet, and a word processing program.

As the design of the product proceeds, the students ask questions related to the anticipated performance. Will it be stable? Will it be strong enough? What is the effect of an impact load? The need to design and to develop something that is real is driving the need for analysis. We respond and begin to present short lecture segments aimed directly at their questions about their design. We do not follow the typical procedure of deriving every equation from first principles. The equations that provide results are presented and used in examples that are related to the design. The solutions to the examples are interpreted to establish the engineering significance of the numerical results. We refrain from extending the theory to situations not directly related to the design analysis that is required to support the product development. During the development of a typical product, the opportunity arises to present material covering from six to ten different basic concepts. For example, in the design of a very simple structure the concepts covered would include: the free body, equilibrium, stresses in axial members, strength, safety factors and failure by rupture.

Finally, the introductory course offers the opportunity to develop personal skills valuable in working together in team arrangements. As this topic will be covered in a separate section later it is only mentioned here as part of the list of opportunities.

There are two primary advantages of offering an introduction to engineering design in the freshman year. The product development process is real, and it motivates the students to accept many concepts and to quickly learn graphics. They appear to understand and retain this knowledge. Finally, the experience is very favorable. Students gain confidence with successful completion of the prototypes and they achieve good grades and the retention rate is high (97 percent).

V. BUILDING THE PRODUCT

There is a great tendency to engage students in design and to stop before the prototype is built. The reasons for not building the product are many and compelling: We do not have the space. We do not have the equipment. We do not have the support staff. It takes too much time and it is dangerous. Some one could get hurt and we will get sued. Some of the faculty cannot operate the required machines or do not have the necessary skills with the machine tools.

It is easy to yield to these pressures and to stop before the students undergo the experience of building the prototype. However, actually building the product is vitally important. The building process provides valuable opportunities for the students to learn new material and to appreciate more fully some of the design concepts that were introduced earlier in the course. For example, they begin to understand the importance of concurrent design and integrated manufacturing even though they are manufacturing simple piece parts on hobby shop equipment. They develop an appreciation of the drawing package and the fact that the drawings are instructions to the person in the shop. When pieces fabricated to the drawing do not fit together, they learn firsthand about tolerances. They learn how their design influences the assembly process. After procuring all of the parts necessary for their build, they understand the importance of minimizing part numbers. When parts do not fit, they understand the impact of rework on the scheduling process. When changes are made in large numbers they appreciate the advantages of CAD in the drawing revision process. When the prototype fails in the first trials, they understand the importance of pretesting subsystems. All of these important learning opportunities are lost if work stops with the drawing package and no product is actually built.

In addition to the significant number of learning opportunities, there is an emotional benefit to the building process. Building is a hands-on activity in a year filled with frustration and abstraction. The students sense ownership as the part on their drawing becomes a real object in their hands. Many students (both men and women) truly enjoy their first opportunity to use a power tool even as simple as a drill motor. The fact that the building process is fun and excites the class should be appreciated by the faculty. Why should learning be considered a boring and painful process?

The costs to build the product can be kept reasonably low for both the student and the institution. We have selected several products that are challenging to the students, yet simple enough to be constructed from materials and components costing about $150 to $250. With teams of five to ten students the cost of the materials per student is less than the cost of a typical textbook in a freshman course or less than the cost of an hour of classroom instruction per student. The costs involved in building the prototype afford yet another learning opportunity. The students begin to understand for the first time the cost-performance tradeoffs of engineering materials and components. If they need to procure a bearing, they quickly recognize the significant difference between the cost of a simple sleeve bearing that exhibits high friction and a roller bearing with much lower friction but at a much higher cost.

The cost to the university is more difficult to assess. If space is available, a good shop suitable for working in wood, plastic, and sheet metal can be developed for less than $10,000. Considering the fact that this shop can accommodate 600 students per year for several years before the tools need to be replaced indicates that costs for the shop are negligible. The faculty and TA costs are more difficult to assess. If an institution handles their first freshman course in a lecture, recitation arrangement, offering the traditional blend of graphics, programming, and orientation, staffing costs are probably lower by a factor of about two. However, about a third of the students in this traditional offering perform unsatisfactorily and either leave the engineering program or repeat the course. If the true costs of these failures are considered, the lower-cost option is probably the one in which students have a better opportunity to succeed by designing and building a real product; the loss of students in this course is less than 5 percent.

VI. TEAM WORK

Product realization in the industrial setting involves teams, often large and interdisciplinary, that are established to design, build, and market the product. For a new model automobile this process involves a team that resembles a small army, a time involvement of about four years, and an investment in excess of $4 billion. Other products require smaller numbers of team members, and less time and money. However, the point is made that teams are established in essentially all cases to design and develop new products and to maintain the production of established product lines. In spite of the importance of the team concept and the ability of employees to work effectively as team members in industry, the educational process is carefully tuned to reward individual efforts. Our methods of lecturing, regardless of class size, and evaluating the students response to exam questions rewards individual accomplishment. We do not have processes in place in any course to develop team work, and we usually measure only individual performance. Indeed, even on homework assignments we seek to discourage cooperation that might taint the individual record. It is only in team sports that students have an opportunity to learn the importance of the team effort that goes into a winning season.

Individual effort and team effort differ in significant ways. In a team the work load is divided and shared. The task division is complex and involves the individual capabilities of the individual team members. While individual efforts are important in moving the team toward its goals, cooperation and the ability to compromise are equally important. Scheduling and checking are important skills for an individual to learn, but as a team member scheduling and checking are vital. The workload is much too heavy for any individual to carry and even too large for a sub-element of the team to manage. All members in the team must work in unison to achieve their goals in a timely manner.

Since many members are working on different segments of the design, it is important that communication between team members be effective. If the communication fails, the interfaces between design tasks will surely fail. The importance of communication becomes apparent to the students as the design evolves and problems develop. These problems rarely lead to serious difficulties, because the students develop a respect for one another. They appreciate the differences between team members, and recognize each other's strengths and weakness. Those strong in one area may be weak in another. As the semester evolves the students working as a team begin to understand each other and self-imposed work division and task assignment are related to these strengths and weakness. The ability to cooperate is also important, and it is interesting to observe team behavior when dealing with an individualistic team member who insists on having only his or her ideas implemented in the design. The process is slow but it is inevitable; the implemented team will isolate the individual not willing to cooperate and to share.

Leadership is another skill that is developed when the students work in a team situation. We do not assign team leaders when the students are assigned to their respective teams. Instead, we watch and observe the evolution of the leadership process. In many cases a student with a strong personality (usually male) will take over the leadership. Female students rarely compete early in the process for the leadership role. With time, changes occur in leadership. If the individual who initially assumes the leadership role cannot command the respect of his or her team members, the team seeks a new leader and deposes the old leader simply by ignoring his or her directions. A new leader emerges that the team members follow and support. The former leader can continue on the team in self-imposed isolation or can accept the change and join in the collective efforts of the team as an ordinary member.

Formation of teams has an additional and important advantage that we rarely consider in the normal engineering classroom situation. A social structure develops that is beneficial to the students in two different ways. First, the teams as social units provide an opportunity for students to bond and to support each other. Remember, most of these students are first semester, freshmen, far from home for the first time with few developed friendships. The team offers the opportunity to make new friends and to work together on common problems. We have noted that the friendships are enduring and that the students appreciate the support system that the team provides.

The team also provides a communication circle. In the traditional lecture course, the communication is on a single channel and is one way—from the lecturer to the student. If the student's attention wavers during the fifty minute lecture, the communication channel fails and no information is transmitted. In a team of five students there are always two or three communicators with even more channels of communication. The information communicated from one student to another is usually accepted more readily than that presented by the lecturer. If questions arise, the teams seek the opinion of the instructor to clarify the issue. When class is dismissed communication ceases until the next class. When class ends for the team, they often decide to meet at another time and place and communication resumes. Even without a team meeting between classes, communication frequently continues as the students interact in the dorms or over the phone. The teams with students working in small friendly groups provide a cooperative learning environment that is far more conducive to prolonged and effective communication than the traditional lecture format.

VII DEMANDS ON THE SYSTEM

Teaching design in an effective manner during the freshman year places many demands on the system. The student, our first concern, is impacted because this course is extremely demanding of their time. They attend class, meet with their team, prepare homework assignments, attend the computer laboratory, and, in the final weeks, go into a crisis mode to build and test their product before the end of the semester. They rarely complain about the significant time involved because they enjoy developing a product, and they know that they are succeeding. Even when we assign evening classes for instruction in CAD or spreadsheets, students attend cheerfully and perform well. They recognize that design is challenging with more unknowns. They also learn that effective use of the team reduces the time required to complete the prototype. When team members fail to complete an assignment on time, they begin to recognize the true meaning of responsibility and dependability.

Teaching design effectively also places new demands on the faculty. Most important is the changing role that they must play in teaching this type of a design course. They must serve more as a coach than an instructor and should severely limit their lectures. They should appreciate that the students can learn by discovery although it sometimes is a painfully slow process. They should offer suggestions rather than instructions and should always listen and observe. Never should they try to lead a team; the design belongs to the team and not the instructor. The team will vigorously defend the design regardless of its deficiencies; therefore, the team must develop and make its own mistakes. It is remarkable how well students can finish a design and how much they can learn on their own, if they are provided with a challenge and an opportunity.

The time required of the instructor in teaching a freshman design course is comparable to any other traditional course. However, the instructor spends less time in preparation for well-defined formal presentations and more time trying to anticipate problems that may develop and in learning ways to diplomatically intercede to mitigate their impact on a team or on the class. In class there is much less structure, less detail in the presentations, and no derivations. There are many questions and these questions lead the direction of the class. The instructor responds to the questions, presents related (and only related material) begins the interpretation of this material and then brings the entire class into the discussion. Basically, the instructor sets the direction, and the students determine through their questions the details that are to be discussed.

The demands on the institution are modest considering the outcome of the course. The institution will be required to devote more faculty time to the freshman class and that will cost more. However, the retention rate is higher and the number of students repeating the class is much lower, which will clearly save more in total cost to society. Most institutions have the faculty in place to markedly improve the quality of undergraduate education. The dual role of the institution that includes graduate education, which has a very significant research element, and undergraduate education, which has a minor research element, is currently out of balance in many of our leading institutions. Institutions must make a major change in their philosophy and emphasize quality undergraduate education as a prime goal. They must abandon their quest for top ranking in the popular press. These rankings are based largely on research volume and Ph.D. production. While we do not suggest that we return the missions of the institutions to those of the 1950s, we do believe that a better balance can be struck regarding faculty efforts devoted to undergraduate versus graduate education. While research is and should remain an important element for every faculty member, it should not become the consuming element.

Most faculty members truly enjoy teaching, and they appreciate their interactions with undergraduate students. The reason that many faculty members are not more devoted to quality undergraduate education is due to the reward system. There are two aspects to the reward system. Tenure is the first aspect. It is well recognized that one earns tenure based primarily on research accomplishments, which include the number of papers published, the amount of funding on an annual basis, and the number of students graduated with a Ph.D. Of course good teaching evaluations are necessary, but certainly not sufficient to merit tenure.

The second aspect is merit-based salary increases. After tenure, is it possible to strike a better balance between the classroom and the research laboratory? No, because salary increments are largely determined on the research accomplishments listed above. Faculty surveys clearly show this statement reflects the faculty perception of the current situation regarding the position of the administration on rewarding faculty effort.

There have been encouraging signs that new emphasis may be placed on undergraduate education [18]. Presidents and provosts make statements to the effect that faculty must be more concerned with undergraduate education. Occasionally a dean of engineering is supportive, but rarely is a chairperson supportive. The chairperson, a first line administrator with a swollen budget of soft money needed to support graduate TAs and RAs, is in a difficult position. If we emphasize undergraduate education, funding from the research grants decreases and budget problems force a reduction in the size of the graduate program. Since size of the graduate program is often equated to quality of the department, chairpersons are only willing to emphasize undergraduate education to the extent that it does not detract from either the size or the rate of growth of the Ph.D. programs.

The situation will change only when the employment opportunities for our graduates with the B.S. reflect the difference in the institutions with quality undergraduate programs and those with inferior offerings.

VIII. ANCHORED INSTRUCTION IN MECHANICS

Instruction in mechanics has not changed significantly over the past thirty years since vector analysis (including the dot and cross product) was added to the course content in statics and dynamics. The most significant difference between the new textbooks published in recent years covering mechanics is in the number of colors employed in the production. The content is nearly the same, although there is some difference in the difficulty of the exercises provided for the homework assignments.

These mechanics courses are very important to the curriculum and to the development of the student. The mechanics courses provide the first opportunities for the transition from a science based delivery to an engineering approach. In the physics course students are exposed to the science presentation where the knowledge of Newton's laws is emphasized. In the mechanics courses, students learn how to apply Newton's laws to structures so that they will be able to understand the principles related to the design of real structures. It is the first from abstract science and mathematics to real objects like trusses, beams, and columns in which students can recognize and begin to understand their engineering significance. The mechanics courses are extremely important in establishing the foundation for most of the upper-level courses in many of the engineering disciplines. The courses are packed with vital concepts that must be clearly understood by the student as soon as possible in their development. The fact that there is little difference in the content of the textbooks is not an accident. There is a very strong consensus about the importance and the significance of every topic included in the coverage on the part of the faculty teaching mechanics. The course content is based on developments that are many centuries old. For example, Archimedes who (287-212 BC) described the principles of the lever, and Newton, (1642-1727) who wrote the laws of motion, was a relative new comer.

There is no argument that we know what the content of mechanics courses should be and that the material that is currently offered is fundamental and important. The problem is with our ability to teach this material to the students. Many students have difficulty in learning the material in the first engineering mechanics course (statics), and they have difficulty in retaining this material as they move to the second course (mechanics of materials). Finally, they are at a loss to use engineering mechanics effectively in analysis of products that are designed in the senior level courses. The magnitude of the problem is quite serious. We typically experience 30 to 35 percent losses (D, F or W) in teaching statics. While most faculty members and even some students consider statics an easy course, about half of our students have serious difficulty in mastering the material and the problem-solving approaches that are introduced. The problem is even more serious than the numbers cited above indicate, because it continues to affect the educational process. Many students who learned enough material to pass statics do not retain that knowledge when undertaking subsequent courses in either dynamics or mechanics of materials. If a statics exam is given on the first day of class in one of these upper level mechanics courses, the results are predictable—over half of the students will fail it. They have not retained the material even when the problems are posed in the traditional manner. A significant amount of time is spent in reviewing statics before new material can be introduced in the upper-level courses.

The "mechanics" problem persists into the senior year in teaching design. In this learning situation the problem is not cast in the traditional format. A drawing of the object with a clearly worded problem statement is not available to the student. It is expected that the student will recognize the problem from a much broader design effort and will define the problem statement in such a way as to provide results that give insight in making decisions regarding the design of some object, system, or process. Most students cannot perform design analysis without significant help. They have difficulty moving from the vague concepts of a design to a model that can be used as a basis for beginning the formulation of a problem statement. Their experience is based on solving clearly stated problems so they are often at a loss when they have to write their own problem statement. When they finally manage a solution, they often fail to relate the solution to the design in such a way as to verify the adequacy of the design or to suggest methods for its improvement. Again, this deficiency should not be a surprise, because very little class time is spent on developing skills in interpreting solutions. Instead, we are too involved in developing students' ability to obtain accurate and unique solutions to precisely stated problems.

We need more educational experiments and different approaches in order to develop new methods of teaching mechanics that are more effective than the current standardized methods. Since little educational research has been accomplished in mechanics, this presentation on new methods based on anchored instruction must be somewhat speculative.

Today, mechanics is offered in a fragmented sequence of course offerings. For example, mechanical engineering students usually take separate courses in statics; mechanics of materials; dynamics; mechanics of deformable solids; dynamics of machinery; and structure and properties of engineering materials. These six courses require eighteen credit hours and are taught over a four-year period with statics in the freshman year and the materials course in the senior year (at the University of Maryland). The students study these segments of knowledge as separate entities. More often than not, the students never put the pieces together and fail to see the continuity and the interrelationships that exist when mechanics and materials is viewed in its entirety.

A similar fragmentation occurs on the thermal science side of mechanical engineering. Courses in fluid mechanics, thermodynamics I and II, heat transfer, and energy conversion are separate. The student often views each course as a separate body of knowledge instead of as a total body of information that is closely coupled.

It appears reasonable to consider some amount of integration of mechanics and materials in the curriculum. However, the thought of integrating all six of the solid mechanics and materials courses into a single course is overwhelming. The amount of material would choke both the faculty and the students. It is possible to consider dividing the material into three areas and integrate it with some of the design offerings normally delayed until the senior year. Several different combinations are possible; it would be helpful to conduct pilot programs to gain experience and to determine student reactions, performance, and reaction to different approaches. We suggest one possible arrangement of content that would include:

1. Mechanics and Materials (six credit hours).
   Include a non-vector approach to statics, mechanics of materials, and properties of engineering materials.
2. Dynamics and Vibrations (six credit hours).
   Include vector statics, dynamics, vibrations, and dynamics of machinery.
3. Design and Analysis of Machines (six credit hours).
   Include advanced mechanics, finite element methods, engineering materials, and machine design.

These three courses would reduce the degree of fragmentation and provide students with the opportunity to learn mechanics, materials, and design as a technical continuum.

In recasting these courses and establishing in detail the course content, it is vitally important to recognize the impact of the computer software available and its effect on the content of the courses. Spreadsheets and math packages with several hundred functions, and user-friendly finite-element programs, all with powerful graphics output, must change our course objectives. First, the availability of these programs dictates that the obsolete methods of how to calculate this or that must be stripped from the new content. Teaching *what* and *why* we perform an analysis and demonstrating methods used in interpreting results is much more important than the old "how to do it" approach. The graphical capabilities must be exploited in the new content, since we can view families of solutions rather than a single numerical value for the solution. The computational abilities give us more latitude in the selection of the problems that we can employ and the types of answers that we seek. Open-ended problems become feasible and some aspects of design optimization can be considered very early in the educational process.

While integration of mechanics and materials should aid the student in the initial process of learning, in retaining the material, and in methods of problem solving, this form of integration does not address the problems students encounter in performing analysis to support a design effort. To enhance student performance in this area will require a much closer coupling of the theory covered in mechanics of materials with significant design experiences for the students. We have done almost nothing to integrate design with engineering mechanics.

In most mechanics courses a recitation period is used to supplement the normal lecture. The recitation is devoted for the most part to problem solving, often with a teaching assistant in charge of the class. In the new context of an integration of mechanics and materials as described above, we are suggesting that the recitation periods be replaced with a studio period. We use the term *studio* in a broad sense in that we would intend that the setting and the time be used for several different activities. The studio could be used for the same function as the traditional recitation—for illustrating solution methods. The studio could be used also as a laboratory where experiments are performed to prove or to illustrate the concepts covered in a recent lecture. On other occasions demonstrations could be conducted to illustrate key concepts. Finally, the studio could be used to accommodate students working in teams to design a structure, a component, a subsystem or a process. The idea here is to change the traditional lecture-recitation period that involves one way communication between the instructor or the TA and the students.

The studio changes the communication pattern. It affords the opportunity for communication between the students and it affords the opportunity to establish groups of students who can work together as teams. In turn, the teams afford the opportunity for the students to bond together into social groups where learning is encouraged. We have found that often the bonding is permanent and that the students continue their team meetings after the regularly scheduled studio hours.

The studio affords the opportunity to pursue learning in many different ways. The traditional blackboard and chalk is a great method of teaching but it is clear that other techniques are needed to teach mechanics. We need better methods to stimulate the interest of the students and to help make the connection between the abstract theories and the design of a real structure.

Examples of the activities to be pursued during the studio are problem solving, experimentation, demonstration, designing, building, and testing. We will describe a few simple ideas in the following paragraphs that are based on forming students into small groups of three to five students to pursue learning opportunities in a student-dominated studio environment. The instructor and the TA serves as coaches and give general directions, but they usually do not lead the activities.

Problem-solving skills are currently taught in recitation by solving one carefully worded problem after another. In all cases these problems have a unique answer that is rarely ever interpreted either by the student or the instructor. This approach teaches the student to select the correct formula and to substitute the correct quantities into the formula and to manipulate the mathematics without making a mistake. The process does not help much in aiding the student in designing even a very simple structure. Some of these problem-solving skills should be taught, but not all of our activity should involve these artificial problems. We must consider more realistic problems that are not completely specified for the student. The open-ended problem permits many solutions that are correct. The key is to interpret the many correct solutions to find the best one.

Consider giving a team of students the following open-ended statics problem. Design and analyze a Warren truss that is to span 100 m and support a load of 65 kN rolling across its span. The members used in the truss will support forces of 50 kN in either compression or tension. Clearly, the teams will generate many solutions to this type of problem. It is the interpretation of the solution that leads to the design process. What criterion do we use in interpreting our solutions in judging the merits of the design? Is the height of the truss important, or is the weight of the members used in its construction a major factor? The freshman year is not too soon to be introducing the fact that weight, cost, and size are important design criteria in engineering.

We lecture endlessly and rarely conduct experiments where measurements of engineering quantities like force, displacement, velocity, momentum, etc. are made. In those few cases where we do have a laboratory that supports a course, we often uncouple the two and permit the students to take them independently of one another. This practice may aid in the registration process, but it is not wise from an educational viewpoint. We lecture about physical quantities that the student should know and understand. If we handed you 100 N would you need one or two hands to hold it? That is a simple experiment but one which permits one to sense the scale of a force.

Several decades ago we embellished the curriculum with many laboratories, and it was common practice for a student to have engineering laboratories scheduled several afternoons each week. For a variety of reasons [19], these laboratories have, with few exceptions, vanished. We are not suggesting that these laboratories all be replaced in the curriculum, but we are indicating that we need to bring laboratory experiences to the student. Moreover, these experiences should be related to the theory being presented and should be presented in a timely manner. The laboratory, as we remember it, is probably not the best way of introducing the realism of the physical quantities. Our laboratory experiments tended with time to become canned, and the students clearly resented the imposition of writing a laboratory report covering each experiment. The students need to participate in experiments that are fun. They need to understand that measurements help them appreciate the size and/or scale of the physical quantities.

The studios provide a better structure in many instances than the laboratories. We need the opportunity to conduct simple experiments. Students need to measure some of the physical quantities rather than thinking that they should calculate everything. To accomplish this we need portable experiments that we can wheel into the studios. The apparatus must be safe and simple to operate. We do not need computer controlled testing machines that require ten hours of instruction before they can be used to conduct a simple tension experiment. Students should conduct the experiments and interpret the test results. We do not need another report to grade and the students certainly do not need to write a report to help them understand the significance of these physical quantities and their scale.

VIII. DISCUSSION

We have examined the educational process employed in the first two years of the engineering curriculum and find that it is seriously flawed. We have identified several problems related to student acceptance and success and to an obsolete curriculum. During the first two years of the engineering program, we give our students very little attention, and, in some instances, even refuse to acknowledge that they are engineering students. We farm our students out to service departments, expose them to a flood of abstract material, and teach them in a hostile environment often with indifferent faculty or poorly trained TAs. The results are predictable—students quit in large numbers. About 30 percent give up in the first year and another 15 to 20 percent leave during the second year.

These are costly failures for the student, the student's family, the educational system, and to society in general. If the student's departure is because of academic failure, it is a very traumatic and unpleasant experience. In some cases the student never completely recovers and the damage imparted is permanent. The potential is lost and is never regained. The educational system has lost a valued customer, and the opportunity to develop an engineer. Furthermore, a career or even worse, a personality, may have been seriously damaged.

The engineering curriculum is very old and significant changes have not been implemented for about forty years. We have fragmented a growing body of knowledge and teach it in narrow and often isolated segments. The students do not retain this material, and we find it necessary to be repetitive. We have not implemented the new technologies in either the renovation of the course content or in the delivery of the course material. While we have examples of innovations in employing new technologies in course offerings, these examples are local and wide spread reform has not occurred. This fact is not surprising because curricula reform in the sciences and mathematics, which has preceded attempted reforms in the engineering curricula has a relatively poor record of achievement. Sheila Tobias [20] indicates that the 300 studies on revitalizing undergraduate science have done little more than add to the growing body of literature. Based on the results of these previous changes in the science and mathematics curricula, it appears that innovations leading to lasting improvements in the educational process for engineers will be slow and difficult.

We have suggested that the process be reversed with design being placed in the curriculum *before* all of the basic science, mathematics and engineering science courses have been completed. Indeed, the most successful example of this reversal is in the offering of an introductory engineering design course to first semester freshman students. Our experience with this course has been extensive and it is now offered to all of the incoming freshman students (about 600 each year) at the University of Maryland. The results have been encouraging. The students give the course excellent ratings in their evaluations. They have fun, build confidence, and learn about the product realization process. More importantly they complete the course with good grades and stay in the engineering program.

Most, but not all, of the faculty enjoy teaching the course. The small class size and the enthusiasm of the freshman is appealing. Developing a free and open dialog with a faculty member serving more as a coach than as an instructor is a challenge. Some faculty enjoy the new role but others cannot cope with this aspect of the course. They prefer the structured class where they have total control over all facets of the educational process. In these instances, we have found that the faculty member completes the semester but does not participate again.

We argue that it is essential that a prototype of the product designed as part of the product realization process be built. While this imposes a significant burden on the facilities and the support required, the learning opportunities afforded by the building process are many and important. In addition, there is an emotional benefit to the building process. Building is a hands-on activity in a year filled with frustration and abstraction. The students sense ownership as the part on their drawing becomes a real object in their hands. Many students (both male and female) truly enjoy their first opportunity to use a power tool, even one as simple as a drill motor. The fact that the building process is fun and excites the class should be appreciated by the faculty. Why should education necessarily be considered a boring and painful process?

Team work is also an important element of the introductory engineering design course. Team leadership, cooperation, and dynamics are taught by requiring the students to work in teams. Using teams has an additional and important advantage that we rarely consider in the normal classroom situations: A social structure develops that is highly beneficial to the students. The teams, as social units, provide an opportunity for students to bond and to support each other. This support goes far beyond the classroom. This support is important since most students are first semester freshmen, far from home for the first time, with few or no friends. The team offers the opportunity to meet new friends and to work together. The friendships are enduring and students appreciate the support system that the team provides.

Offering an introductory design course is demanding for students, faculty, and the institution. The student invests a significant amount of time and energy. The faculty member copes with a new style of teaching and the institution is required to devote more resources to the freshman class. We contend that although the direct costs are higher, the overall costs to society are lower because of the high success ratio and the improved student confidence.

Anchored instruction and design, manufacture, and assembly are closely coupled. However, it is possible to use anchored instruction in many other courses in the engineering curriculum. We indicate supplementing statics with some aspects of anchored instruction. We suggest the use of a studio experience where students have the opportunity to work as teams to perform a variety of tasks including problem solving, simple experiments, and model building. A closer coupling of the abstract theory of mechanics with realism of experiments and models will enhance the learning process.

REFERENCES

1. **Creese, R.C., and Pedersen, E.**, Project based education at Allborg University in *Innovation in Undergraduate Engineering Education II*, Engineering Foundation Conference, Potosi, Missouri, 51-60, 1990.
2. **Fruensgaard, N.O.**, Project work as a study technique at the University of Allborg, World Conference on Education Technology, Cologne, April 16-19, 1984.
3. **Dally, J. W., and Zhang, G. M.**, Experiences in offering a freshman course in engineering, *New Approaches to Undergraduate Engineering Education*, Proceedings of the Engineering Foundation Conference, Banff, Alberta, Canada, 211-226, August 1991.
4. **Dally, J. W.**, and Zhang, G. M., A freshman engineering design course, *Journal of Engineering Education*, 83-91, April, 1993.
5. **Zhang, G. M., and Dally, J. W.**, Teaching design to high school women, engineering education: curriculum innovation and integration, *Proceedings of the Engineering Education Foundation Conference*, Santa Barbara, California, 307-314, January, 1992.
6. **Hoover, C. W., and Jones, J.B.**, *Improving Engineering Design: Design for Competitive Advantage*, National Academy Press, Washington, DC, 1991.
7. **Viets, H.**, Restructuring the undergraduate engineering curriculum, *Proceedings of the NSF Conference on Innovation in Undergraduate Engineering Education*, Mercersburg, Pennsylvania, The Engineering Foundation, August, 1989.
8. **Viets, H.**, A modular approach to engineering design, *Innovation in Undergraduate Engineering Education II*, Engineering Foundation Conference, Potosi, Missouri, 133-138, August 5-10, 1990.
9. **Vagners, J.**, Undergraduate control systems engineering design revisited, *Innovation in Undergraduate Engineering Education II*, Engineering Foundation Conference, Potosi, Missouri, 39-50, August 5-10, 1993.
10. **Sheppard, S. S.**, Mechanical dissection: an experience in how things work, *Proceedings of the Engineering Education Foundation Conference*, Santa Barbara, California, 327-332, January, 1992.
11. **Regan, T.M.**, The extended class room: plant trips as design, *Proceedings of the Engineering Education Foundation Conference*, Santa Barbara, California, 301-306, January, 1992.
12. **Hagler, M., and Marcy, W. M.**, Simulation implementations of peripatetic learning, *New Approaches to Undergraduate Engineering Education*, Proceedings of the Engineering Foundation Conference, Banff, Alberta, Canada, 125-140, August 1991.
13. **Finlayson, B. A.**, Graphic computer aids for chemical engineering, *Innovation in Undergraduate Engineering Education II*, Engineering Foundation Conference, Potosi, Missouri, 105-122, August 5-10, 1993.

14. **Bordogna, J.**, Future of engineering education, *Proceedings Engineering Education: Curriculum Innovation and Integration,* Aung, W., and Carmi, S., Eds., Engineering Foundation Conference, Santa Barbara, CA, 33-38, January 5-10, 1992.
15. **Anderson, C.W., Fine, J.M., Froyd, J.E., McLean, H.L., Maloney, M.J., Mottel, E.A., Rogers, G.J., and Winkle, B.J.**, Engineering education renovation: Do we go it alone?, *Proceedings Engineering Education: Curriculum Innovation and Integration,* Aung, W., and Carmi, S., Eds., Engineering Foundation Conference, Santa Barbara, California, 115-120, January 5-10, 1992.
16. **Anon**, Has ours standard of living slipped, *Consumer Reports*, 392-393, June 1992.
17. **Conner, F.L.**, *The Student Edition of Autosketch-Version 3,* Addison Wesley, Reading, MA, 1991.
18. **Kennedy**, Stanford Address.
19. **Hilborn, H.**, Can we afford our grandfathers' laboratories, *Proceedings Innovation in Undergraduate Education II,* the Engineering Foundation Conference, Potosi, Missouri, 1-6, 1990.
20. **Tobias, S.**, *Revitalizing Undergraduate Science: Why Some things Work and Others Don't,* Research Corporation, Tucson, AZ 1992.

Chapter 6

THE FUTURE OF MATHEMATICS EDUCATION FOR ENGINEERING STUDENTS

Malcolm Getz, Vanderbilt University

Students who want to become engineers, as well as practicing engineers themselves, see a creative tension between the practical and the theoretical, between the scientific method and conventional wisdom, between formal constructs that yield optimal designs, on the one hand, and on the other, the incremental process of trial and error, the commitment to getting a workable result and improving on it. Few would trust an engineer who lacks knowledge of the basic sciences, because such a person will have little scope for challenging standard practice when it needs to be challenged. Likewise, few would trust an engineer who lacks experience with the tools, materials, and people of the trade, because turning a bright idea into a marketable result is a path with many steps.

The same tension that characterizes engineering is mirrored within mathematics. On the one hand, students of mathematics and their teachers are challenged by the elegance of mathematics, by the desire to prove relationships, the details of form and process, the search to make a statement that is unambiguously true. On the other hand, students of mathematics and their teachers gain a sense of mathematical intuition, an appreciation for the shapes and textures of mathematical objects and their uses. Just as one can learn a language by studying its grammar and vocabulary to the point of perfection and find it useless without its cultural context, so a student can study the formal constructs of mathematics and find it pointless unless it is grounded in application. At the same time, a student who drills on applications without grasping the elegance of the underlying structures will find mathematics tiresome.

Computer-based mathematics affords the possibility of assuaging this tension. The computer can demonstrate proofs; indeed, some proofs may be constructed with computer math that are not possible in any other way. In addition, the computer allows users to apply powerful, elegant mathematics to complex, interesting problems. Moreover, the methods may be stored for easy, repeated use. Thus, computer math addresses both points of view.

When computer math allows a user a cross-ruff between the two points of view, the experience may be transformational. Because interesting problems may be addressed relatively easily with computer math, even a beginning student may adopt a discovery mode to work a variety of problems and so trace the underlying mathematical object. By combining the numeric, symbolic, and graphing capabilities of computer math, a student may gain deeper insight into the nature of the underlying mathematical ideas. With such deeper insight and the substantially reduced effort required to apply even quite sophisticated mathematics, mathematicians, engineers, and their students are able to use more mathematics more creatively for more purposes. Computer math has considerable potential.

Of course, not all tools are well designed or well applied, just as not all serrated metal blades are equally good at crosscutting wood nor are all sawyers equally adept with identical saws. Computer math is evolving rapidly, and so its limitations today may be overcome tomorrow. Moreover, as users gain experience, we can expect new approaches to the practice of engineering and to the education of engineers.

Computer math will be used by practicing professionals as part of their trade because the productivity gains from using such tools are substantial. Indeed, the productivity gains from computer math seem likely to be several times greater than the productivity gains from word processing. Equation systems can be manipulated, solved, and graphed in three dimensions with animation, and cross-checked in various ways in a matter of minutes. Calculators, even computers with conventional programming languages, require days or weeks to achieve similar results when working from scratch. The productivity gains from computer math in practice may be a sufficient reason for computer math to be learned in pre-professional education.

However, computer math may have even deeper import for education because it will allow quite different pedagogical strategies.

I. COMPUTER MATH

To set the stage for the discussion of educational issues an overview of the features of software for computer mathematics is in order. From a mathematical point of view, computer math software has several distinguishing features we might refer to as back-end issues. Computer math, sometimes called computer algebra, accepts algebraic expressions as inputs and yields algebraic expressions as outputs. Some algebraic software also functions as a programming language so that one expression can be used as input to others. Computer math includes a set of functions commonly used in mathematics, science, and engineering. Although some functions are typically found in spreadsheets or in conventional programming languages, better computer math software has a more extensive set of functions. Indeed, the quality of computer math software will be judged, in part, by the variety of the functions defined within the software. A popular computer algebra system, *Maple*, is written in a conventional, procedural language, while its main competitor, *Mathematica*, is built from rules and so offers more flexible definition of mathematical ideas. Neither *Maple* nor *Mathematica* provides automatic limitations of domains, a feature found in *Axiom*. The algorithms embedded in computer mathematics software may be judged for their logical soundness, accuracy, and computational efficiency.

On the front end, users of computer math are interested in the interface. First, better software supports a "notebook" interface that allows a user to write text, display graphics, and manipulate equations without knowledge of the underlying programming language. This feature is especially valuable for instruction because it lowers the threshold of knowledge required of first time users. Second, better software also supports three-dimensional color graphics with animation. The ability to display mathematical ideas graphically also has important consequences for teaching, both for instructors presenting demonstrations and for students working problems or developing projects. Third, better software supports a typesetting interface, allowing users to read and write with the same symbol set that mathematicians have developed over the centuries. Seeing Σ instead of seeing "sum" allows easier connection with conventional texts. Computer math differs in its computational environment as well. First, a significant feature is the availability of a programming language because the scope of the software can be extended by the addition of new functions written in the software's own language. The most popular computer mathematics systems grow as their communities of users grow, and the software accrues commands and procedures specific to various disciplines. *MATLAB, Mathematica,* and *Maple* are leading examples. Second, computer math software differs in the variety of computer platforms on which it will run;

better software runs on more platforms. Some software allows processing to be spread among multiple computers, with interfaces on personal computers and the computationally intensive tasks supported on remote, faster, shared machinery. Some software provides linkages between the core computational processing and other, dissimilar software such as spreadsheets. Such software can then be more easily integrated into existing tools. Third, computer math software differs in the demands it makes on hardware. More sophisticated software often occupies more disk storage, requires more memory, and needs faster processors to achieve reasonable performance. As hardware becomes less expensive, we may expect more sophisticated software to become accessible to more people. In the meantime, initial experiences for students are likely to be with less sophisticated and less expensive tools.

Software for computer mathematics is evolving. With better interfaces, it will become accessible to more people. As it becomes the norm in more disciplines, we can expect mathematics to become fully embedded in software, whether it is used by mathematical theorists, practicing engineers, or applied economists. As the cost of hardware and software continues to fall, we may expect it to be used at lower levels in the curriculum, even to a degree in secondary schools. Computer mathematics, then, will become an increasingly essential tool in all intellectual disciplines where mathematics is a language of discourse.

II. TEACHING WITH COMPUTER MATH

Much recent discussion about how best to teach collegiate mathematics relates to a conference on teaching calculus that took place at Tulane in 1986. [Rodi,1991] The resulting calculus reform movement has influenced instruction of calculus on many campuses, and the pedagogical ideas at play in the calculus reform movement also have wide implications for instruction of other subjects in mathematics and beyond. A calculus curriculum developed at Harvard is embodied in a text. [Hughes-Hallett, 1992] Duke faculty developed Project CALC, Calculus As a Laboratory Course.[Smith, 1991] The Harvard and Duke schemes use computers. A very computer-intensive calculus course was developed by Jerry Uhl at Illinois. [Uhl, 1991] Non-computer-based courses that use projects, writing assignments, and groups have come to Grinnell, New Mexico State, and elsewhere. [Alvarez, 1993] Some institutions like Baylor depend on graphing calculators as tools in calculus and beyond. The National Science Foundation has sponsored a number of initiatives.

At root, those seeking to become better teachers of calculus ask fundamental questions both about how students learn and about the kinds of learning that is useful. [Ferrini-Mundy, 1991] Studies of how students learn suggest that students learn better when they are active participants, formulating their own questions, exploring concepts for themselves, and organizing their discoveries around their own preconceptions and experiences. More broadly, students (like all of us) are social creatures. Their behavior, including the amount of time spent studying, is influenced by the expectations of their peers and their personal relationship to their instructors. Good teaching will have several ways of influencing student behavior and thinking and, thereby, affecting outcomes. We know that different students learn differently not simply because some have better backgrounds or more native intelligence, but rather became different people have different learning styles. Different students respond differently to print, image, or speech, and so organize new knowledge in different ways. More deeply, we know that the human brain stores information about verbs in different places than it stores nouns, visual images are stored as shapes, colors, and motion separately, and still other regions of the brain store faces, sounds, and numbers. The brain actively parses experiences for storage and reintegrates the stored matter for use. Good teaching then can provide rich experiences that stimulate several kinds of mental processes in an integrated manner. From these ideas, one can conceive a variety of methods of enhancing teaching.

A. MOTIVATION

One such idea is an old one, namely, that good teachers motivate their students. Motivation has multiple dimensions. On one level, the teacher provides a familiar context for each new idea that is understandable to students. On another level, the instructor shows enthusiasm for each concept and so captures the students' imagination. On yet another level, the instructor establishes a social context for learning and an expectation that each student will learn and find the learning to be meaningful. In these ways, the teacher can induce the student to engage the material at hand. A textbook writer may motivate an idea by giving an example embedded in a familiar story, and then draw the generalization from the example. A teacher may go further by inducing a student to provide a suitable story line for herself.

To put it a little differently, what counts is what the student does. What the teacher does matters only in so far as it has a consequence for students. Motivation, then, is the range of teacher behaviors that has a positive effect on student behavior, including study time, attitude toward learning, speaking, writing, performance on tests, and subsequent use of the ideas.

If one believes that student behavior matters most, then one avenue to influencing student behavior is to induce students to talk. One such approach is the Socratic method, wherein the teacher asks questions and students talk. One might think of measuring the amount of class time during which students talk as a percentage of the total class time. As long as the talk is productive, the higher the percentage of air time provided by students, the better. The teacher will set the agenda, define problems, and most especially, understand exactly what dimensions of a problem trouble students and guide the discussion accordingly. Moreover, students who know they will be expected to talk and that their performance will be judged in a significant way by their talk, will put effort into preparing to talk. Thus, the students will put effort into preparing for class, thinking of issues in advance, and organizing their thought so as to be effective as speakers. They will engage the material actively and learn more. Many of the ideas afoot in the calculus reform movement involve ways of having students do more talking.

A second way to ensure that students are engaging a subject is to induce them to write. An effective writer will have thought about what is to be said, internalized the concepts to the point of organizing them on the writer's own terms, and generally will reveal the writer's own thought process. Even more surely than with speech, the writer will understand a concept on the writer's own terms. Many of the reform efforts in calculus instruction involve students writing about calculus. Typically, the writing assignment derives from a problem or project wherein the student describes each step of the process used to reach a conclusion about the issue at hand.

A third method of inducing students to engage a subject is to involve them in addressing the subject in small groups, typically three to five students. If most of the value of the Socratic method derives from the student's own speech and the class meets as a group of thirty-five people in fifty minute sessions, then the average amount of air time per student is severely constrained. If the teacher consumes fifty percent of the air time, each student will have less than one minute of air time per class. By organizing as small groups, say groups of five, to meet for an hour, each person might average twelve minutes of air time per session.

Moreover, students relate differently to their peers than to their instructor. They may have less inhibition, fear of talking about what they don't understand, and a heightened sense of responsibility. Because the simple questions are often the most helpful, each member of a group can contribute to the group. The effort put to responding to a question can be as helpful in inducing the respondent to organize thoughts as to the questioner in hearing an answer quickly. Most students enjoy working together and so the opportunity to be a part of a group improves how students feel about study.[Watkins, 1989 describes the success of Philip Uri Treisman with groups of black students at Berkeley.]

Many calculus reform initiatives (indeed many efforts to improve collegiate instruction in all disciplines) involve increased use of student teams. The instructor organizes the teams and assigns projects to the teams. Typically, the team submits one product and all members of the team receive the same grade. In some cases, a team may work on a common project but each member submits his or her own write-up.

Team work has risks, however. Some students may shirk their responsibility to the their teammates, and take the average grade while contributing little to the effort. As counterweight, some instructors have each student grade the contribution of each member of the team and use this information in formulating individual grades. Some students with strong personalities may dominate a group even when their own understanding of the concepts is vague or incorrect. As counterweight, some instructors assign each team member to a different role so that dominant personalities may be more easily contained. Many instructors say that learning to work in a team is a valuable skill in its own right, and so accept some of the problems inherent in inviting students to work as teams.

B. LABORATORY

A fourth method for inducing students to engage calculus is to provide laboratory experiences. In calculus, a laboratory means that students meet in small groups at a specific time with the instructor present as advisor but not leader. The students are assigned a problem or project and typically must submit a written report of procedure and result. The laboratory combines the first three methods of engaging students. It provides more student air time, encourages group work, and requires written products. All three of these dimensions of the laboratory tend to engage students with calculus.

One of the ways in which computer math may be used in instruction is as the basis for laboratories. To the extent that computer math laboratories are more successful than laboratories without computer math, we may say that computer math enhances learning in mathematics. It may offer other benefits as well, but its role in the laboratory highlights its appeal.

Computer math brings several advantages to a laboratory. A principal advantage of computer math for a mathematics laboratory is its integration of numbers, equations, and graphs. Because each of the three forms of mathematics can be manipulated in the same medium, indeed because computer math can present a single idea in any of the three forms at the student's command, a student can quickly learn to tackle a given lab problem on any of the three fronts. Moreover, a student can easily learn to cross check an algebraic solution by examining a plot. With computer math, a student can see that a mathematical idea can be expressed as numbers, symbols, and graphs, and that each one of the three has advantages and disadvantages. Without computer math, the interrelationships of numbers, equations, and graphs are more difficult for students to comprehend. By exploring the interactions for themselves in a lab setting, students have a basis for focused conversations with one another and something to write about. Moreover, the integration of numbers, equations, and graphs stimulates more pathways to memory and will appeal to diverse styles of learning, which is a bonus.

A second advantage of computer math in a laboratory setting is that it can allow students to address more meaningful problems. Assume that the amount of time a student has available is limited and that the time to complete a project is divided between time spent formulating a strategy for solution and time spent in executing the strategy. If computation is by hand or with a limited calculator, problems must be kept relatively simple so that some balance is maintained between strategy time and execution time. Because computer math speeds execution, especially for more complex problems, laboratory projects may be much more complex and still allow students to devote a majority of their time to formulating strategies. Indeed, if a laboratory encourages trial and error as a way to improve students skills at formulating strategies, then the execution by computer math is extraordinarily important. By gaining experience with more realistic and therefore more complex problems and by putting relatively more effort into formulating strategies than to execution, students may gain more intuition about problem solving than otherwise. Because intuition derives from deeper insights into mathematical phenomena, such experiences will generally have positive effects on future student use of mathematical thinking.

A third advantage of computer math for a math laboratory is that it can be made to be interactive. That is, with a notebook interface, a student can be given clues and a variety of numbers, graphs, and functions can be readily available with suitable explanations. In this way, a student can move through a multi-step process with computer-based support at each step. A well developed laboratory may need less support from the instructor.

C. CONSTRAINTS

Up to this point in our discussion of the problem of teaching mathematics we have defined as a goal the engaging of students and we have reviewed several strategies that show promise in moving toward that goal. To understand the environment in which mathematics is generally taught, however, we need to understand the constraints under which students, teachers, and their institutions operate. The three major constraints are 1) the student's time, 2) the teacher's time, and 3) the institution's budget.

The student's time is limited. Full-time students have several courses to attend to, and have varied interests, and so they typically commit only so much time to their studies. Let's suppose a student spends forty to fifty hours per week for five courses, so typically students commit eight to ten hours per week to a given course including time spent in class. Although a few students may have their own intellectual agendas, most focus on achieving the best grades they can for the effort expended. They will get the best information they can about what will count in their grade and optimize their time commitment so as to get the best grade. If grades are based exclusively on tests, then students will want to drill on practice test questions. If the test questions are the same as in conventional calculus courses, for example, "evaluate the following six integrals," then students will understandably allocate their time to drill and practice on such questions much as they always have, regardless of teams, labs, or the availability of computer math. [Gale, 1993] If talking, writing, teams, and labs are to get student attention, then student performance in these activities must weigh in the course grade. More broadly, if deeper conceptual understanding is the goal, then teachers must find appropriate ways to test whether students have gained deeper conceptual understanding.

Computer math may help. If students are allowed to use computer math during tests, either take home or in class, then the tests may include richer problems just as open book tests generally do. Because basic computational skill is also valued, some reformed calculus courses include mastery tests on such skills. For example, a perfect score on a test on computing derivatives may be required to pass the course. The letter grade may, then, be based on more conceptual tests and written work, without sacrificing assurance that students can compute.

Use of computer math involves its own time cost for students. Although computer math software continues to improve in ways that reduce the effort required for first successful use, still all computer math requires learning syntax and learning to work on a computer. Some instructors who are committed to computer math devote two weeks or so at the beginning of their courses to having the students learn to do arithmetic, algebra, and plotting in a computer math system before turning to new ideas in mathematics. These instructors contend that the gains in learning later in the course more than make up for the two weeks devoted to learning computer math. Of course, the amount of time needed to get to first base with computer math will depend on the students' prior experience with computers and their willingness to learn some new computer skills. For computer- and math-phobic students, this threshold may be high. Better students may use their time more efficiently and so find the extra time demands of using computer math more acceptable.

The second constraint is the teacher's time. If the average section of a calculus course has thirty-five students and meets 150 minutes per week for fourteen weeks, the time an instructor will devote to each student in class is limited. Instructors typically teach three or more courses at a time and have other responsibilities and interests. The amount of time they are willing to devote to class preparation, to extra class counseling, and to grading is limited. At this point, most of the calculus reform movement is concentrated among enthusiasts who commit extraordinary amounts of time to experimenting with new pedagogical methods. Whether the reform ideas will become routine will depend in part on the time commitments required of faculty on average. To the extent that computer math is a valuable professional tool for practicing mathematicians, we may believe that the time and cost of learning computer math will have its own rewards for faculty. However, if the use of computer math for instruction or the other methods of engaging students mentioned above come at the expense of greater time commitment by faculty, then the extent of use will be limited.

The financial constraint limits class sizes. Expenditures per student are limited by the revenues that come from government and philanthropy and from charges to students. The financial constraint limits the size of the faculty and so the student-faculty ratio. If faculty's teaching loads and the students' course loads are given, then average class size is constrained as well [Getz, 1991]; which is to say, average class size is linked to expenditure per student. If successful use of computer math depends on smaller class sizes, the cost of higher education on a per-student basis will likely be higher as a consequence.

An important issue for the widespread deployment of computer math in collegiate mathematics instruction is its effect on class size and thus the faculty-student ratio and cost per student. If computer math yields dramatically improved results with smaller classes, then such higher education will be more expensive. Unless other costs can be trimmed and state legislatures are convinced of the payoff of the more expensive education, students will likely be expected to pay more for the higher quality educational service. Given that the cost per student of providing higher education is generally rising faster than the consumer price index, the gains from smaller class size and use of computer math would have to be quite substantial to be funded on a permanent and universal basis.

There is a trend toward increased use of part-time faculty because they are less expensive and, generally, can be hired or fired more quickly in light of changing demands. If use of computer math requires training of faculty and investment in learning new teaching strategies, then use may be more difficult with part-time faculty.

On the other hand, should instruction that makes extensive use of computer math be as effective with larger class sizes as with smaller ones, then use of the computer math based instruction would tend to lower costs. At this point, experimentation with computer math in instruction is too scant to provide clear evidence about implications for class sizes.

Generally, if an instructor is to be responsible for more students without loss of effect, and the computer is to carry more of the burden, then more than computer math will be required. One might imagine computer math embedded in interactive tutorial materials that anticipates a variety of student difficulties and provides diagnostics and review materials for students needing more help. Such an intensive computer-based curriculum might be called Computer Aided Instruction. Laying computer-based tutorials and problems on top of computer math software may create text-like electronic products that students can use to very good effect in supervised lab settings such that a faculty member can be responsible for substantially more students. The development, testing, and marketing of such computer-aided instruction in mathematics seems likely to require a significant investment. At this point, the concept is not proven.

The computer tools themselves entail an expense that might be borne by the institution or by the individual. A few institutions now expect nearly every student to own a personal computer. A number of institutions find that more than half of their student bodies own computers. Many students say they would be at a competitive disadvantage if they did not own a computer. Today, the primary purpose students find for their computers is as an aid to writing. Given that the productivity advantage to use of computer math is proportionately even higher than word processing is for writing, we expect more students with computers to invest in computer math tools as well.

Although computer math software today is relatively expensive and the hardware requirements in processor speed, memory, and disk storage are above the level of equipment most students buy, the cost of both hardware and software seem likely to continue their downward course so that they may be well affordable by students later in this decade. The requirements for instructional computing for the majority of students in mathematics are then likely to be met by student owned machines. Students are likely to encounter productivity gains with other software appearing in other disciplines so that the total productivity gains from using personal computers will be substantial. Only a fraction of the computers' cost then need be attributed to its value in mathematics instruction.

Until the costs of full function computer math software and the hardware necessary to run it fall to a level students find affordable, institutions will likely choose less capable devices. Graphing calculators are enjoying some success now. Some computer software that supports some algebra but without programming languages or support for a full range of functions may be sufficient, particularly for large numbers of beginning students. In more advanced courses, faculty will likely choose more advanced software. As the cost of the best tools fall, however, we might expect a migration of more sophisticated tools into lower-level courses.

Institutionally owned equipment generally entails more expense because of the higher level of wear and tear on publicly used equipment, because the institution must dedicate space to labs, and because staffing and other measures are required to maintain security. Institutionally owned systems will likely be used for more sophisticated work than can be conveniently performed on systems owned by students. Of course, institutions who serve part-time and less affluent students may wish to provide institutionally owned machines for student use so as to allow instruction using computer math because student ownership is less likely.

Just as students are likely to own computers, so are their computers likely to be connected to a network that is used for a variety of purposes. Among the purposes may be collaboration in mathematics instruction. Indeed, as networks allow ready sharing of documents, group work may be easier. In particular, written work can be reviewed by peers. The network may also provide access to higher speed processing and more sophisticated software for particular activities. When the network is used for a variety of purposes, only a fraction of its cost need be attributed to instruction in mathematics.

D. EVALUATION

In making decisions about how to teach mathematics, math faculties and their deans will make judgments about how to do the most good within the constraints of time and finances. In making such judgments, they will be interested in evaluations of the performance of various strategies. What is the evidence on the question of whether students learn more when using computer math? What is the evidence of the use of computer math with the assignment of written projects, teams, and laboratories? What is the evidence on the effectiveness of computer math with classes of differing sizes?

To answer these questions in a formal way, one would like to have controlled experiments wherein students are randomly assigned to multiple treatments groups and suitable control groups. Pre- and post-tests of skills in mathematics would be needed in which other attributes of the students are measured, including abilities like spatial reasoning and quantitative aptitude, personality attributes and study styles, time devoted to study, prior experience with computers and keyboards, and more. One might also track attributes of teachers, their level of experience, skills with computer math, education, teaching styles, and personality factors so as to discover what kinds of teacher attributes are most successful with computer-math based courses with particular kinds of students. One would then hope to infer what kinds of pedagogical strategies used by what kinds of teachers yield the biggest gains in achievement for specific kinds of students.

The research task is compounded in several ways. Alternative pedagogical strategies may aim at different goals. For example, one group might place an emphasis on conceptual understanding as embodied in written lab exercises and projects as opposed to a conventional course's emphasis on computational skills. Performance testing might then measure computational tasks across groups. Measurement of conceptual skills should be measured as well. That is to say, the performance measurement should measure a variety of outcomes. Designers of educational programs can then hope to keep separate the issue of what goals to pursue from the issue of what pedagogical strategies are best used to achieve particular goals.

The possibility of such research is further compounded by the fact that students will typically object to being randomly assigned to different teaching regimens. If one regimen is computer-based and emphasizes conceptual skills while another uses calculators and emphasizes computation, students will want to choose their regimen for themselves. They will sort themselves out in distinctly non-random ways that make inference about the consequences of the difference in pedagogy difficult. Without random assignment to control groups, it will be nearly impossible to distinguish the influence of difference in pedagogy from differences in which students select each regimen.

Two studies compare the performance of students in college calculus where computer math was available to students in a control group where computer math was not available. Robert Cunningham conducted such a study for his thesis at Temple in 1991. [Cunningham, 1991] He taught two sections of calculus at a college in New Jersey. One group was allowed to use Kemeny's calculus from True Basic, Inc. The control group saw computer demonstrations but did not have access to the software themselves. The group using the software out performed the control group on the manipulative or computational test. The group using the computer software did somewhat better on the conceptual test. The two groups had the same teacher, used the same text, and the courses were otherwise similar. The experiment reveals little about how pedagogical strategies might change to make better use of computer math and does not address the possible roles of different teacher and student attributes.

A study by Jeanette R. Palmiter involved engineering students. [Palmiter, 1991] Students who volunteered to participate in an experiment were randomly assigned to experimental and control groups. The control group was taught the material in four chapters of a text in a conventional way for ten weeks. The experimental group covered the same material in five weeks, except that the students had access to MACSYMA (computer math software) on a DEC-20 to compute integrals instead of being taught techniques for integration. Identical computational and conceptual exams were administered at the end of the ten and five week periods, respectively. The group using MACSYMA used the software during the examination. The group with computer math achieved higher mean scores on both the computational and the conceptual parts of the examination. The experimental group also got somewhat higher grades in subsequent calculus courses. The higher performance might reflect a Hawthorne effect, the effect of getting attention in an experiment. Also, the difference between the five and ten week sequences may have had an effect. The study does not seem to have included a pre-test, so one cannot be sure of the skill levels of students entering the two groups. They may differ even though they were randomly assigned.

An appendix to this essay by Bruce Black reports on the use of computer algebra at Rose-Hulman Institute. He comments on an analysis of the grades in subsequent courses of students who used Maple in calculus and differential equations compared to the grades of similar students who took a conventional sequence in mathematics. [Rogers, 1993] The experience at Rose-Hulman at least suggests that students using computer math in their early years are not at a disadvantage in later years in upper level courses that use conventional mathematics.

In summary, a small number of studies tend to suggest that computer math may have a beneficial role in student learning. However, an investigation of sufficient breadth to control for all the possible sources of variation would seem to be an expensive undertaking. Moreover, the nature of computer math continues to change as computers become more powerful and capable of better graphics and as the software becomes more sophisticated. Thus, experiments designed to demonstrate the gains in student performance from pedagogies using computer math are likely to become out of date relatively quickly.

III. COMPUTERS AND THE CURRICULUM

Let's suppose that the case for computer math in teaching is convincing. How, then, should it enter the curriculum?

As noted above, a principal focus has been on the calculus course. Large numbers of students take calculus and the success rate is low. Improvement in instruction in calculus would have wide effect. Yet, calculus students are often first-year students adjusting to college, uncertain of their intellectual ability and engaging as much in learning how to study as they are in learning specific skills in mathematics. Reforming calculus may therefore be the most daunting challenge.

As a tactical matter, one might want to begin computer math in upper-level courses so that faculty get experience with the tools with students who may be better able to cope with a new teaching method. In addition, as upper-level students gain positive experience, they may be able to be of counsel to beginning students. Indeed, where graduate students play a teaching role, introducing graduate students to the tools first may make the most sense. One might then imagine use of computer math moving backward in the curriculum ending with the biggest challenge of all, calculus. Of course, once computer math is used in calculus, teachers in more advanced courses can use it more easily because they can assume students know the rudiments of computer math.

The curriculum designer faces a decision about whether to have courses devoted to computer math itself or to rely on integrating computer math into the conventional course sequence. A one- or two-credit course in computer math might be focused on creative use of the computer math and its facilities for manipulating numbers, solving equations, and plotting and animating diagrams. If an instructor with high enthusiasm for computer math had success in teaching such a course, the burden on other faculty to use computer math in their courses might be relieved.

Although pioneering faculty may teach themselves the computer math and develop their own pedagogical techniques for using it, most teachers will need support. Moreover, the use of computer math and the pedagogical strategies it makes possible may be so sufficiently different from conventional approaches, that departments of mathematics may need to organize formal training sessions for their teachers. This may be especially important for adjunct faculty, teaching assistants, and newcomers. Training will be as important for laboratories, assigning and grading written projects, and for team work, as for using computer math. Training will likely be a continuing necessity.

Training costs will be lower when published materials support computer math and other new pedagogical techniques more effectively. The time required of a teacher to adopt a particular pedagogical approach is much reduced if suitable reading, exercises, problems, and projects may be used off the shelf. As new pedagogy succeeds, the market for published tools that support the pedagogy will grow.

Authors of textbooks and associated electronic materials want to meet a market as well. Such materials will sell better the easier they are to use and the more successful the instructors who use them find themselves to be. As pedagogical techniques using computer math succeed, we may expect them to be distributed in published printed and electronic materials.

A critical issue for the author and publisher of electronic materials is what particular computer math software to use. The choice of a computer math program will be difficult not only because of the diversity of higher education, but also because of the rapidity with which the computing world changes. Even if one chooses a brand, one may expect new versions to appear every year or two that may make obsolete (to some degree) materials focused on the older version of the software. If text materials take two years to develop and bring to market and new versions of the software that substantially makes old versions obsolete appear every two years, the text material will be perpetually out of date. The fall back position for authors and publishers is to support only electronic products with very wide markets running on common denominator computer systems. We might expect a delay in published products to support pedagogy using computer math until computer math because a less volatile, more standard software market.

The issue of what software students will be using has some connection to the hardware platform they will use. Microcomputers are a likely first choice because they are relatively low in cost and can be used for a variety of purposes. However, better computer math requires larger, faster systems than most students purchase. This fact raises several possibilities. First, as mentioned above, the curriculum designer might choose to use less capable, less expensive software that runs on less expensive hardware for the lower level courses where the numbers of students are large. In Vanderbilt's experience, however, it is relatively easy for students with modest sophistication to encounter problems that are computationally intensive, say in physics or biology. The value of more sophisticated software may not be closely correlated with the sophistication of the student. Second, one might consider using servers on the network when high-speed computation is needed and allowing students to call on the faster server when faster computation is required. If use of faster computation means using different software, use of a network server may be relatively inconvenient. Third, one might think of providing specialized computational labs for mathematics instruction. Provision of such facilities, however, is substantially more expensive than depending primarily on student-owned machines. Fourth, one may suppose that the continued rapid decline in the cost of computers and software will allow today's more sophisticated computer math to become available at prices students can afford that operates on computers students ordinarily buy within a few years.

Designing a mathematics curriculum to take advantage of computer math will continue to be a challenge. A number of processes are in motion and so the optimum curricular strategy will also likely continue to evolve.

IV. COMPUTER MATH IN ENGINEERING

For engineering education, an important issue will be how engineering faculty come to use computer mathematics in their professional work. The dominant software in engineering now are packages written in conventional programming languages that are highly optimized for a particular problem—for example, the simulation of integrated circuits or the optimization of a combustion chamber. It seems possible, however, that in time, some of these tools will be recast into the environment of computer mathematics because of the greater flexibility and extensibility such an environment provides. The tables of values that have played a role in practical engineering will give way to functions in computer mathematics. Computer simulations will likely be constructed from objects defined in computer math. The ease of migration from equation, to numbers, to graphics will become more important to practicing engineers and so seems likely to become central in engineering education. The changes occurring in mathematics education may then be mirrored in engineering education.

REFERENCES

1. **Alvarez, L., Finston, D., Gehrke, M., and Morandi, P.**, Calculus Instruction at New Mexico State University through Weekly Themes and Cooperative Learning, *Primus* III #1, March, pp. 83-98.
2. **Cunningham**, R.F. 1991. The Effects on Achievement of Using Computer Software to Reduce Hand-generated Symbolic Manipulation in Freshman Calculus, Dissertation at Temple University, 1991.[1] Palmiter, Jeanette R., Effects of Computer Algebra Systems on Conceptual and Skill Acquisition in Calculus, Journal of Research in Mathematics Education, XXII #2, pp. 151-156.
3. **Ferrini-Mundy, Joan, and Karen Guether Graham**. 1991. An Overview of the Calculus Reform Effort: Issues for Learning, Teaching, and Curriculum Development, *The Teaching of Mathematics*, (August-September) pp. 627-635
4. **Gale, David**, 1993. Calculus Reform: Past, Present, -Future(?), *Notices of the American Mathematical Society*, XXXIX #8, October, pp. 822-4.
5. **Getz, Malcolm and John J. Siegfried**. 1991."Costs and Productivity in American Colleges and Universities, in Charles T. Clotfelter, et al, *Economics Challenges in Higher Education*, University of Chicago Press, Chicago, pp. 357-392.
6. **Hughes-Hallett, D. A., Am. M. Gleason, et. al,**. 1992. *Calculus, Preliminary Edition*, (New York: John Wiley & Sons, Inc.).
7. **Rodi, Stephen B**. 1993. Preface: Special Issue on Teaching Calculus, *Primus*, III#1 March, p. 1.
8. **Rogers, Gloria and Elton Graves**. 1993. Using a Computer Algebra System in Calculus and Differential Equations: Does It Help or Hinder in Later Courses? Proceedings of the ASEE Annual Conference, Urbana-Champaign, IL, June 20-24, pp. 1991-1994.
9. **Smith, D. and L. C. Moore**, 1991. The Calculus Reader, Project CALC, (Durham, N. C., Duke University).
10. **Uhl, J. Jerry et al.**, 1991, Calculus & Mathematica, Reading, Mass: Addison Wesley. [reference to be completed]
11. **Watkins, Beverly T.**, 1989. Berkeley Mathematician Strives "to Help People
12. Get Moving". *Chronicle of Higher Education*. June 14 v35n40, p. A16

COMPUTER ALGEBRA SYSTEMS AT ROSE-HULMAN INSTITUTE OF TECHNOLOGY

Bruce A. Black,
Professor of Electrical and Computer Engineering

Rose-Hulman Institute of Technology is an independent, fully-accredited engineering and science college. Originally founded as Rose Polytechnic Institute in 1874, it is now located on a 130 acre campus near Terre Haute, Indiana. The Rose-Hulman mission is to provide a superior undergraduate and master's level engineering and science education that includes the broad humanities and fine arts background required for a fully developed professional person. Rose-Hulman is committed to the undergraduate and master's level of education and therefore offers no doctoral degrees in any discipline.

The student body consists of 1,300 academically superior students. Of the 360 incoming freshmen, ten percent are National Merit scholars. Over ninety percent of the student body ranked in the top fifth of their high school class.

Rose-Hulman provides rigorous academic programs in Applied Optics, Chemical Engineering, Chemistry, Civil Engineering, Computer Engineering, Computer Science, Electrical Engineering, Mathematics, Mathematical Economics, Mechanical Engineering, and Physics. All classes are taught by faculty members. More than 90 percent of the faculty have Ph.D. degrees. Many of the faculty have extensive industrial experience, are active consultants in their fields, and are registered professional engineers.

Because of Rose-Hulman's special orientation as an engineering and science college, all entering students complete three quarters of calculus in their freshman year and one or two quarters of differential equations as sophomores. Changes made in the teaching of calculus and differential equations are significant therefore, as they affect the entire student body. Further, the preparation that the students receive in these basic mathematics subjects have a profound influence on the technical education in scientific and engineering subjects that they receive later.

Rose-Hulman's active use of computer algebra systems began in the spring of 1988 when Professors Robert Lopez and Elton Graves of the Department of Mathematics obtained a grant from the National Science Foundation [1] to equip a classroom with VAXStations running the *Maple* computer algebra system. Thirty VAXStations were purchased; a sufficient number so that each student could be seated at a workstation during a class session. In the Fall of 1988, Dr. Lopez taught two of the ten calculus sections in the computer-equipped classrooms using the *Maple* Computer Algebra System (CAS) as an integral part of the instruction. An illustration of the way in which the CAS is used to teach engineering mathematics is given in [2]. The use of computer algebra systems has subsequently expanded, as will be described below, to involve all sections of the calculus and differential equations courses.

In the summer of 1988 a group of five faculty from the Department of Mathematics and several science and engineering departments received a grant from the Lilly Endowment, Inc. to develop a syllabus for an integrated first-year program in science, engineering, and mathematics [3]. Subsequently, funding to purchase computer equipment in support of the proposed curriculum was sought and received from the National Science Foundation. Seventy NeXT workstations were purchased and networked. This was sufficient to equip two 30-machine classrooms and several faculty offices. The Integrated First Year Curriculum in Science, Engineering, and Mathematics (IFYCSEM) was offered for the first time in the fall of 1990 to a group of 60 volunteer first-year students. The program comprises a sequence of three 12-credit hour courses that includes and integrates calculus, physics, chemistry, engineering graphics, and computer programming; that is all of the first-year subjects except for humanities and social sciences. The NeXT workstations were equipped with the *Mathematica* computer algebra system. The NeXT version of *Mathematica* includes a "notebook" interface, which has proved to be a useful tool for organizing and presenting instructional material even apart from mathematical applications. The IFYCSEM has made extensive use of *Mathematica* in teaching calculus and basic science, and also as a first programming language.

The IFYCSEM has been offered, still on an experimental basis, each year since its inception. In 1991-92, 120 volunteers enrolled in the program. The number of NeXT computers on campus has increased to nearly 200, including five 30-station classrooms. To encourage a degree of uniformity, *Mathematica* was adopted as the Institute "standard" CAS for the teaching of calculus. In the fall of 1991 all sections of calculus, including those in the "traditional" first-year program were taught in the computer-equipped classrooms. In the fall of 1992 all of the sections of differential equations were moved to the computer-equipped classrooms as well.

As the students who learned calculus using the CAS move on into their junior and senior years, there has been considerable interest (and some anxiety) in seeing how their mathematical skills stand up in the upper level programs. Recognizing a widespread faculty concern that students who had learned to depend on CAS to perform algebraic manipulation would be at a disadvantage in upper-level courses where workstations would not be available in the classrooms, Gloria Rogers, Dean for Academic Services, and Professor Elton Graves carried out a study to evaluate the performance of the first group of students taught calculus using the CAS [4]. The performance in upper-level mathematics-dependent courses of 33 students who had completed all five quarters of calculus and differential equations in sections that used the CAS was compared with a control group of students of similar mathematical ability and with the student body as a whole. Drs. Rogers and Graves found no significant difference between the performance in later courses of students who learned their calculus using CAS and students who learned their calculus in traditional sections. What makes this finding interesting is that the mathematical ability, measured by SAT scores, predicted indexes, and high school rank percentile, of the 33 students who completed five quarters of calculus and differential equations in CAS-based sections was significantly ($p < 0.05$) lower than the ability of the student body as a whole! At present, it is too early to draw clear conclusions about improvements or declines in the student's mathematical skills resulting from exposure to CAS. At least the evidence to date is that the experiment is doing no harm.

As the entire student body is now learning calculus and differential equations in computer-equipped classrooms using a CAS, it is inevitable that pressures have developed for changes in upper-level courses in response to the students' new computer-related skills. In the Electrical and Computer Engineering Department, although computer-based circuit analysis packages such as SPICE have been in common use for more than a decade, computer usage in classrooms and laboratories has increased dramatically in recent years. At present the Advanced Circuits Laboratory, the Digital Laboratory, the Controls Laboratory, the Power Laboratory, the Image Processing Laboratory, and the Digital Signal Processing Laboratory are each equipped with computers at each student work bench. All of these computers have access to some variety of computer-math system. In response to the students familiarity with *Mathematica* and CAS-based computation, one section of the required Circuits and Systems course sequence offered to sophomores has been taught for several years in one of the computer-equipped classrooms. For several years faculty from the Departments of Electrical and Computer Engineering and Mathematics have offered summer workshops under the auspices of the ASEE to faculty from schools across the country interested in learning to use CAS in their engineering courses. It is not clear, however, whether the *Mathematica* and *Maple* computer algebra systems used to teach mathematics will turn out to be the systems of choice for teaching electrical and computer engineering subjects. At present, there is a growing use of MATLAB in both classes and laboratories in the signals and systems, communications, controls, and electromagnetics areas.

As the use of computer algebra systems as a classroom tool has become more prevalent in the upper division courses, the pressure on the computer-equipped classrooms has increased. It became apparent during the 1992-93 academic year that the Institute would not be able to continue to equip computer-outfitted classrooms on an ever-increasing scale indefinitely. Replacing workstations as they age was also seen as a challenge. Consequently President Samuel Hulbert convened an ad-hoc committee to review the role of computers in the curricula. In December of 1992, the committee proposed that future incoming students be required to purchase a laptop computer. The laptop computers will be equipped with a standard software package that will include a computer algebra system. Students will be asked to bring their computers to class for "computer equipped classroom" instruction. At the present time, standards are being developed for the laptop computers, and networking of the residence halls is in progress. The plan is to require purchase of laptop computers by the class that enters in the fall of 1995. In anticipation of an even greater demand for the availability of computer mathematics packages throughout the campus, the Institute has recently acquired site licenses for *Mathematica*, *Maple*, and *MATLAB*.

REFERENCES

1. **Robert Lopez and Elton Graves**, A Symbolic, Numerical and Graphical Computations Lab to Support a Revised Curriculum in Calculus and Differential Equations. A proposal submitted to the National Science Foundation Instrumentation and Laboratory Improvement Program, Rose-Hulman Institute of Technology, November, 1987.
2. **Robert J. Lopez**, Computer Algebra as the Instrument for Engineering Analysis, Proceedings of the ASEE Annual Conference, Toronto, Ontario, June 24-29, 1990, pp. 713-716.
3. **Brian J. Winkel and Jeffrey E. Froyd**, An Integrated First-Year Curriculum in Science, Engineering, and Mathematics: Progress, Pitfalls, and Promise, in Lawrence P. Grayson, ed. Proceedings of the 22nd Annual Frontiers in Education Conference, IEEE, November 11-14, 1992, pp. 557-562.
4. **Gloria Rogers and Elton Graves**, Using a Computer Algebra System in Calculus and Differential Equations: Does It Help or Hinder in Later Courses? Proceedings of the ASEE Annual Conference, Urbana-Champaign, IL, June 20-24, 1993, pp. 1991-1994.

Chapter 7

THE ROLE OF DISTANCE LEARNING IN ENGINEERING EDUCATION

R. Mattson
National Technological University

I. INTRODUCTION

A. DEFINITIONS AND CONCEPTS

Learning is an individual accomplishment. No one can learn for another, but learning can be enhanced in group efforts or with the guidance of an instructor. Learning can be realized in a variety of ways and under different conditions. Individualized learning matched to a person's desires and abilities is often a stated goal of educators.

Education and training are organized efforts to implement learning. Training provides the learner with job specific capabilities. Education provides the theoretical knowledge and critical thinking skills needed to understand, learn, and apply new skills and competencies as well as to create new designs.

Engineering education is an organized activity designed by engineering educators to implement learning accomplishments for individuals who will perform creative engineering tasks. Distance engineering education is an organized effort to realize engineering learning, using remote delivery of the educational experiences. Distance learning removes time and place constraints from education, which is especially useful for lifelong learning. Lifelong learning enhances professional competencies.

Educational technology provides a process for implementing learning, especially distance learning. It includes the hardware and software that define an educational delivery system. It also includes the human organization that operates the delivery system as well the persons and methods used to present the educational materials. Thus, educational technology includes people and their organizational structure as part of the total remote educational delivery system. It is this total structure that is used to accomplish distance learning.

Education consists of two major thrusts: the transfer of a body of knowledge to a student and the development of that individual to the highest level attainable. The bodies of knowledge in engineering continue to grow and expand. The skills desired by engineering graduates continue to grow in number and complexity. Educational technology must be used to deal with the explosion of knowledge. There are not enough experts in all areas to teach using traditional lecture methods, all those desiring the knowledge.

B. PRESSURES IMPACTING ENGINEERING EDUCATION

Competition has caused manufacturing industries to change drastically and permanently. Restructuring within the industry has led to improved worker productivity and product quality, with downsizing accounting for a major part of the effort. Service industries are responding to competitive pressures with similar results; improved worker productivity and quality has resulted in improved customer satisfaction. Competitive pressures are now appearing in education, and engineering education is not immune to these pressures. Restructuring and downsizing are now happening in institutions of higher education, and this thrust is expected to grow.

The local, state, and federal levels of government, driven by taxpayer concerns, are seeking methods to improve worker productivity, improve the quality of the services, and respond to the needs of society. The pressure is clearly focused on increasing and improving the efforts of workers. Reinventing government is the thrust, and it is happening at all levels.

The educational system in the U.S. is being forced by its funding sources, students, and society, to make changes. Publicly funded education, including engineering education, faces decreasing budget support. Brewster C. Denny of the University of Washington [1] makes the following observations: Higher education's tuitions have been rising rapidly. State legislative support is decreasing. Higher education's share of the states' budgets has been dropping steadily, now averaging about 10 percent as compared to 20 percent a few years ago. Faculty ranks at the University of California have shrunk, with physics and civil engineering at Berkeley losing 26 percent of the senior faculty. In 1993, faculty salaries were to be cut 5 percent and programs 9 percent. At the University of Washington, the percentage of the budget provided by state appropriations has dropped from 50 percent in the early 1960s to 25 percent in 1993. A 4 percent cut is effective this fall, and two "tax revolt" measures on the ballot in the fall would cut higher education budgets sharply throughout the state. These cuts would impact on faculty, enrollment, and student aid, with closure of whole departments, schools, and colleges a real possibility. Other schools face similar belt-tightening with the tax payer revolt of California spreading across the country. The academic culture that rewards research has been challenged, and most colleges and universities are re-evaluating their commitment to undergraduate education.

The economic realities will force faculty to improve productivity, provide higher quality education and be responsive to customer (student) needs. Engineering educators must recognize that they must become more productive, providing better quality education for their students. The question is where to start and how to proceed.

C. RESPONSES TO PRESSURES

NSF has recently created coalitions of institutions to address undergraduate engineering curriculum revisions with a view that major changes are needed. A recent study [2] published by the National Society of Professional Engineers' Professional Engineers in Education practice division shows that a major change in engineering education is not the top priority with U.S. industry. Appendix A of the document shows that 67 percent of engineering deans felt that the current four-year baccalaureate program is adequate or can be made adequate for graduates. Approximately 100 engineering deans and 1,000 employers in industry and government participated in the study. Nearly two-thirds of industry engineers feel that the current four-year B.S. engineering curricula meet their companies' basic needs. Employers did agree that new graduates lacked crucial knowledge and skill in the areas of teamwork, product/system design, leadership, integrative thinking, social and environmental areas, and ethics.

The study message is clear. Employers and engineering deans agree that undergraduate engineering programs need little change, but practicing engineers need additional skills. Lifelong learning is the answer. Distance learning can provide cost effective access to the needed skills. This means that engineers who wish to be productive and successful throughout their careers need to devote time to continuing education. This creates pressures on engineering education to provide lifelong learning opportunities.

Wrenching changes forced by competition have led to the development and implementation of Total Quality Management (TQM) concepts in manufacturing and service organizations. TQM concepts are simple common-sense engineering concepts: 1. Develop metrics to measure outcomes; 2. Compare the outcome measurement to a desired result; 3. Feedback the difference between the outcome and the desired result to adjust the process to reduce the difference; and 4. Provide a means to repeat the steps for continuous improvement. Most engineers have been exposed to control system concepts, so the TQM concepts are easy to accept.

A concerted effort on the part of industry has resulted in the initiation of programs to implement TQM concepts in engineering education. This is creating another pressure impacting on engineering education in general, and engineering faculty specifically. This response to pressures has promise, but there is much work to be done. Engineering education should lead the way in this response because the practitioners understand the concept.

The cost of engineering education is tied closely to the productivity of engineering faculty. There is no question that costs for engineering education in the future will be reduced. Engineering educators must find means to become more productive if they are to retain their standard of living. Distance education provides the opportunity for substantial increases in faculty productivity. This response has promise, but needs much more work. The educational technology applied to distance learning can be used effectively in undergraduate as well as on-campus graduate engineering education resulting in reduced costs as well as improved quality.

The quality of engineering education must be retained and improved if possible. Presently, there are few efforts to measure that quality and use the measurements for improvement. A very good way to measure the quality in courses is to ask the student. If that student is a practitioner, it would appear to be the very best measure of quality possible. Distance education provides the opportunity for making such measurements and using them to improve the quality of engineering education.

D. EDUCATIONAL TECHNOLOGY AS A RESPONSE TO PRESSURES

The goals of improved faculty productivity and course quality not only can, but must be implemented using educational technology. There are no other responses that will relieve the pressures on education as completely as properly applied educational technology. The unique characteristics of engineering education make it the appropriate testing ground for experimenting with educational technologies and applying TQM concepts to education. Engineering practitioners and students are technologically advanced, well suited to be the leaders in evaluating educational technologies. The evaluations can be used to identify the best from the many options available. The lessons learned in engineering education can be easily transferred to other educational fields and levels with appropriate alterations. This is the message to be developed in this chapter.

Distance education can be used to address the real problems of productivity and quality directly. It is not possible to accurately predict the part distance education will play in the future of engineering education, but it is possible to initiate an experimental environment which will allow distance education technologies to compete in a way that will allow the best to succeed.

The purpose of this chapter is to make an argument for implementing a process which will allow the many emerging forms of distance education to compete intelligently for acceptance by all involved. The role of distance learning in engineering education will be substantial in the future. Distance engineering education is already important, and will become much more important in the future. Revolutionary engineering curriculum design is not the road to an improved future, but the implementation of educational technologies is. Engineering educators must engineer education.

This chapter has nine sections. The history of distance learning section precedes a section describing the present status of distance learning. The next three sections explore the effectiveness of, the benefits of and the barriers to distance learning. Section VII describes various visions of the future of education. Section VIII provides conclusions and Section IX recommendations. Distance learning can provide a very important mechanism for engineering education to alter the way it functions. Those who read this chapter are encouraged to contact the author with reactions and rebuttal.

II. HISTORY

A. OVERVIEW

Story telling has been a means of teaching for centuries. It continues today, especially in some societies which have had no written language until very recently, like the aborigines and some American Indian tribes. Story telling involves no educational technology. Much present delivery of engineering education is based on the lecture form of story telling. Writing and printing were the first educational technologies, and they still provide one of the most used technologies for distance learning. Correspondence programs based on printed materials have been, and continue to be, used world wide. Correspondence based engineering education has not been used much, mainly because of the unique nature of engineering education such as a hands-on design thrust, small numbers, difficulty to implement and the availability of other mechanisms. The pervasive use of printed materials has impeded the implementation of new technologies mainly because of the inertia in the educational system.

Self-paced instruction can be considered a form of correspondence instruction. There are self-paced instructional packages available for engineering education. This interesting form of instruction has not been widely accepted by engineering educators. It does have advantages which will find application in distance learning based on telecommunications systems. These applications have not emerged as yet, but may do so in a few decades.

Instructional radio appeared in the 1920s as a distance learning technology. This technology continues to be used in parts of the world, such as Australia, where operating systems have proved effective. Engineering education via instructional radio has been used in China to serve large numbers of students. This technology has not been effectively applied to engineering education in the U.S. or Europe.

With the advent of television, a few visionary educators hailed the beginning of a new era for education. Many universities experimented in the 50s and 60s with the use of television, especially for teaching large classes. Master teachers were asked to prepare materials for delivery via television to on-campus students with large lecture sections learning via television and small discussion sections led by graduate students. The results of these efforts were mixed. The efforts were driven by university administrators, not faculty nor the students. Anticipated savings were not realized. Most universities abandoned the efforts by the 70s. Engineering education did not experiment with this application of television because very large classes did not exist. These efforts preceded distance learning via television. They were useful because the measurements of learning outcomes showed that television education was effective. The section of this chapter devoted to the effectiveness issue provides additional details regarding the evaluation results.

About the same time—1960s—a large-scale educational effort based on computer technology was established. The PLATO project at the University of Illinois was funded by NSF and the Control Data Corporation. Many important lessons were learned and forgotten. The program was not a great success, although there are still enthusiastic supporters of this pioneering effort. There was an attempt to implement the basic undergraduate engineering course materials on a large main frame computer. Basic physics and engineering courses were created so the course materials were available. The major problem was that engineering faculty would not accept the materials for their courses. The project faded away for a number of reasons, including lack of faculty acceptance, availability of new improved computer technology, lack of student access because of costs, and reduced external funding for the project.

Both the early television and computer educational efforts were designed to serve the on-campus student body. Few large-scale efforts were mounted before 1970 to serve distance learners because of the costs. Both educational technologies—computer and television—were compatible with distance learning, but not applied extensively to the challenge.

During the 1970s, U.S. high technology firms pressured universities to provide practicing engineers with access to graduate courses. They wanted this education because they realized the future of their firms relied on technical expertise. They wanted the education delivered to the engineer at the work site because of the costs of requiring the practicing engineer/student to attend an institution. In some cases there were no local institutions available.

Large corporations were willing to commit resources to establish distance learning centers and to reimburse the practicing engineers for educational costs. This was the major reason for the development and growth of distance learning in the U.S. Distance learning applied to engineering education continues to lead in the development of educational technology in the U.S.

During the 1970s, distance learning used for engineering education started in a number of places. Engineering education, especially graduate and continuing education, is well matched to remote delivery methods. There is a well-defined body of knowledge to be delivered. Graduate engineers are mobile and mature, requiring little personal attention. The first distance learning systems for graduate engineering education were established. Pioneers included Dr. Thomas Martin, who launched systems at Southern Methodist University (TAGER) and the University of Florida (GENESYS), and Dr. Lionel V. Baldwin, who established the SURGE system at Colorado State University. The systems were successful, and continue to serve the needs of practicing engineers today, although in altered form in some cases.

The format used was called Instructional Television (ITV), because it was based on televising the lecture classes. This allowed instructors to continue with their basic teaching methods. They did not have to learn how to create learning materials in new formats.

During the 1970s, federal government action allowed institutions to establish instructional television fixed service systems (ITFS) in the microwave band for the use of educational institutions to deliver distance education. A number of schools in large metropolitan areas established outreach programs based on course delivery via microwave transmissions. Engineering education was a primary aspect of most of these deliveries.

Even more important than the availability of ITFS frequencies, was the development of low cost video cassette tape recorder/playback equipment. This device allowed courses to be taped and shipped using mail delivery to remote sites. The technology improved and as the cost decreased with time, more institutions joined the effort.

The candid classroom teaching techniques used by engineering educators for distance learning initially were the same as they had been using—traditional lecture methods. The instructor did not have to change delivery or preparation methodology. The candid classroom was designed to make the technology invisible. The instructor was recorded in normal delivery, so faculty acceptance was greatly enhanced. This was a key element in the growth of ITV and video tape delivered engineering education. The existing educational system, the faculty preparation, and delivery methods were not disturbed by this implementation of distance learning. It was a small-impact add-on which responded to the demands of industry and at the same time increased the income of the institutions and the productivity of faculty.

Since the 1970s, universities have invested in television technologies for distance learning without large-scale government support. They have developed markets, many in engineering fields. They have been able to grow this market because faculty acceptance provided a source of programming, and industry provided the resources for implementation and growth.

Computer-based educational technology for engineering education has not developed as rapidly as television based educational technology. The lack of faculty acceptance is the major reason. Engineering faculty do not normally use educational materials developed by others, except for books. They have not accepted computer-based course materials for their use, nor have they changed their course development methods. Therefore, computer technology has not grown like the ITV technology for distance learning uses. The federal government is supplying start-up funding to support the development and implementation of computer-based educational systems. This technology will have an impact after course programming becomes available and faculty resistance is reduced.

By the late 1970s, there were enough ITV systems for engineering education outreach, so a consortium of universities was formed. Dr. Lionel V. Baldwin was a leader in the effort to create the Association for Media-Based Continuing Education for Engineers (AMCEE). Each member university had a College of Engineering and a local or regional ITV or video tape delivery system. AMCEE's role was to market the video tapes created by members in a cooperative arrangement. AMCEE continues in this role, but has not developed into a large endeavor. The reason it has not been successful seems to be embedded in the way it is organized. The human organizational aspects of the AMCEE form of educational technology is not arranged to encourage growth.

The National Technological University (NTU) was established by Dr. Baldwin in 1984. It is a private, non-profit, accredited institution founded to serve the advanced educational needs of practicing engineers, scientists and technical managers. It consists of three parts; suppliers, customers, and an advanced technology satellite network. The program suppliers are mainly member universities; the customers are subscribing government, industry, and university organizations. The delivery system is based on digital compressed video technology, and presently operates on the GSTAR 1 satellite. NTU will be described in some detail in a later section. It is the largest and most advanced engineering education delivery network for distance learning in the U.S.

History teaches that large scale distance learning in engineering education based on technology will happen only if faculty accept the technology. They must learn to create educational materials with the technology and learn to be innovative when using them. The growth in the use of ITV technology happened because the technology is transparent. Computer-based technologies have great potential for distance learning in engineering education but they have not been accepted by the vast majority of the engineering faculty. If this form of educational technology is to be implemented, the technology must become more transparent and engineering faculty must be motivated to learn how to use the technology.

B. BALDWIN/DOWN REPORT

The 1981 National Academy of Engineering report, Educational Technology in Engineering [3] is out of print. It reviews the literature and 1981 level of knowledge on the use of modern technological methods in engineering education, with an emphasis on video tape systems. The authors evaluate the practices as of 1981, and offer recommendations for future activities. A brief review of the report will be presented here. It provides a basis for comparing 1981 and 1993 distance learning activities.

The report addresses the role of distance learning in engineering education and provides suggested actions for supporting the implementation of educational technologies. It has been 12 years since the report appeared. The prioritized principle suggestions and their impact are reviewed below. The objective is to show what actions resulted from this National Academy of Engineering report, and to show how responsive government agencies were to the recommendations of the document. The report recommendations and their impacts are:

1. Government agencies should design funding programs which encourage broad-scale, diversified development and testing of Engineering/Technology (ET) courseware in various media at the local course level. Although some basic instructional research in artificial intelligence and other similar areas should be funded, major support should be given to improving local courses and to productivity studies. The purpose of these programs would be to involve many more faculty, nationwide, in meaningful ET activities.

   This top priority suggestion designed to entice engineering faculty to become active in educational technology activities had no discernible impact on government funding actions.

2. At least twelve prototype computer graphics centers should be funded in the next two years at selected engineering colleges. These centers would serve as regional models for the organization and operation of interdisciplinary facilities, as clearinghouses for vendors and educators, and as developers of outreach materials for engineering educators nationwide. External funding of these pioneering centers might require one-third to one-quarter local matching funds. A second round of facility grants to ensure the general availability of computer graphic equipment might involve as much as a 50 percent match by colleges and universities.

   This suggestion to create computer graphics centers to facilitate engineering faculty creativity and the use of computer graphics in engineering education had no discernible impact on either government funding or engineering college priorities.

3. Several university video presses should be endowed at universities having both a strong commitment to continuing education of professionals and the media expertise to set high

standards of quality. These new enterprises would provide a creative outlet for faculty everywhere and encourage the development of modern instructional programs under the guidance of skilled staff.

> Although no endowed university video presses have been created, many universities have established high quality instructional television systems. Relatively little government or foundation funding has been allocated to help establish these systems. The universities have chosen to establish the systems, often at the request of state government and/or local industries. Most of the systems have been established for engineering education to provide distance learning at the work site for practicing engineers.

4. An operational plan to test the cost effectiveness of communications satellites should be funded to provide engineers with the following educational services at their job sites: national delivery of graduate credit courses in a coordinated program initially involving 10 AMCEE universities; delivery of continuing education short courses; and teleconferencing between originating sites to provide for direct faculty exchanges, special seminars, and technical meetings. Studies should be undertaken to involve ethnic minority colleges of engineering in activities of their choosing.

> This suggestion has been implemented by Dr. Baldwin with the establishment of the National Technological University (NTU) in 1984. Little government or foundation funding was available to support this dramatic creative effort. The present status of NTU will be discussed in a latter section. In the opinion of the author of this chapter, the creation of NTU was the most important thing to happen in U.S. engineering education since World War II. The fact that this was accomplished without major foundation or government support is a tribute to those involved in creating NTU.

5. Post-graduate study programs offered by U.S. engineering colleges might be transmitted overseas via satellite. The potential advantages of high-quality educational opportunities available in the homeland of the student may warrant the investment necessary to employ modern communications technology.

> This suggestion has been implemented on an experimental basis by NTU. It can be done, and expansion of this outreach for engineering education is being pursued. Government funding has not been available for this development. Central and Eastern Europe are prime targets for such outreach, but resources have not been made available to implement the effort.

6. An in-depth study of the equipment problems of colleges of engineering should be undertaken, perhaps by an industrial team sponsored by the National Research Council. Recommendations might include government incentives to industry, as well as emergency programs to overcome what has become a massive problem of hardware acquisition.

> This recommendation continues to surface at many meetings and conferences. Engineering colleges need laboratory equipment as well as educational technology equipment. The problem is not limited to engineering colleges, but impacts on the hard sciences as well. Industry has contributed laboratory hardware and software to engineering colleges in significant amounts, but to selected institutions only. This

action has created larger gaps in the quality of engineering education between first and third tier engineering colleges.

Government funding for educational technology has been focused on the National Research and Education Network (NREN) project which is part of the NSFNET, an advanced technology telecommunications superhighway which can be used for engineering education. This is not educational technology because it does not include the human organization and the presentation of educational materials aspects of educational technology. Although these investments are highly desirable, they do not address the human organization problems.

7. Federal agencies should design a number of new programs to expedite the transition to decentralize computing. These diverse efforts might include sponsorship of the following: consumers' unions to evaluate machines and courseware; visiting lectureships; workshops and summer institutes; and media-based, packaged, short courses for in-service training of teachers.

   The NSF approach to this suggestion has been to support the development of NSFNET and NREN with funds for hardware and software to implement the telecommunications network. Again, the human factor is neglected. There is a need to encourage the educational infrastructure with funding for the types of programs listed.

8. Foundations should explore the creation of working consortia or nonprofit organizations to produce high quality films and video tapes in support of case study instruction, especially in engineering design and in professional ethics.

   Foundations have not stepped forward to support the types of efforts noted above.

9. The use of ET in attracting, motivating, and preparing ethnic minority students and others who are educationally disadvantaged in grades 9 through 12 should be extended. A few university-based centers could supply materials and information to the many programs which are now sponsored by businessmen and educators regionally.

   No organized effort to implement this suggestion has been noted.

10. The National Academy of Engineering (NAE) should sponsor a feasibility study on the creation of several regional computation centers for engineering research.

    This suggestion is being developed by NSF, with emphasis is on super computing. There is a need to provide centers based on engineering design workstations and personal computers.

11. The NAE should update its 1974 study on issues and public policies in ET in the light of current trends and forecasts in hardware delivery systems.

    This suggestion is focused on the National Academy of Engineering. Clearly, government agencies, including the National Science Foundation, have not supported the development of educational technology, especially the human organizational aspects. The funding to support education has been devoted mainly to creating Engineering Coalitions, Engineering Research Center, and NREN. The Coalitions are supposed to reform engineering education. The recent NSPE report

[2] shows that employers, engineering deans and practicing engineers do not support such reform. They claim that the present basic undergraduate engineering program is adequate.

The NSF funding to develop an educational infrastructure, NREN, has supported the development of the hardware and software to provide access to a superhighway based on advanced telecommunications. This is desirable, but does not address the implementation and use of educational technologies. Educational technologies provide the basis for improved faculty productivity and course quality. They also are essential for distance learning.

The need for educational technology is apparent for distance learners. It is not so obvious for undergraduate education. Yet the application of educational technology to undergraduate education will happen, and the lessons learned in distance learning will be very important in developing the application of educational technologies to undergraduate engineering education.

C. ECONOMIC CONSIDERATIONS

Public funding of education is standard procedure in the U.S. as well as most of the world. Private institutions exist, but the funding of public institutions dominates education. The U.S. taxpayer revolt continues to reduce the resources available for education, including engineering education.

The ASEE 1993 Directory of Engineering and Engineering Technology Undergraduate Programs shows that there are 274 schools offering 1,382 accredited engineering programs with 23,574.5 full time equivalent faculty. These schools granted 63,094 undergraduate degrees, 23,659 masters degree's and 5,963 doctorates. In addition, 76 schools with 231 ABET accredited engineering technology programs and 1,591.4 full time equivalent faculty granted 14,774 bachelor's degrees, 3,846 master's degrees and 967 doctorates. Clearly, engineering education is big business.

Faculty productivity is a parameter of major interest in today's environment. It is interesting to compare the degrees awarded per faculty member in engineering technology and engineering as a gross measure of faculty productivity as shown in Table 1.

DEGREES PER FULL TIME EQUIVALENT FACULTY MEMBER

Program	Bachelor	Master	Doctorate
Engineering	2 .68	1.0	0.25
Technology	9.26	2.4	0.6

Table 1.

There is a message here that engineering educators must recognize. They argue that the engineering degree is of better quality, but if it is twice as good or ten times as good is not known. The point of this observation is that it is essential that proper metrics be developed to deal with this kind of data. Historically, engineering educators have not paid much attention to measuring faculty productivity on a financial basis. Most of the measurements were related to the production of research papers, the number of graduate students, the number of students in a class, presentations and public service. Most universities are required to maintain their financial records in a fashion that is not compatible with making accurate measurements of faculty or staff productivity. This should change.

The TQM thrust in engineering education requires accurate metrics for implementation. Industry has implemented new financial metrics called Activity Based Costing (ABC). The ABC process relates costs to activities and provides a basis for making faculty productivity measurements. NSF should encourage this development by funding conferences devoted to this concept. NSF should require principle investigators to report on their research with some common metrics so research productivity can be measured.

D. SUMMARY

This section reviewed the history of distance learning as used in engineering education. The 1981 report [3] recommended that a number of actions be initiated by government agencies to encourage the development of distance learning and educational technologies. Very little was done in response to the recommendations, many of which should now be funded. Finally, a brief comparison of the productivity of engineering and engineering technology faculty showed that the latter were much more productive if degrees were the only measure of such productivity.

III. PRESENT STATUS

A. OVERVIEW

Distance learning via educational technologies is no longer an option which engineering educators can ignore. It is now present and growing. The major application is in graduate and continuing education for practicing engineers delivered by universities to remote sites. The Baldwin/Down report [3] showed televised off-campus engineering continuing education program data for the 1979-80 academic year. Thirty-one institutions reported 10,627 enrollments in 1,098 ITV courses at 545 remote locations.

Some of these institutions offered academic engineering courses for credit. Twenty-four institutions awarded 2,837 M.S. engineering or M.B.A. degrees to distance learners from when their distance learning programs were initiated until 1980. No data is available from 1980 until 1993.

Peterson's Guides Inc. published THE ELECTRONIC UNIVERSITY: A Guide to Distance Learning Programs [4] in cooperation with the National University Continuing Education Association in 1993. It lists 136 subjects in which a student can seek a remotely delivered post secondary degree. There were thirty-eight different bachelor's degree programs as well as sixty-three master's and one Ph.D. degree program available. These programs are from eighty-eight institutions in forty states, and one in the District of Columbia. In addition, there were five institutions listed in four Canadian provinces which offer degrees using remote delivery. The states with the most programs are California, with nine schools offering programs at remote sites and New York and Texas listing six institutions offering distance learning programs in each state.

The Guide lists nine distance degree consortia and networks. Of these nine consortia and networks, only the Electronic University Network offers credit courses and degrees via computers and modems. The member institutions of that network include the California Institute of Integral Studies offering a Ph.D. in Integral Studies, Heriot-Watt University offering M.B.A. degrees, and Rogers State College offering A.A. and A.S. degrees. Credit earned by examination at these institutions are accepted by 2,500 U.S. colleges and universities. There are nearly 12,000 students participating in this computer-based engineering related distance learning activity. All the other consortia and networks listed in the Peterson's guide are based on instructional television (ITV) systems. Clearly, computer-based instruction has not been accepted by the major universities.

Other consortia listed in the Guide include the Agricultural Satellite Corporation (AG*SAT), which is a consortium involving forty-seven U.S. universities, business, government and industry. It specializes in the delivery of programs related to agriculture. The International University Consortium (IUC) is a cooperative world wide organization which emphasizes general management, health-care management, liberal studies and American studies. The Michigan Information Technology Network Inc. (MITN) is a nonprofit partnership offering degrees in 15 disciplines to Michigan citizens. It does offer engineering courses and degrees, and some of its member institutions are members of the National Technological University (NTU) consortium. NTU is a member of the MITN partnership and it is listed in the guide. The Mind Extension University (ME/U) offers credit courses and degree programs from 20 institutions via cable and satellite. Both bachelor and master degree programs are available. Engineering courses are not part of their offerings at this time. The Tribal College Consortium (TCC) serve the Native American community in Montana, and it has three participating colleges with no engineering programming involved. The Western Cooperative for Educational Telecommunications (WICHE) is a 150 member cooperative partnership of states, institutions, networks, public broadcast stations, school districts, and telecommunications concerns. WICHE has actively supported the development of distance learning programs that meet quality standards.

Table 2 lists the forty-one institutions extracted from the Peterson's Guide that offer engineering degrees to distance learners. They show 60,000 enrollments with 1,051 degrees granted in 1993. The guide does not specify enrollment and degrees by discipline, and some of the schools also offer degrees in non-engineering fields. Therefore, these numbers may be a bit inflated. However there are institutions offering engineering degrees that are not listed, such as the University of South Florida with over 500 enrollments per semester and eight undergraduate electrical engineering degrees granted to distance learners. Many of the programs are relatively new, and the growth of these enrollments will continue.

Engineering or engineering technology bachelor degrees are offered to distance learners by Embry-Riddle Aeronautical University (B.S.P.A.), Rochester Institute of Technology (B.S.E.T., B.S.M.T.), University of Maryland (B.S.N.S.) and the University of North Dakota (B.S.E.E., B.S.M.E., B.S.Ch.E.). All forty-one institutions offer master degrees. This is compatible with the concept that the distance learning programs serve the needs of practicing graduate engineers. The University of South Florida offers a BSEE degree, but is not listed in the Guide.

There are thirty-nine institutions that offer undergraduate engineering courses for credit remotely via ITV educational technology systems, according to Dr. Arnold E. Seigel and Cynthia Davis of the University of Maryland, College Park. Their survey results were reported at an NTU meeting. The Delivery of Televised Engineering Courses: A Survey of Engineering Schools with ABET-Accredited Programs was an unpublished 1992 report [5]. It showed that 222 engineering deans out of 269 surveyed (83 percent) responded. Thirty-nine respondents, of which thirty-three are public institutions, offer televised engineering courses to undergraduate students for credit. Eighteen of the thirty-nine reported that they would conceivably offer all junior and senior level engineering courses other than the laboratory courses via television. The authors point out that the reasons for such programs include efficiency, cost-effectiveness, remote delivery in response to political pressures and student performance which equals on campus learning. Four of the 39 schools that offer undergraduate engineering courses also offer degree programs. These results are compatible with the data included in the Peterson's Guide. There is no similar data available for computer-based educational technology systems.

The schools which offer a bachelor degree in engineering do so with the help of other educational institutions. For example, California State University at Chico offers a bachelor degree in Computer Science to distance learners. The remote students work up a study plan with Chico faculty. The computer science courses are delivered via satellite and the student gets support courses in humanities, mathematics, etc., from a local institution.

The University of Missouri-Rolla has proposed a televised distance education program that would originate on the UMR campus, to be delivered to community colleges and four-year institutions within the state not now offering engineering. Clearly, institutions are considering and implementing distance learning for undergraduate engineering degree programs. This thrust is expected to expand in the future.

The Association for Media-Based Continuing Education for Engineers (AMCEE) offers 463 video taped courses in fourteen disciplines from its thirty-four member institutions. Many of the schools are also members of NTU. The courses vary in length from short courses to full 3 semester credit hour courses. They are offered in video tape format. The prices vary depending on the material and the length of the course. Although some of the courses were offered for on-campus credit, AMCEE does not include credit as part of the arrangement. In spite of many observations that engineering education is in high demand, the AMCEE organization has not been able to build a large business from its offerings. The excellence of the course and the prestige of the source institutions do not seem to impact on course sales. It is not clear why this source of good engineering education is not expanding.

Some of the NSF funded Engineering Coalitions, especially the Synthesis Coalition, have developed courses using computer technology, but large scale usage has not been implemented. Large scale NSF support has not generated applications of educational technology as was expected.

The Adult Learning Satellite Service (ALSS) of the Public Broadcasting System Education's Adult Learning Service serves distance learners. The 344 local PBS stations own and operate the satellite delivery service. ALSS programs include telecourses, audiovisual resource programs and live interactive videoconferences. The telecourse is similar to a traditional college course, except the students learn at home from television programs, textbooks, and study guides. Over 2.5 million students have earned college credit by enrolling in such courses at over 1,900 colleges nationwide. Over 60 telecourses are available, and the College by Cassette is also available. Although very active and effective, ALSS has not offered engineering courses or degrees.

THE ELECTRONIC UNIVERSITY
A Guide to Distance Learning Programs

Published by Peterson's Guides Inc.
Princeton, New Jersey
in cooperation with NUCEA, National University Continuing Education Association
1993

Abbreviated Program Profiles
(See the publication for details)

ENGINEERING PROGRAMS ONLY

Institution	1992-93 Enrollment	Degree/Certificate Programs
Arizona State University Tempe, AZ	3,290	M.S. in Engineering. Graduate Degree in E.E.
California State University - Chico Chico, CA	2,000	B.S. and M.S. in Computer Science - 30 graduates
California State University, Northridge Northridge, CA	150	M.S. in Engineering; Teacher Certificate for California
Colorado State University Fort Collins, CO	2,700	M.S. in Agricultural, Chemical, Civil, Electrical, Mechanical, and Interdisciplinary Engineering; M.B.A.; M.S. in Computer Science; M.S. in Management; M.S. in Statistics; Ph.D. - 8 graduates
Embry-Riddle Aeronautical University Daytona Beach, FL	2,500	B.S. in Professional Aeronautics; Master of Aeronautical Science - 21 graduates
Florida Atlantic University Boca Raton, FL	new	M.S.-M.E. in Civil, Electrical, Manufacturing Systems, Mechanical and Ocean Engineering as well as Computer Science and Engineering

Table 2.

Florida State University Tallahassee, FL	35	M.S. in Electrical and Mechanical Engineering - 17 graduates
George Washington University Washington, DC	332	M.S. in Electrical Engineering; Computer Science; M.A. in Educational Technology
Georgia Institute of Technology Atlanta, GA	300	M.S. in Electrical, Environmental, Industrial, Mechanical, and Nuclear Engineering as well as Health Physics and Engineering - 7 graduates
Iowa State University of Science and Technology Ames, IA	1,000	M.S. in Engineering, Agriculture and Biosystems, Chemical, Civil, Computer, Electrical, Industrial, Mechanical and Nuclear; B.S. and M.S. in Agriculture - 20 graduates
Lehigh University Bethlehem, PA	100	M.S. in Chemical Engineering; M.S. in Chemistry
Mary Washington College Fredricksburg, VA	108	Commonwealth Graduate Engineering Program; M.E. in Electrical, Chemical, Nuclear, Aerospace and Ocean, Systems, Civil, Chemical and Administration and Management; M.B.A. - 2 graduates
National Technological University Fort Collins, CO	5,077	M.S. in C.E., C.S., E.E., E.M., H.W.M., H.P., M.O.T., M.S.E., M.A.T., S.E. and Special Majors. M.S.E. - 118 graduates
Oklahoma State University Stillwater, OK	373	M.S. in E.E., M.E., and ChE; M.B.A. - 7 graduates
Old Dominion University Norfolk, VA	2,100	M.S. in E.M., C.E.E., E.C.E., E.E., and M.E.; M.S. in Nursing - 44 graduates
Purdue University West Lafayette, IN	1,131	M.S. in Engineering, I.E., E.E., M.E., and C.E.; Master's Pharmacy - 127 graduates
Rensselaer Polytechnic Institute Troy, NY	600	M.S. in Materials, Mechanical, Manufacturing Systems, and Management Engineering as well

Table 2. (continued)

		as M.S. in Computer Science and Microelectronics Manufacturing; - 40 graduates
Rice University Houston, TX	12	M.E.E. or M.C.S. - 2 graduates
Rochester Institute of Technology Rochester, NY	2,000	M.S. in Software Engineering; B.S. in Electrical/Mechanical Technology; M.S. in Computer Science - 12 graduates
Southern Methodist University Dallas, TX	1,031	M.S. in Computer, Mechanical, and Electrical Engineering as well as Operations Research, Computer Science and Engineering - 87 degrees
Stanford University Stanford, CA	3,352	M.S. in Electrical Engineering and M.S. in Computer Science
Texas Tech University Lubbock, TX	280	Undifferentiated Master of Engineering
University of Alabama Tuscaloosa, AL	170	M.S. Aerospace Engineering; Master of Library Studies
University of Arizona Tucson, AZ	1,650	Master of Science in Engineering with a major in Reliability and Quality Engineering, Electrical Engineering and Optical Sciences; Doctor of Pharmacy; M.A. in Library Science
University of California, Santa Barbara Santa Barbara, CA	112	M.S.E.E., M.S.C.S. - 28 graduates
University of Colorado at Boulder Boulder, CO	790	M.E./M.S. Aerospace, Civil and Environmental, Electrical and Computer, and Mechanical Engineering; M.E. in Engineering Management, Software Engineering, and Telecommunications; M.E. M.S. in Computer Science - 20 graduates
University of Houston Houston, TX	816	Master in Electrical Engineering with Computer Specialization; Master's degree in Industrial Engineering

Table 2. (continued)

University of Idaho Moscow, ID	975	M.S. in Engineering - 33 graduates
University of Illinois at Urbana-Champaign Champaign, IL	300	M.S. in Mechanical, General, Electrical, and Nuclear Engineering as well as Theoretical and Applied Mechanics - 18 graduates
University of Kentucky Lexington, KY	311	Master of Engineering and Master in Mining Engineering; M.S. in Education, Family Studies and Special Education
University of Maryland University College College Park, MD	13,377	B.S. in Nuclear Science Engineering; B.S., B.A. in Business and Management - 137 graduates
University of Massachusetts at Amherst Amherst, MA	1,206	M.S. in Engineering - 22 graduates
University of Missouri- Rolla Rolla, MO	220	M.S. in Engineering Management, M.S. in Engineering/ Computer Science - 30 graduates
University of Nebraska- Lincoln Lincoln, NE	189	M.S. in Industrial and Management Systems, Manufacturing Systems and Mechanical Engineering; M.S. Computer Science; M.B.A. - 2 graduates
University of North Dakota Grand Forks, ND	179	B.S.E.E., B.S.M.E., and B.S. ChE.; Degree programs in nursing, medical technology, public administration, counseling, social work, education - 5 graduate
University of South Carolina Columbia, SC	4,000	M.E., M.S. in Chemical, Civil Mechanical, Electrical and Computer Engineering; degree programs in Library and Information Science and
University of Tennessee, Knoxville Knoxville, TN	4,000	M.S. in Aerospace, Chemical, Electrical, Industrial, Mechanical and Computer Engineering a well as Mechanics, Mathematics, Physics and Engineering Management - 12 graduates
University of Virginia Charlottesville, VA	1,000	M.E. in Chemical, Civil Electrical, Materials Science,

Table 2. (continued)

		Mechanical and Aerospace, Nuclear and Systems Engineering - 30 graduates
University of Wisconsin-Madison Madison, WI	1,137	M.S.E.E. Power Electronics, M.S.M.E. Controls, M.S.N.E.E.P. Nuclear Engineering - 4 graduates
Virginia Polytechnic Institute and State University Blacksburg, VA	1,374	M.S., M.E. in Electrical, Mechanical, Systems, Civil or Industrial Engineering - 103 graduates

Table 2. (continued)

B. THE NATIONAL TECHNOLOGICAL UNIVERSITY

The National Technological University (NTU) is a private, non-profit, accredited institution founded in 1984. It consists of three parts; member universities, subscribing organizations, and an educational technology infrastructure. The infrastructure includes a sophisticated Digital Compressed Video (DCV) telecommunications delivery network which is now delivering via a satellite transponder. The infrastructure also includes a non-traditional human organizational structure which is a paradigm shift in engineering education. Since NTU represents the leading initiative in distance learning applied to engineering education, it will be discussed in some detail.

NTU is pioneering innovative and cost effective methods to serve the educational needs of engineers, scientists and technical managers. The NTU vision statement is:

"Enabling technical professionals and managers to share premier educational resources globally via telecommunications"

An engineering educator, Dr. Lionel V. Baldwin is the entrepreneur/visionary who has led the development of the institution. Many engineering educators also contributed significantly. The Association for Media-Based Continuing Education for Engineers (AMCEE) played an important role in the establishment of NTU. NTU is headquartered in Fort Collins, Colorado. It is accredited by the North Central Association of Colleges and Schools. The accreditation became effective November 19, 1986.

The NTU educational mission is to:

- Serve the advanced educational needs of graduate engineers, technical professionals and managers thereby enabling them to reach their professional and organizational goals
- Award degrees at the master's level to qualified candidates

The NTU technology mission is to:

- Explore, develop and use advanced educational and telecommunications technologies in delivering instructional programs to graduate engineers and technical professionals at their employment locations

- Provide a satellite network infrastructure linking technical professionals and managers nationally and potentially internationally, in research seminars, technology transfer activities and related technical exchanges

NTU uses instructional television (ITV) video based instruction because it is flexible and convenient. It can be delivered on site and most important—there is a large mass of course materials available.

There are forty-five universities that are members of NTU. They all have a local or regional ITV program to deliver their courses for credit to local subscribing sites. Thus they have a local educational technology infrastructure with equipment, candid classrooms and a human organizational structure that is accepted and used by local faculty.

There are over 600 NTU subscribing sites, one in Mexico and six in Canada. The Monterey Institute of Technology in Mexico purchases credit courses for faculty professional development at their twenty-six satellite connected sites. The Canadian sites are at corporate subscribers such as IBM and HP. There are over twenty European and a dozen Asian corporate subscriber sites which receive the NTU Advanced Technology and Management Programs which are not-for-credit short courses, seminars and workshops. About 200 of the 600 NTU sites register for credit courses and degree programs. They have NTU Subscriber Units which allow them to receive four TV channels simultaneously.

The Ku band satellite broadcasts are uplinked live from thirty-eight of the forty-five NTU member institutions that have the equipment. The other schools' programs are broadcast tape delayed from an NTU encoder equipped uplink school. The sites can view the broadcast live if appropriate or video tape the program for later use.

The NTU Digital Compressed Video (DCV) system operates at a 3 megabit rate, and transmits a robust full motion color picture of improved audio and video as compared to an analog signal. The DCV allows NTU to operate up to ten channels on one 55 megahertz transponder on the GSTAR 1 satellite. In early 1994, NTU will move to TELSTAR 401, which will allow more channels on one transponder, up to sixteen. This reduces the per channel cost of broadcast substantially.

The NTU proprietary DCV system allows NTU to activate the subscriber's receiver as well as their VCRs. Facsimile transmission of class notes via the DCV system will reduce operating costs. In addition, the DCV allows for the transmission of computer information, data, graphics, etc. Therefore the network is ready to include computer-based education. NTU views its mission as education delivered via telecommunications, and with the DCV system NTU is positioned to deliver via land lines and the information superhighway.

NTU offers M.S. degrees in Computer Engineering, Computer Science, Electrical Engineering, Engineering Management, Hazardous Waste Management, Health Physics, Management of Technology, Manufacturing Systems Engineering, Material Science and Engineering, Software Engineering and Special Majors. The Special Majors Program allows an individual to create a proposed program of Study using courses listed in the NTU Bulletin. The proposed program along with the individual's past transcripts, brief resume and career plan are sent to an NTU faculty expert for evaluation. If the program is allowed, the expert becomes the student's advisor for the academic program. NTU plans to offer an M.S. in Chemical Engineering starting in 1994.

Table 3 displays the admissions statistics for NTU. With over 1,700 matriculated students, NTU's graduation output will increase. It is expected that NTU will be in the top ten grantors of engineering and technical management M.S. degrees by the end of the decade.

The human organizational aspects of the NTU educational technology effort are very important. Table 4 presents the organizational structure for NTU. The Board of Trustees is made up of corporate executives, academicians and one graduate. The Executive Advisors are from Corporate Subscribers. They help NTU identify new programs and courses for development. The University Administrative Contacts are from member schools. They provide the NTU President with advice and guidance as to how to operate and improve the system. The Administrative Contacts are either staff or faculty at the schools.

The Graduate Programs are administered by the Academic Vice President. The Director of Admissions and Records and the Academic Coordinator are NTU employees. The Director is the Registrar, overseeing all the functions of a traditional university. The Coordinator deals with the member schools while creating the Bulletin and the Class Schedules for NTU. An Evaluation Consultant, an expert in instructional systems assessment especially for engineering instruction is also employed.

All of the programs except for the Computing Systems and Software (CSS) program and the Special Majors program offer only one degree. The CSS program offers three degrees: Computer Engineering, Computer Science and Software Engineering, because of the similarity of the topics. The Special Majors Program offers the M.S. degree with a specialty chosen by the student. This allows a student to pursue a specialty in mechanical engineering as well as reliability engineering. The Special Majors program is managed by the Academic Executive Committee. The Management of Technology M.S. degree program is the NTU executive management program. It operates as a lock-step two-year program requiring forty-five semester credit hours for completion. It is managed by the program chair and the teaching faculty.

	Students Admitted to Major Degree Programs	Students Still in Admission Process	Students Denied Admission	Degrees Conferred	Students Withdrawn
Computer Engineering	334	17	58	117	69
Computer Science	322	39	85	115	95
Electrical Engineering	386	42	39	122	75
Engineering Management	279	23	94	95	61
Manufacturing Systems Eng.	92	4	5	17	30
Management of Technology	82	1		61	46
Materials Science	44	7	0	3	5
Special Majors	66	19	5	6	5
Health Physics	24	7	9	2	4
Hazardous Waste Management	86	11	8	1	2
Software Engineering	47	22	5		
Unspecified Major		2			
Totals	1762	194	308	539	392

Degrees Conferred by Year:

1986 - 1	1991 FEB - 30	1993 JAN - 19
1987 - 6	1991 AUG - 40	1993 FEB - 52
1988 - 14	1992 JAN - 18	1993 AUG - 63
1989 - 37	1992 FEB - 53	1994 JAN - 24
1990 - 56	1992 AUG - 47	1994 FEB - 79

Table 3. Admissions Statistics By Major - February 9, 1994

BOARD OF TRUSTEES: Consists of industrial executives and academic administrators who make NTU policy. **PRESIDENT**: Dr. Lionel V. Baldwin reports to the Board of Trustees which meets quarterly. **EXECUTIVE ADVISORS**: Representative of NTU Corporate Subscribers who advise the President concerning employer needs and meets three times a year. **ADMINISTRATIVE CONTACTS**: Member university representatives meet two times a year to advise the President. They have a Steering Committee which provides continuing advise to the President.

Table 4. NTU Organization

NTU has a Corporate Secretary, Treasurer, Vice President of Marketing, Director of the Satellite Network, Director of Advanced Technology and Management Programs, and an Academic Vice President. The Academic Vice President is responsible for the graduate programs which are organized as outlined below:

- ACADEMIC EXECUTIVE COMMITTEE: The 10 academic program chairmen plus one curriculum committee chairman constitute the Academic Executive Committee which makes academic policies. They are members of the faculty at NTU member institutions, and are paid as consultants. This committee also administers the NTU Special Majors Programs.
- PROGRAM CHAIRMEN: They organize the management of their programs with curriculum, advising and admissions committees consisting of faculty from member institutions. They also have faculty representatives from member schools who meet annually to discuss the curriculum and other academic matters.
- ADVISORS: Each admitted student has an NTU advisor who works with the students. The advisor can overrule general policies to meet student career education goals.
- NTU ALUMNI ASSOCIATION: Graduates can join the association, and one member is a member of the NTU Board of Trustees. NTU STAFF: Staff members include an Academic Coordinator and a Registrar. Additional staff support all the academic functions.

For the usual programs each interested member school is asked to identify a faculty representative for each program. The representatives meet annually to discuss the curriculum, identify acceptable courses and make decisions regarding the admissions and completion requirements. Each degree program has a curriculum committee, and all admitted students have an assigned faculty advisor. All participating faculty are paid consultants to NTU.

The Academic Executive Committee meets three times a year, to make academic policy. This academic organization is unique, and it works quite well. This is the core of the human organization part of the educational technology implementation known as NTU. About 100 engineering faculty are involved in this organization, in addition to hundreds of faculty who have taught on NTU since inception. It is a tribute to Dr. Baldwin and the others who envisioned NTU that there have been no major changes in the human organization part of NTU.

Table 5 shows how the academic information flows in the system with guidelines and policies stated. Table 6 shows how the examinations are handled. The Site Coordinator is responsible for the academic integrity of the effort at the site. Tables 7 and 8 show how homework and projects are handled along with guidelines and policies.

There are many unique and non-traditional aspects to NTU. One of the most significant is related to the fact that member institutions submit more academic courses for broadcast than can be broadcast via the network. Thus, NTU must select courses from those submitted for inclusion in each class schedule. This allows NTU to establish a course selection process. Table 9 shows the process that is used.

Each NTU credit student receives a course evaluation questionnaire upon completion of a course. The results of the student evaluations are used to identify NTU Outstanding Instructors as well as help to select courses. Site Coordinator evaluation surveys as well as NTU instructor surveys are also used to improve the quality of the entire system. Early in each semester NTU asks students to complete "Squawk Cards" which are designed to identify problems in the technology and shipment of course materials. This use of evaluations will lead to continued improvement of the NTU educational system. This approach should be encouraged as the new paradigms evolve for providing engineering education to distance learners. It improves course quality in a systematic fashion.

Since NTU is a consortium, there are many opportunities for sharing academic resources among member institutions. Presently NTU offers a number of such opportunities. The most successful is National Engineering Faculty Forum. The Forum was established in February, 1993 with a small grant from the Department of Energy. It is a one hour program broadcast on the third Tuesday of most months from 4 to 5 PM Eastern time. It is devoted to topics of interest to engineering faculty, and a C band turn around is implemented so engineering faculty at non-NTU institutions can receive the program. Table 10 shows that the number of viewers has been extraordinary. Seldom do 3,000 engineering faculty choose to view the same program. Clearly, this Forum is a means of sharing concerns with engineering faculty in a realistic cost-effective manner. This aspect should be incorporated into any new paradigms that evolve for engineering educators. It provides for faculty professional development as well as a forum for discussing questions of concern to engineering educators.

1. Instructor presentations to on-campus students are either video taped or uplinked live from the home university.

2. Fifty percent of the 200+ courses are broadcast live and the rest broadcast a day or two later.

3. NTU Broadcast Control activates subscriber units for individual courses, switches between uplinking institutions, and monitors and controls all broadcasts for quality.

4. NTU tapes all broadcasts so subscribers can order back-up tapes if there are technical difficulties. The tapes are erased after two weeks.

5. Subscribing site downlinks are turned on by NTU, and one subscriber unit provides up to four receive channels.

6. Site Coordinators provide for live viewing with telephone call in capability at the site, or they video tape the presentations.

7. Students can view the presentation as downlinked or via a video tape. Most of them choose to view the tapes because it provides time and place flexibility.

8. GUIDELINES: Students should view the presentations within four working days of the on-campus session.

9. POLICY: Students must view the presentation within fourteen days of the on-campus session. If this is not possible, the student must make special arrangements with the instructor.

Table 5. Flow Of Academic Instruction

1. Instructors send the examination materials along with instructions for administering the examination to the site coordinator. The site coordinator is responsible for proctoring the examination as requested by the instructor.

2. The completed examination is returned by the site coordinator to the institution, usually to the instructor.

3. The instructor grades are required at NTU before payment to the professor is initiated.

4. Any grade appeals must be made by the student through the source institution's appeal system, and the decisions are final.

5. GUIDELINES: Because of the distributed nature of the NTU system, instructors are urged to be innovative in examination procedures to ensure academic integrity.

Table 6. NTU Examination Flow Chart

1. Homework is assigned by the instructor who realizes the time delay inherent in the system.

2. The source university sends all course materials to the site coordinator who passes it on to the student.

3. The student returns the homework to the institution, sometimes through the site coordinator. The instructor tells the student how to handle homework mailings. Many instructors are now using electronic mail for communicating with students.

4. Instructors return graded homework in time for it to be useful to the student.

5. GUIDELINES: Students are responsible to meet instructor imposed deadlines. Special arrangements can be made between students and the instructor.

Table 7. NTU Homework Flow

All projects are individually arranged between the student, the involved faculty member and NTU.

POLICY: NTU must be notified in advance of any M.S. thesis projects. Direct payments to involved faculty are part of the arrangements.

Table 8. NTU Projects Flow

1. The Administrative Contact at each member university submits courses each academic period creating a pool of courses for broadcast consideration.
2. NTU Outstanding Instructor taught courses are automatically included in the NTU Class Schedule.
3. NTU Program Chairmen make their recommendations based on past enrollment histories, instructor evaluations and student needs for all courses in their technical areas.
4. NTU staff make their recommendations for all courses not assigned to a technical area controlled by a program chair.
5. NTU staff introduce student request information, and a preliminary course list is produced for the class schedule.
6. This course list is shared with the Administrative Contacts, and adjustments are made to accommodate the member schools.
7. NTU staff then prepares the Class Schedule which is sent to all NTU students and site coordinators.
8. Student enrollments determine the courses which will be broadcast.

Table 9. Course Selection Process

NTU NATIONAL ENGINEERING FACULTY FORUM
PARTICIPATION & SCHEDULE FOR UPCOMING PROGRAMS

Program Date **# Sites** **(Speakers)**	**Title** ***Viewers**
Feb. 16, 1993 (Raymond Landis, California State/LA) 221	Facilitating Collaborative Learning Among Minority Students 7,061
Mar. 16, 1993 (Eleanor Baum, Cooper Union, NYC) 90	Undergraduate Women in Engineering 11,215
Apr. 20, 1993 (Joseph Bordogna, NSF, & Paul Herer, NSF) 1	Engineering & Change at NSF 5959
May 18, 1993 (Karl Reid, Okla. State Univ.) 92	Continuous Quality Improvement: Will the Principles Work in the University? 945
Jun. 22, 1993 (J. Ray Bowen, Univ. of Washington) 3	Engineering Education in ASEE's Second Century 8,911
Sep. 21, 1993 (John Gibbons & Edward David) 111	New Opportunities for Engineering Under the Clinton Administration 2442

Table 10.

Oct. 19, 1993 (Sheri Sheppard, Stanford Univ.) 78	Teaching Students Hands-On Engineering 14,630
Nov. 16, 1993 (NACME Conference, George Campbell, President) 115 *3:30pm - 5:25pm ET.*	New Institutional Vision: Change from the Inside Out 1,989
Dec 7, 1993 (Linda Enghagen, Univ. of Massachusetts) 24	Intellectual Property Concerns for Faculty 12,120
Dec 10, 1993 (National Academy of Engineering; Robert M. White, Pres.) 13 *11:00am - 5:00pm ET*	NAE Symposium: New Directions in Science and Technology Policy 125
Jan 18, 1994 (Joseph Wujek, Univ. of California, Berkeley) 109	First-Year Students Emulating Industry: Berkeley's Engineering Design Studio 1,822
Feb 15, 1994 (Richard Felder, North Carolina State University) 28	How Students Learn, How Teacher's Teach: And What Usually Goes Wrong With the Process 14,325
Mar 15, 1994 (Frank Wayno, Cornell University)	When Engineering Skills Are Not Enough: Teams, Concurrent Engineering and Time-Based Competition
Apr 19, 1994 (Robert White, Carnegie Mellon University)	CMU's New Electrical and Computer Engineering Curriculum
May 10, 1994 (J. Ray Bowen, Univ. of Washington)	Persons with Disabilities in Engineering Education and the Engineering Profession

All times 4-5pm ET, unless listed otherwise (adjourn for June,July & August)

*Indicates minimum number of persons expected to view at time of broadcast. It does not include the number of persons who actually attended, or those who viewed video tapes after the broadcast.

Table 10. (continued)

The NTU Course Exchange Program (see Table 11) provides a realistic opportunity to improve faculty productivity by sharing academic resources. A course which is being broadcast via NTU can be purchased for internal use by a member institution. Reasons for purchasing such courses include response to budget pressures, response to faculty shortages, to improve faculty productivity, supplement on-campus courses, provide local students with courses that are not available on campus and increase cooperation between institutions. Agreement of the source school and instructor is required. The purchasing school must provide their own instructor who manages the course, deals with the local students, monitors and grades examinations and homework, and allocates grades. The local instructor does not have to prepare lectures, homework or examinations, and therefore can become more productive in other duties. The cost for this exchange is $2,500 with $2,000 going to the source institution and instructor and $500 retained by NTU. If the purchasing school has a course desired by the source institution the course exchange would cost each school only $500. Innovative use of such exchanges will increase faculty productivity and course quality. Such exchanges will require new engineering faculty attitudes. Many faculty will become managers of education using materials prepared by master teachers with technical topic competence. This policy will allow outstanding instructors from member institutions to be shared nationally.

NTU COURSE EXCHANGE
Live at the source institution Agreement of instructor and source institution User institution offers their credit for their students
OUTCOMES
Respond to faculty shortage problems Respond to departmental budget problems Improve faculty productivity Increase cooperation between institutions Supplement on campus courses Peer review of academic courses
COSTS
$2,500-$500 NTU - $2,000 Source institution and instructor $5,000 Monterey Institute of Technology for faculty and staff

Table 11.

The NTU ATMP courses are short courses and seminars in advanced technical fields taught by experts. These non-credit courses are available to the faculty and staff at member institutions at a fee that is 60 percent lower than the NTU site fees. The purpose of this policy is to encourage faculty professional development.

The NTU Material Science and Engineering faculty offer a seminar course, MS 799-P Materials Science and Engineering Seminar. This course is a series of presentations by national and international materials scientist experts on topics at the forefront of science and technology, for one credit via the NTU network. This course is available at no cost to member institutions. The MSE faculty at member schools share this course with their students. Another example of sharing academic resources nationwide.

Academic resource sharing via NTU can be expanded extensively. New plans are now being implemented to create a power engineering consortium. Power engineering at many institutions has been neglected. Few institutions have all aspects of power engineering in-house. NTU member schools with the interest are identifying courses that will be shared with the most expert instructor in various technical areas teaching via NTU and other schools using the course materials. This type of use of the NTU network can be implemented to improve the cooperation within and between Engineering Coalitions, Engineering Research Centers, member universities, industry and government organizations.

NTU is a pioneer in the area of engineering education and distance learning. It provides an accepted educational technology for investigating and implementing new paradigms in engineering education. Although the emphasis is on distance learning, the applications of the technology to on-campus engineering education is obvious.

C. THE OPEN UNIVERSITY

NTU has been created by sharing existing engineering educational resources via telecommunications. The Open University (OU), Walton Hall Milton Keynes UK, is a very different model for distance learning. It is the largest and most innovative university in the United Kingdom, with a world-wide reputation for the quality of its courses and the effectiveness of its teaching methods. Since this report is focused on new paradigms for engineering education the OU model will be discussed in some detail. The information quoted here was obtained from brochures describing the Open University programs.

OU was created about twenty years ago and two million people have studied with it. The OU undergraduate programs have enabled 115,000 people to earn a degree. The OU engineering courses provide students with access to associate, undergraduate, masters, and Ph.D. degrees as well as associated short courses. During 1992, the engineering courses had 80,000 undergraduate, 8,000 M.Sc./Ph.D., and 15,000 associate degree enrollments. OU granted over 7,000 undergraduate engineering degrees, 1,700 honors engineering degrees and 1,200 masters and Ph.D. degrees. Clearly, the OU model is successful.

The OU mission is:

- Open as to people and will play a leading role in the transition to mass higher education by serving an increasingly large and diverse student body.
- Open as to places and will contribute to a widening of educational opportunities by making its programs, courses and services available throughout the UK and more widely in Europe and the world.
- Open as to methods and will use distance teaching methods and new learning technologies and teaching techniques to serve home and work-based students.
- Open as to ideas and will be a vibrant academic community dedicated to the expansion, refinement and sharing of knowledge.

The similarity of the NTU and OU missions is obvious. The financial support for the creation of the two universities was very different. OU was initiated with substantial government funding. NTU was initiated without government support and continues to grow without any government subsidies or support. These differences have a major impact on the evolution of the organizations.

OU was able to study distance learning concepts and design a system with relatively little constraint imposed by the lack of resources. At the time of the creation of OU, telecommunications was not at the point of development as it was at the time NTU was designed. Therefore OU has not incorporated the advanced telecommunications technology that forms the basis for NTU.

OU provides specially written textbooks or workbooks as the heart of the educational programs. Each student has a tutor-counselor content expert to work with on a course by course basis. The tutor-counselor is available via telephone, correspondence or electronic mail. Student self-help groups are encouraged and there are 250 study centers located in colleges, schools, libraries and community centers throughout the UK as well as in Ireland and Europe. Many courses have regular television and radio broadcasts transmitted on national BBC network and available on tapes elsewhere. Full time residencies during the summer provide tutorials, seminars, workshops and practical experience. Thirteen regional centers oversee the 6,700 tutors and counselors and link them to groups; of students. Thus the OU educational program is planned, created and implemented within the OU system as opposed to the NTU model which shares existing resources via telecommunications.

D. OTHER PARADIGMS

The National Technological University and Open University models discussed above are the two major new paradigms for engineering education that are successful and presently exist. They are not the only new paradigms that exist, Motorola University and Skill Dynamics are examples of other models that offer engineering related education to students.

The Calvin Center for Continuing Education, located in Schaumburg, Illinois, is the largest Motorola training facility with classrooms, laboratories, and administrative offices for Motorola University and the Lake Forest Graduate School of Management. The Center has several state-of-the-art technologies available, including personal computer, engineering, manufacturing, and Programmable Automation Literacy laboratories; an advanced audiovisual system; and a wellness-oriented food service. Motorola University has facilities in Mesa, Arizona; Mansfield, Massachusetts; Austin, Texas; Singapore; Edinburgh, Scotland; Slough, United Kingdom; and is developing training facilities in China and Japan. Most Motorola University programs are offered at most Motorola locations in the U.S., Asia, Europe, Eastern Europe, and Latin America. This industry-based model is extensive and effective. Although presently serving the educational needs of Motorola and its customers, it is apparent that Motorola University can compete with U.S. engineering education program offerings.

E. SUMMARY

Distance learning is an established part of engineering education in the U.S. and the U.K. Many U.S. engineering colleges are offering their engineering courses and in some cases, their engineering degrees to distance learners. The educational technology used is based on television and telecommunications for the hardware and software, and on an extension of the usual institutional human organizational structure to outreach programs. This model represents very little change in the education of engineers except for the distance delivery of the materials. Because there is little change, faculty participate extensively.

The National Technological University represents a new paradigm for engineering education. It is a consortium of U.S. universities. NTU strives to deliver the best courses with the best instructors from the member institutions to the distance learners. NTU has established a course selection process which uses student evaluations to make decisions. NTU is supported financially by student payments which are usually reimbursed by student employers. The NTU digital compressed video telecommunications network provides cost effective distribution via satellite. The network is also capable of transmitting computer data and computer-based instruction to distance learners.

The Open University of the United Kingdom is also a new paradigm for engineering education. It uses course materials such as books, handouts, etc., and tutor counselors as well as regional centers throughout the U.K. It has been very successful in delivering quality education to distance learners. Each course is costly to develop, but is used for large numbers of students, and so can be cost effective. The major difference in the two paradigms is central control of the educational materials as in the OU, and selection of the best of the existing courses as in NTU. Both institutions can expand course delivery systems to include computer-based instruction. Both paradigms will evolve during the next decade to include computer-based engineering education, and will compete for students internationally.

IV. EFFECTIVENESS OF DISTANCE EDUCATION

A. INTRODUCTION

Distance education is realized with the use of technology. The printing press provided the first such technology, and printed materials still provide the most used distance education media. Billions of people use printed or written materials for educational purposes. This forces virtually all supporters of new technologies to face the reality that most learning involves written materials. Humans have most likely limited their capabilities to learn using other technologies because of the emphasis on written information. Learning based on this technology, coupled with face to face verbal communication is the basis against which other learning methodologies are measured.

Most academicians automatically assume that the standard lecture method with student interaction is the norm. The idea of sitting on a log with Plato for discussion and education is not the norm, even if it might be a better learning experience. Thus, the resistance to change and the inertia of the educational system must be recognized and planned for if major changes in education are to be realized. It takes about 3 generations to make major changes in an imbedded educational system. Thus if engineering education is to embrace distance learning technologies a planned process is needed to help speed up the acceptance rate. In any event, competition will drive the new technologies to eventually replace the old.

B. GENERAL EFFECTIVENESS

Maureen P. Gibbins of the Center for Communications at San Diego State University authored a report [6] in January, 1989 titled The Effectiveness of Technology Applied to Instruction: A Summary of Research Literature. This report reviews 80 articles which deal with the evaluation of the effectiveness of distance education. A quote from a 1977 publication, Big Media Little Media by W. Schramm summarizes the matter "...over and over again the answer has come back: of course, students can learn effectively from the media, from any medium...It depends on the performance of the teacher, the content of the media, what is being taught, and to whom."

Ms. Gibbins' conclusion states: "Any instruction depends, as it always has, on the skill of the teacher, the content of the media, what is being taught, and to whom. Wilbur Schramm was right. The ongoing question is not whether media will be used effectively in teaching, but how best to utilize all our instructional resources in order to benefit from the unique capabilities of each." This conclusion applies to engineering education. A process which will implement the application of new technologies and evaluate the results in terms of cost, quality and productivity would provide guidance for the use of instructional resources.

An article in the March, 1977, issue of *Science* by James Gibbons et al [7] showed that off-campus graduate engineering students taught remotely using instructional television (ITV) delivered materials did not differ statistically in their learning accomplishment from on-campus students based on grades. These results have been supported by unpublished studies done by the National Technological University. Grade analyses show that off-campus students did better than on-campus students.

Dr. Harvey P. Stone authored A Comparative Analysis of Interactive and Non-Interactive Video Delivery of Off-Campus Graduate Engineering Education [8]. This 135-page document concludes that "off-campus graduate engineering students do not suffer from the inability to 'talk back' to faculty in real time." This result, coupled with the others, documents the effectiveness of distance education and remote learning, especially for graduate engineering students.

Distance education has not been applied extensively to undergraduate engineering education because laboratory experiences are costly and difficult to provide. There have been discussions, especially in legislative bodies, aimed at providing access to engineering education via distance education systems. Many states have initiated efforts to implement state-wide distance education systems. The principle motivation is to reduce the costs of education by means of technology. Such systems tie the state universities together so that education can become less costly and more responsive to student needs.

Computer-based remote education is now developing, and there have been few studies of effectiveness. Based on previous evaluations of distance learning, there is no reason to believe that such technology will be less effective than other media. There is also no reason to believe that the technology will be any more or less effective than other media. The media does not add to the learning accomplishment, it provides a means for reaching the learner. What computer technology can bring to the student is a broader learning experience with mediated interaction between students, experts, and tutors.

Again, it is apparent that there is a need to establish a process in which educational technologies can be implemented and their impact evaluated so potential users can make intelligent decisions about acceptance.

In summary, distance education can be as effective as on-campus education, especially for graduate engineering education. Faculty will argue that undergraduate students require on-campus experiences, but legislators will continue to press for reduced costs and more faculty productivity. As the state-wide technology based distance education systems are implemented, all faculty will have to reevaluate their roles in education, and the role of distance learning will become more important.

C. LEARNING ACCOMPLISHMENTS FOR NTU STUDENTS

Assessing learning outcomes has become a major interest of educators. Measuring return on investment in engineering education is of great interest to industry, but it is a difficult and complex measurement to make. A few companies have established a three-layer evaluation system. The first level evaluation is done by the student using a student questionnaire tool. Most universities have established this level of evaluation. The second level of assessment is made by asking the supervisor of the student whether the learning accomplishment has improved the performance of the student. The third level is a follow-up assessment performed about a year after the learning effort, again trying to assess the impact of the learning experience on the productivity of the learner. This three-layer effort is expensive and not standardized, so little evidence as to return on investment is available based on such an effort.

Most assessments that are used to compare student learning accomplishments are related to grades. This is a reasonable means of making comparisons as long as the grades are based on similar experiences. The NTU educational system is well-positioned to make measurements based on grades because the course is usually taught live on campus to local students while at the same time it is broadcast to local ITV students as well as NTU students. The homework, tests, and final examinations are identical, so grade evaluations are based on the same assessment tools.

Information is available from NTU that shows grade data for NTU, ITV and on-campus students from 1982 to 1990. Admitted NTU students performed the best. All NTU students who are not admitted into NTU degree programs were next best, performing as well as ITV students. The on-campus students were at the bottom, about 0.25 grade points below the admitted NTU students. The data shows that learning accomplishments as measured by grades is best for NTU degree-seeking students. There are other aspects that must be considered including maturity, desire to get reimbursed, and the industry selection process involved in hiring the NTU student. Yet the results are conclusive.

D. CONCLUSIONS

Distance learning is an effective means to educate engineers. Distance learning technology is effective as a medium for engineering education. Educational technology as represented by local ITV systems and NTU are effective learning systems for engineers. Distance learning technology and methods have not been extensively used for undergraduate engineering education, but they are now being implemented in many places.

V. BENEFITS OF DISTANCE EDUCATION

A. INTRODUCTION

The benefits available with distance education include improved learning accomplishments, improved quality of education, removal of time and place constraints for students and instructors, expanded access to education for the home-bound and disabled, improved faculty productivity, and lower education costs. Engineering education should provide leadership in developing the methods for realizing these benefits. Other educational fields will use the methods developed and apply them to their educational systems. Engineering education needs an organized nationwide cooperative effort to implement the new education paradigms.

Learning accomplishments as measured by grades have been shown to be improved with distance learning systems as shown in Sections IV B and IV C. The quality of instruction can also be improved as shown in Section III B by using student evaluations for course selection. Distance education will provide the home bound and disabled with improved access opportunities. Lower education costs will accrue as the use of distance learning systems becomes more prevalent.

Since distance education is now dominated by television, the benefits of this paradigm are apparent. The promise of distance education delivered via an information superhighway has not been realized because there are so few examples and experience with this delivery mode. Little effort has been expended in a systematic way to implement computer-based distance learning on a large scale basis. Since digital compressed video can be transmitted via the information superhighway, a cooperative effort to deliver instructional television and computer based engineering education via the superhighway is needed.

Many educators believe that interaction is an essential part of learning. Interaction provides the learners with assessment, recognition and answers to questions. Therefore a brief discussion of interactive education is provided before describing the benefits of distance education, both realized and anticipated.

B. INTERACTION

Distance education provides opportunities for both live interaction between the presenter and the learner during the presentation as well as delayed interaction. Live interaction does not provide opportunities for improved faculty productivity, since the presenter must be directly available to the student, and the number of students that can actually interact is limited. Although useful, this form of interaction in distance learning lacks great impact potential. It also tends to be more costly since the interaction requires telecommunications equipment, and interconnect time charges from each learner to the presenter, which are added to the cost. The presenter cannot handle more than a standard lecture enrollment. Some distance engineering education institutions, such as the University of South Florida, require their distance undergraduate learners to be present at the remote site during class.

Delayed interaction can provide improved faculty productivity with other benefits. Delayed interaction can occur with the presenter, a tutor/faculty facilitator person, other students in collaborative learning, or with a computer program. Delayed interaction can be accomplished via telephone, electronic mail, and facsimile, as well as computer or personal communication. Engineering education relies on interaction for many purposes, including assessment and responding to inquiries. Therefore, educational technologies applied to engineering education must include some method which will allow students to get access to people or information for question/answer and discussion interactions. There is also a need to provide laboratory experiences for undergraduates which can be done remotely via simulations or remote laboratory centers. With properly prepared programs, the interaction can provide high quality learning. All of these options provide opportunities for innovative experimentation to explore the benefits available with various forms of interactions.

C. REMOVAL OF TIME AND PLACE CONSTRAINTS

Practicing engineers are very busy and mobile. They have difficulty attending traditional institutions because of business and personal demands on their time. The mundane problem of travel time to the school, if one is close by, coupled with the time needed to find a parking place and sit through the lecture, deters practitioners from pursuing advanced engineering education or continuing education in traditional programs. This is a major reason distance education has been implemented for practicing engineers. This is a real benefit for the student.

Distance education is delivered to the student, the student can view the materials at a convenient time and perform the needed study, homework, etc. with some degree of flexibility. Deadlines are important to ensure progress toward completion. However, if a practicing engineer has to move to a new location, even overseas, it is possible to continue the educational effort because of the place and time flexibility.

Some engineering practitioners aspire to Ph.D. degrees. Distance education has provided a new approach to satisfy this aspiration. NTU students have made arrangements with NTU schools to get Ph.D. course materials delivered to a subscribing site to complete the courses needed for the degree. The student then goes to the institution to complete a dissertation. This is a benefit to the institution because there is no need to support the candidate while completing course work, and to the student by allowing completion of degree requirements cost effectively. Some NTU subscribers support their employees while completing this type of program.

Removing the time and place constraints is a benefit to the practitioner's firm. Each firm has its own policies concerning support of employee education, but most engineering firms reimburse for engineering educational efforts. Tax laws have encouraged such actions by making reimbursements non-taxable to the student. Large firms have developed large-scale distance learning programs for their employees in conjunction with local schools as well as NTU.

High technology firms have supported these efforts, but lately have reassessed their positions. Firms have moved away from a policy that employee education is a benefit to be used as desired by the employee to a position that employee education must be in line with company goals to be reimbursed. Although more costly than most on-campus education, distance education is still considered a good investment by most major engineering firms. It represents an investment in the intellectual capitol of the firm, a most important investment for engineering talent, while at the same time allowing the firm to utilize the capitol while it is improving.

The time and place constraints are also imposed on engineering educators. Instruction in traditional programs is on campus at specified times. Both of these limiting constraints are removed with distance education. Instructors can develop learning materials when and where it is convenient. Students can study the educational materials at their convenience and on demand. Such flexibility is highly prized and leads to additional advantages.

Top quality researchers are often limited in the amount of time they choose to allocate to teaching. Teaching in distance education systems provides video materials which the instructor can control. The instructor can determine whether to develop new lectures every term, or use old ones. Obviously this can improve faculty productivity and is a benefit to the faculty.

Innovative use of the video materials with tutors and at other schools represents additional benefits which are evolving. The NTU Course Exchange Program provides member universities with access to course offerings from other member institutions for on-campus use as discussed in Section III.

Removing the constraints of time and place are also of interest to undergraduate students. A major fraction of undergraduate engineering education can be implemented using distance learning methods. This is now being initiated at a few schools as noted in Section III.

Engineering educators are seeking to encourage women and minorities to enter the profession. These individuals, especially minorities, hesitate to enter the white male dominated engineering education environment. Distance education, coupled with innovative implementation can provide women and minorities access to engineering education in a friendly environment. For example, traditional black and women colleges will be able to use distance learning engineering course materials to teach their students on their campuses. Some of this thrust is now being implemented through the engineering education coalitions.

D. IMPROVED FACULTY PRODUCTIVITY

Distance education can bring an outstanding instructor to the student. This person can be an excellent teacher as well as an expert in the course materials based on up-to-date research efforts. Such faculty are normally extremely busy and distance education allows their expertise to be shared. This expert becomes the presenter who teaches the course on campus to live students. The experience is video taped and shared with other schools, and/or saved for later use on campus. This improves the educational productivity of the presenter.

A disadvantage of this arrangement is that the presenter may not be available to the distance learners. Faculty using the tapes of another presenter must be retrained to become managers and facilitators of engineering education. Using distance education, the best experts/teachers can be shared as presenters, but their improved productivity limits their availability to students. Engineering faculty can provide the personal interaction needed to best serve the student. As distance education processes evolve, course developers will develop mechanisms to provide interaction via materials similar to self-paced instruction and technologies such as interactive video conferencing.

Another advantage of educational technology is improved ITV and CBI course quality for faculty use. By using the course materials prepared by others, the faculty can improve productivity and increase personal interaction with their students. Undergraduate teaching faculty will have access to a number of learning materials developed by the expert/teacher presenter. This will allow the faculty to reduce lecture preparation and delivery time. The teaching faculty can use the released time to interact personally with undergraduate students. This will enhance the quality of the student's learning experience and improve the teaching faculty productivity. The Software Engineering Institute at Carnegie Mellon University has developed and video taped a series of graduate courses in software engineering. These course tapes are used by faculty at a number of colleges and universities as part of the institutions academic degree program.

Student access to learning materials will be enhanced since the time and place limitations are removed. Students will be able to access specialties that their home institution does not offer. They will have access to remedial materials as needed. Thus quality, flexibility and productivity are anticipated results of the intelligent application of distance education.

Although distance education requires hardware and telecommunications technology, the opportunities for reduced costs are real. The ability for many students to share the expertise of the best will give lower costs if properly developed. There is a need to develop a process for monitoring and evaluating the best of the many distance education options. Such a process should start with graduate education since this is where the majority of distance learning efforts are now emerging, and extend the best results into the undergraduate programs.

VI. BARRIERS TO DISTANCE LEARNING

A. INTRODUCTION

The major barrier to the acceptance of distance education for engineering education is the resistance and inertia of the present system, especially on the part of faculty. The system requirements of students include entrance requirements, lock-step pursuit of a degree, inflexible course participation rules, and increasing tuition and fees. The faculty culture emphasizes competition between students with little emphasis on group efforts, freedom to control class room and course materials, student assessments based on individual performance, research over teaching, consulting over teaching, service only if desired, and devotion to specialized fields. A major cultural change will occur among engineering faculty over the next few decades.

The Accreditation Board for Engineering and Technology (ABET) accredits undergraduate programs. The process tends to focus on the degree requirements as opposed to assessing the learning outcomes. The NSPE report [2] showed that practitioners and engineering deans believed that undergraduate engineering education programs do a good job. ABET is proposing changes in the accreditation process. The outcome is not clear, especially as it impinges on distance learning for undergraduate engineering education, but the accreditation process is seen as a barrier by many engineering educators.

The cost of implementing distance learning systems is not a major barrier because the cost often does not impact on engineering education directly. In the past, the systems were created with the help of companies. Now they are being installed by institutions and government agencies in the hope of reducing the cost and improving the quality of education as well as improving the productivity of teachers.

Most major universities have installed instructional television systems for the generation and delivery of course materials. The systems include candid classrooms, television equipment including transmitters and appropriate space as well as technical and managerial talent to operate the system. Many of these systems are used by engineering faculty, and some of them are managed by personnel in the engineering college. All of these institutions have developed payment policies which reward the participating faculty and protect the intellectual property rights of the lecturer.

Most of these systems were developed in response to the continuing education needs of practicing engineers. The employers of such engineers were willing to pay for cost-effective continuing engineering education delivered to the work site, so local and regional instructional television systems were created. As a result, engineering education led the way in distance education based on television technologies. Television technology is the most-used distance education method with delivery accomplished by microwave, land lines, and mail delivery of video tapes and satellite. No other technology has become as pervasive as television for educational purposes.

Section III shows that there were 60,000 enrollments in engineering credit courses delivered to distance learners in 1993 in the U.S. There were over 1,000 degrees granted in engineering fields to distance learners, most of them graduate degrees. It is clear that the number of engineering graduate students who use distance education has grown drastically in the last decades. This data, coupled with the 1992 data for the Open University of the UK (80,000 undergraduate, 8,000 M.Sc./Ph.D. and 15,000 associate degree enrollments with 7,000 engineering undergraduate, 1,700 honors engineering, and 1,200 masters and Ph.D. engineering degrees), indicates that the U.S. is lagging in its distance engineering education efforts.

The ASEE-collected engineering enrollment data that is published annually does not identify or tabulate the distance education data as distinct from the regular on-campus data. Gathering this data is a useful task, and the ASEE should do so.

B. FACULTY CULTURE

There are a great many dedicated and productive engineering faculty. Their productivity is constrained by the existing faculty culture. The culture is quite well-defined ranging from the reward structure to interaction with students. A major aspect is the lack of accountability. Faculty tend to act like volunteers, and feel free to decide what work they will do and when.

Milton Greenberg, past provost of American University, wrote an interesting article, Accounting for Faculty Members' Time, in *The Chronicle of Higher Education* [9]. He points out that there is a difference between the academic world and the rest of the world of work, where regular schedules and accounting for time are normal expectations. He concludes that the way that faculty members choose how and when to perform their academic functions most closely resembles the behaviors of volunteers. For most faculty members, all work other than teaching classes is optional, and even teaching sometimes yields to other concerns.

Dr. Greenberg points out that universities, unlike any other employers, allow faculty to consult for pay, and often do not document the activities. Often faculty manage to be on campus only two or three days a week, and there are documented cases of individuals who have worked full time at two institutions at the same time. He notes that research is tediously hard work, and academe pays homage to good scholarly publications, but few faculty members perish from failure to publish.

He writes that faculty members' service obligations are the most voluntary of all. They know they will not be evaluated based on service activities, and many choose not to attend departmental faculty meetings, much less college or university faculty meetings. He notes that academic administrators must persuade faculty members to participate in the same fashion that administrators persuade volunteers to participate.

Dr. Greenberg writes that small, private colleges with strong traditions of faculty-student interaction do not subscribe to this culture, but the larger colleges and universities provide this culture to faculty. Henry Rosovsky, Harvard University's noted former dean of arts and sciences, called faculty a "society largely without rules", and he cited a need to redo the social contract so as to redefine the meaning of campus citizenship. Miron Tribus of MIT has described the management of faculty as "herding cats." The concerns of Drs. Greenberg, Tribus and Rosovsky will be addressed in the next decade. The financial pressures, student needs, and society as a whole will impose a new culture on higher education, including engineering education. Competition will play a major role in this reorientation of faculty culture.

C. BARRIERS ERECTED BY ENGINEERING FACULTY

Engineering education is demanding. Engineering faculty like to work with the very best students available. Engineering schools publicize the average grade point average of the entering student body, and rejoice if it is very high. This represents an entrance barrier to a prospective engineering student. Proper preparation is needed for entering students. Often, high school graduates have not received proper preparation, which results in low rankings, although they can perform well if given the chance with good remedial educational materials. Engineering faculty must become sensitive to this barrier and work to reduce it. Distance learning can provide well designed review packages which can be used in advance of specific courses or in high school to better prepare entering students.

The engineering education on-campus environment is unique. It stresses competition, apprenticeship, a high level of stress and a great deal of work. Students entering this environment usually take some time to adjust. If engineering faculty want to reduce this barrier they must provide improved counseling and pay more attention to the new students. Distance education can allow those students who have problems adjusting to large busy campuses, especially women and minorities, to have access to the engineering part of their programs via distance education while living and adjusting to life after high school at a small local campus which is not threatening.

Most undergraduate engineering education is highly competitive. Team work is discouraged, and performance is evaluated individually, with cooperation discouraged. Practicing engineers seldom work alone. They are part of teams, and performance is measured on the success of the team project. This real world environment must be included in undergraduate engineering education, and made a significant part of the experience.

Engineering faculty erect contact barriers to students in the form of office hours. Electronic technology can remove this barrier through e-mail bulletin boards where students can communicate with one another as well as the instructor. Student questions and instructor responses can be shared by all the students. This form of interaction which is being implemented with distance learners can work well for on-campus students as well.

Engineering faculty members use traditional content-oriented examinations for evaluation of learning accomplishments of students. This represents a barrier to some students. New assessment techniques are needed, especially if the team work concept is implemented in engineering education. These techniques are being implemented in distance learning environments because the students are often not on campus for examinations and instructors learn to create examinations that assess competence without encouraging cheating. Project reports and essay exams are often used. With the development of the telecommunications superhighway, personalized exams with direct interaction with remote students will evolve. On-campus faculty members should lead the way in these developments.

Considerable progress has been made in the role model barrier for minorities and women because of the support of NSF and the realization by engineering education administrators of the need to hire women and minorities as faculty. Still, engineering educators must remain sensitive to these realities.

D. SUMMARY

A number of barriers to engineering students have been identified. They impact mainly on entering students with an interest or inclination toward engineering. The biggest problem is the engineering culture, which is the result of decades of development. This culture will change, and the faculty members should be aware of the needs, and lead in developing new, more responsive environments for students. Otherwise, the faculty members will be out of work, and certainly their rewards will be reduced if they do not change.

VII. VISIONS

A. INTRODUCTION

The question is, what will be the impact on engineering education of the new educational technologies? Section III dwells on the two major existing models, the National Technological University and the Open University of the U.K. Both offer engineering education to remote learners via an educational technology system. This section will present some additional information concerning the impact of technology on the future of education in general, and engineering education specifically.

Engineering education includes continuing education. Leenamaija Otala [10] points out that there are different approaches to lifelong learning for technical professionals in Europe, the U.S., and Japan. She notes that traditionally, companies in Western Europe have taken the responsibility for training and educating the working population with three types of approaches. The German approach is industry driven and job-oriented. The Latin European approach is similar, but elitist-oriented, emphasizing lifelong learning for management. The U.K. and small European countries try to meet the educational needs of working adults with the Open University approach. The European Community tries to enhance Europe-wide cooperation in the lifelong learning environment for engineers. Extensive use of distance learning is now under study and development in Europe.

Otala notes that the U.S. has two paths for lifelong learning: special university and college programs, and programs provided by companies. Companies are developing a longer-term focus based on strategic planning for their continuing education needs. Yet practicing engineers are learning that life-long employment is a declining trend in the U.S., and responsibility for lifelong learning must be assumed by the individual. Distance learning is becoming more available, with NTU as a leading developer of opportunities. Individual institutions are extending their lifelong educational opportunities for engineers via distance learning technology.

In Japan, industry has the main responsibility for adult education and training. Engineers get shop floor and sometimes product development experience, and they are rotated within the company to expand their expertise. There is little cooperation between companies and universities. Recently, there is concern in Japan about the need for more public lifelong learning opportunities, especially for engineers. There are few distance learning opportunities available in Japan.

Major areas of the world lack any access to engineering continuing education. The underdeveloped countries and the nations of Central Europe do not have well developed continuing engineering education programs. There is little access to distance learning for engineering education in these areas. There is a real need to develop an educational technology infrastructure to serve the engineering education needs world wide. This vision would result in a world-wide sharing of both academic and continuing engineering education.

Technologies and Lifelong Learning, a report [11] prepared for the Canadian Prosperity Secretariat, summarizes key successes and barriers to technologies in lifelong learning. It proposes initiatives to create an infrastructure for the realization of lifelong learning in Canada. Key issues identified as impeding infrastructure development include: government and education institutions will lose their traditional control as commercial interests compete, the learners will force changes to allow for flexibility and choices in lifelong learning options, institutions will change drastically to use learning technologies effectively, educational professionals will need to get access to technologies to learn how to use them effectively, educators and institutions will have to work out difficult issues, including funding structures, and jurisdictional and structural issues will dominate the debate over technologies for lifelong learning based on distance learning technologies. The issues are very similar to those in the U.S. A critical need identified in the report is to establish partnerships between public educators and private interests. The same need exists in the U.S.

B. VISIONS FOR HARDWARE AND SOFTWARE

Most computer manufacturers are designing computers that are more and more powerful, with more dramatic interfaces. Mark Weiser and other researchers at the Xerox Palo Alto Research Center want to develop devices that are as unobtrusive as possible [12]. They envision an environment in which computers are as available as telephones. They note that individuals don't have personal telephones. Individuals use whatever phone is available. The same should be true in the researcher's view for computers, which should be ubiquitous and interconnected in a wireless network. The network would allow anyone to use any computer to seek any information, including specialized personal information. Such a vision would not try to create supercomputers for the use of an elite class, but rather very user-friendly low-cost "scratch computers" available to all, like telephones. An interesting vision from a research center renowned for its contributions to new technologies.

Dr. John H. Gibbons, Director, Office of Science and Technology Policy, appeared before the U.S. House Committee on Science, Space, and Technology April 27, 1993. His statement dealt with the information infrastructure, the High Performance Computing and High Speed Networking Applications Act of 1993. The information infrastructure consists of computers, computer data banks, fax machines, telephones, and video displays linked by high-speed telecommunication links capable of transmitting billions of bits of information in a second. He noted that this infrastructure will provide Americans the information they need when they need it, and it will revolutionize the way Americans work, learn, shop, and live. The strategy he presented for building the National Information Infrastructure included: implementation of the High-Performance Computing and Communications Program which encourages cooperative work between industry, universities and Federal Labs to develop the technologies needed; provide funding for networking pilot projects; promotes dissemination of Federal information; and reforms telecommunications policies. The scope of the proposed program is so large that many government agencies are involved including: ARPA, NSF, DOE, NASA, NIH, NSA, NOAA, EPA, NIST, and ED. Obviously such a large effort requires substantial funding. The administration is proceeding with the effort to realize the goal of an information superhighway by funding five components: High Performance Computing Systems; National Research and Education Network; Advance Software Technology and Algorithms; Basic Research and Human Resources; and Information Infrastructure Technology and Applications. This effort will provide engineering educators with the opportunity to use the resulting hardware and software to improve engineering education. What is lacking at this time is a plan to intelligently implement utilization of the resulting information superhighway.

C. VISIONS FOR THE ACADEMIC WORKPLACE

The 1994 National Conference of the American Association for Higher Education, "The 21st Century Academic Workplace" will address many visions for the future of education. The September, 1993 AAHE Bulletin [13] included a conversation with Charles Handy talking to Russell Edgerton concerning the conference topic. Dr. Handy will address the conference in March, 1994.

Dr. Handy, noted author of a half a dozen books on the management of change, noted that the nature of work has changed. Large organizations are getting rid of jobs, and small organizations provide the job growth seen in industrialized nations. The small organizations are often working for larger organizations without pensions, careers commitments or other aspects of traditional work environments, like lifelong work. Only 55 percent of the U.K. workforce are in conventional full-time employment, and this percentage is dropping everywhere.

Modern organizations have three kinds of workforces, the professional core full-time, the contractual organizational, and the independent workforce. The efficient company is cutting its core people and out-sourcing for workers and continuing engineering education. This same pattern is apparent in the U.S.

Dr. Handy notes that universities are faced with a strategic decision regarding the change in workforces. Should faculty be core employees or independent contractors like doctors who work for hospitals? Clearly there are faculty who are essential to the core, but others might be independents. He noted that academics sit in little offices, and often are not present. He suggests that institutions spend more effort on activity space and reduce the individual spaces to encourage interactive learning.

Dr. Handy commented on student needs. He claims that educators force-feed students by telling them what to do, where to go, what to learn, how to learn it, what to read, and what tests to take. Students learn to react, but the work life they are entering is one where they will have to invent their own careers. He notes that students are taught to analyze data that is provided for them, and the more difficult task is to find the data that is needed to solve their problems. Students have to learn to look at a mess, and come up with initiatives that are going to work through the mess. This is threatening to the instructor ,because of the unknown aspects, as well as being chaotic. Students must learn that they are responsible for their careers, and to take control of their futures.

Dr. Handy suggests that educators move to the concept of "student as a worker", rather than student as an observer. He believes the student should be actually doing, solving problems, and having creative initiatives. He identifies the core task of education "to improve people's capacity to make a difference in the situations in which they find themselves, using a whole range of skills." That turns professors into coaches and mentors rather than subject experts. He recommends contracting with the best professors and grouping them into small research units. The core function of the educating organization would be coaching and mentoring people in actual action. And as people grow older, he sees more and more of that action taking place in the workplace, using the university as a helper in the learning process. The university becomes a place to encourage and develop learning rather than a place of knowledge.

Dr. Handy predicts that the individualistic tone of education would be replaced by students working in teams that will be hard to organize and evaluate. Therefore different evaluation of learning achievements methods must be developed. Learning will be much more individualistic, based on new technologies rather than learning from the expert directly. Experts will be outside the organization and brought in through the use of media. The new educational organization will be a honeycomb of small groups of people working on problems, and it will be hard to continue the traditional semester structure. Universities will be working all the time, like other organizations, with everyone using all the facilities all the time—electronic communication, video communication, televisual communication—to coach the students. He summarizes by noting that devising real projects to maximize student learning becomes the educational task. The key people in the university become the educational designers and the educational managers. The faculty experts are in the wings brought into the picture via technology. This will cause tremendous upheavals because it will change the power structure of higher education.

Dr. Handy presents a vision of the university of the future that is certainly different. The Open University of the U.K. is a start in this direction, as is the National Technological University. Yet these two institutions are quite different, so there is a range of future options for engineering education. How the U.S. engineering education system will evolve is not clear, but the observations of Dr. Handy are based on what has already happened in many industries.

D. SCHOOL'S OUT

Lewis J. Perlman's book, School's Out: Hyperlearning, the New Technology, and the End of Education [14] provides another vision of the future for education. Most of the ideas that Dr. Handy mentioned are developed in much detail in Mr. Perlman's book. Perlman points out that the General Motors Computer Aided Maintenance System (CAMS) is a good example of a hyperlearning system which incorporates expert systems with maintenance efforts. GM mechanics can connect the microprocessor of a sick GM car via a communications system to a mainframe computer in Flint, Michigan. The mainframe, microchip, and mechanic all work together to identify problems and fix them. The mechanic learns, the mainframe adds to its understanding of problems, and the microchip provides data. CAMS provides an on-the-job learning environment which involves no schools, instructors or credentialing.

Perlman argues persuasively that hyperlearning will replace teachers and classrooms with experts, expert systems and on-demand training delivered via modern technologies. Learning will be in everything, everywhere; school buildings will be replaced by hyperlearning channels; expertise will be more in the network and less in the person; learning will span the human life cycle; learning and work will merge; capitol expenditures on robots will replace labor; and handicapped will be redefined. He maintains that over reduction is now leading to job loss for individuals as companies downsize. Hyperlearning will provide skills as needed by workers at lower costs to employers. As a result productivity will improve, and replace the need for overeducated individuals.

Education as it presently exists must expand and provide the necessary improvements for educating adults. This can be done three ways. Two of them are unacceptable, doubling budgets or taking resources from K-12 education. That leaves only one realistic way to create a sensibly balanced national plan of human capitol investment. It requires an increase in the productivity of education and training in general by using computer-based, multimedia, and telelearning systems. It has been shown that education resource inputs such as expenditures per student, teacher/pupil ratio, teacher education, teacher experience, teacher salary, administrative inputs, and facilities have very little impact on student performance. Perlman argues that hyperlearning provides a new and realistic approach to improved quality and cost effectiveness of education in general.

He lists a number of myths: people learn best in school, school is preparation for working in the real world, the teacher is the fountain and the learner is the bowl, more or less academic achievement means more or less learning, you have to learn to walk before you can learn to run, education is different from training, some people are smarter than others, facts are more important than skills, learning is solitaire, and schooling is good for socialization. These myths are debunked with telling accuracy.

He quotes research results [15] which show that college science and engineering majors claim that the faculty are poor teachers and unapproachable, leading to either changes in major studies or “grit the teeth completion” in spite of poor faculty support. Many science departments intentionally and mistakenly scrimp on the kind of course and career advising that helps students survive because of a “weed out” mentality that assumes attrition is a good thing, because it leaves the best and most dedicated students.

Perlman claims a lack of spending on educational R&D and that innovation is lacking due to entrenched bureaucracies. He calls for metrics to measure the value of all learning, especially hyperlearning, and standardization so comparisons can be made. He calls for a new type of accountability based on dealing with the customers. He relates how Regis College in Denver transformed itself into a college that realizes a “profit” from one that was failing, by listening to the customers. Registration materials and books are delivered to the students, and no student needs to travel more than 10 minutes to get to a learning center. One fourth of the old faculty have been laid off, and 85 percent of the adult, business- oriented courses are presented by adjunct faculty with real world practical experiences. Perlman claims this is a move in the right direction, but not complete by any means.

Perlman's vision is a system where accountability will be: to the customer—not academics; education will come from competent vendors—not teachers, schools or public bureaucracies; it will be delivered in real time—not in a few years or months; accomplishments will be recognized by certification of competency—not credentials of attendance. He notes "There is no job in this economy that requires an academic diploma for its successful performance. None. Nada. Zippo.". He argues against credentialing and for performance. Return on investment in education must be measured, and total quality concepts implemented in the education process.

Perlman provides an action plan to liberate learning. The steps include: outlawing credentialism; bypass the establishment with distance learning and the national information infrastructure; privatize and encourage competition; capitalize by reallocating, reorganizing and set-aside funding for technology development and implementation; replace education schools; create a National Institute for Learning; and create a bank for innovation and redevelopment. He advocates action in political, business and family environments by making learning a normal, daily activity for the family and at work by being a proponent in the political arena. He concludes: "In the words of the immortal philosopher Pogo, we seem to be confronted with an insurmountable opportunity."

E. OECD STUDIES

The Organization for Economic Co-operation and Development (OCED, 2, rue Andre-Pascal, 75775 PARIS CEDEX 16, Paris, France) has established a Center of Educational Research and Innovation. The Center has established a number of study activities such as "The Impact of Information and Communication Technologies on Post-Secondary Education", and "The Future of Face-to-Face and Distance Education in Dual Mode Institutions." A number of experts from around the world met in April, 1993 to study the latter topic. Dr. William Renwick, Honorary Senior Research Fellow, Stout Research Centre, Victoria University of Wellington, New Zealand, prepared a background paper for the meeting with the same title as the study topic. This unpublished paper provides another vision for the future of engineering education, and its contents are discussed here.

Dr. Renwick clarifies a number of issues and provides a vision of the future of post-secondary education. He notes that a new mind set has evolved recently, impacting education, that experts no longer control the way professional services are conceived and managed. Three changes have led to this mind set; economic as well as social life have been transformed by computer technology which has created new forms of interdependence between knowledge and economic activity; better understanding of the essentially global nature of economic competitiveness; and economic libertarianism has become the guiding light for government policies for economic development. Education systems in several countries are being deregulated and educational institutions are becoming competitors. Governments are telling educators that education is crucially important and more is needed, but not more of the same. Post-secondary education is no longer for 18-25 year old individuals, but must be extended to adults. Just as advances in communication and information technology are transforming economies so, too, do they have the potential to transform teaching and learning.

Dr. Renwick identifies three types of post-graduate educational institutions: face-to-face, distance, and dual mode. Face-to-face teachers equate learning with the effectiveness of their teaching performance. Distance teachers equate learning with the quality of the learning materials and the support services for their students. The nature of their work forces them to make a distinction between what they want to teach and ways by which it can be learned. This is where the new technologies that break down the barriers of distance while allowing interaction enter the discussion. These innovations are taking the distance out of distance education, and focus attention on the concept of self-guided studies. These innovations call for a paradigm shift in the way that post-secondary teachers conceive their role as teachers. Teachers must now consider the advantages that communication and information technology can offer all students.

Dr. Renwick identifies two types of distance learning institutions: open and dual mode. Open institutions, like the Open University of the U.K., are devoted totally to distance education. They alone, develop and deliver their learning experiences, and have been created for the almost exclusive use of adults. The dual mode institutions feature primarily face-to-face teaching efforts with distance education as an add-on effort. Most of these institutions are now facing the necessity of deciding how to reorganize and respond to the resource needs of the distance education effort. The budget pressures impacting all education make this decision more difficult to make.

Dr. Renwick points out that open institutions do a good job when there are large numbers of students requiring basic knowledge, usually at the bachelor level. Large numbers are needed because of the cost of course development. Dual mode institutions can respond better to small numbers of students with specialized needs because the institution has the programs due to their face-to-face teaching activities. The most successful dual mode institutions are those whose mission includes serving the needs of part-time students and providing continuing education opportunities. The disadvantages of a dual mode institution is that the distance aspect has grown like Topsy, with the planning, budgeting, facilities, organization and management functions controlled by the face-to-face teaching effort. Dual mode institutions offer advantages when compared with open institutions, including that they can offer a wider range of degrees and qualifications, they can diversify at less cost, would not require large enrollments, and could offer a wider range of options to students. If dual mode institutions work together cooperatively in consortia, they could well become very competitive in their distance offerings.

Dr. Renwick describes a vision for the future based on learning rather than teaching. He writes about problem-based learning, teachers becoming an expert resource, devising and organizing the courses of study, posing the questions their students are to wrestle with, identifying the texts and resources to be used, acting as guides and facilitators, and monitoring and assessing progress. Computers, telephones, and fiber optics are providing new possibilities for campus teaching as well as distance learning. He discusses a pedagogy of learning and identifies seven good practices: encourage student-teacher contact; encourage active learning; encourage cooperation among students; provide prompt feedback; emphasize time on task; communicate high expectations; and respect diverse talents and ways of learning. He recommends funding more research on pedagogy and learning.

Dr. Renwick predicts that most face-to-face institutions will choose to become dual mode institutions which will bring distance learning into the main stream of post-secondary education. Strategic alliances facilitated by telecommunications will become more and more important, and will involve cooperation between public and private sector activities. He envisions a merging of face-to-face teaching and distance learning methods to the benefit of the students. He quoted a paragraph from the Report of the Commission of Inquiry on Canadian University Education which is especially applicable to engineering educators:

"Still it is a general feature of universities that, the vast majority of the time, a person is standing in front of a room of students and lecturing to them. In this respect, things have not changed in a century or more. Given the inventive, creative minds that constitute University faculties, The Commission has asked why it is that those minds rarely apply themselves to creating changes in their own industry as revolutionary as those they instigate in most other sectors of the economy."

Grenville Rumble's article, The Competitive Vulnerability of Distance Teaching Universities [16] provides another vision of the future. He notes that there are twenty-six distance teaching universities in twenty-two countries, most of them established during the 1970s and 80s. He lists the strengths of these institutions and compares them with the strengths of dual mode universities. He concludes that the dual mode universities have an advantage, and notes that they both serve large markets. In the autumn of 1988, forty-three percent or 5.5 out of 12.8 million degree credit students in the U.S. were enrolled as part-time students. The usually commuting, almost always older students reflected a dramatic shift from first-time, full-time residential students of the past.

Private sector companies are marketing sophisticated training packages. Motorola offers an extensive catalogue of courses that are now being taught under franchise in a growing number of countries. Apple computer and Microsoft have an increasing number of computer programs which can be used for self-instruction. As a result, public funded institutions are being forced to compete with private sector educational organizations.

F. SUMMARY

There are many individuals and groups who are convinced that technology will bring a new educational environment as noted above. The aspect that is needed is implementation of a process that will result in the "best" results in a well-defined sense. This best must be the learning of the student, and the application of the learning to an improved life for that student and society. Engineering education in conjunction with other professions can and should be the leaders in defining and implementing the process.

VIII. CONCLUSIONS AND IMPLICATIONS

A. INTRODUCTION

This section draws conclusions from the preceding materials. These conclusions have been chosen by the author, and focus on changes that are expected to happen in engineering education. The changes are so basic that they impact on the missions and organization of universities. It is concluded that all institutions of higher education will change, but they will have choices as to which will be the best direction to pursue. The choices will impact on universities, engineering colleges, engineering faculty and engineering students. They will be presented and discussed in the order noted.

B. INSTITUTIONS OF HIGHER EDUCATION

Conclusions include:

1. These institutions will have three choices for future directions in teaching: to function as face-to-face, dual mode, or open institutions.

2. These institutions will face competition from private organizations.
3. The institutions must reconsider their faculty governance structure.
4. They must reconsider the faculty reward system.
5. They must revise their resource allocation structure.

Discussion:

The conclusions are based on the materials presented. All readers may not agree with the conclusions, and all universities will have different views regarding the statements. Yet all of the conclusions are reasonable in view of the impact of technology on education. All major industrial organizations have already made decisions based on similar conclusions impacting their organizations. Service organizations have also acted, and government organizations are now undergoing examinations that will provide similar conclusions.

The traditional face-to-face teaching university will be forced to change its mode of operation or go out of business. Distance learning will provide their students with access to education delivered from other sources at competitive costs with high quality pertinent courses. Industrial organizations will provide workers with on-site access to work related distance learning opportunities.

Most U.S. universities are now, or will become, dual mode teaching institutions with on-campus face-to-face teaching, along with the educational technology to deliver their courses to distance learners. These institutions will serve the educational needs of their local or regional customers. They will form nationwide consortia. These consortia will learn how to work together to provide the best education at reasonable prices. The institutions will share courses so they can focus on what they can do best. Research institutions will provide the latest research in graduate courses which will be shared and the material will migrate into the undergraduate programs. Dual mode institutions will be the dominant U.S. type of university.

The open university is another type of institution which will be present in the future. They will not be dominant, because of their limited present mission. They are aimed at adults, and they have a limited range of materials that they can offer, not nearly as large a range as the present U.S. research university. Their courses are costly, and can be created only if there is enough demand. Yet they will be a major force, especially in regions where there are few educational institutions and a large unmet demand for education. The U.S. does not represent such a region, even for engineering education.

The private universities face the same choices as public-funded institutions. They tend to be more responsive because of a need to meet costs to survive. A new type of educational organization has appeared which is private and for-profit. They will grow with time, and will become competitors to institutions of higher education. The major obstacle facing such institutions is credentialing of students involving the licensing and accreditation aspect of education. This aspect is under attack, and the future is uncertain. The consumers of education are demanding fast response focused education. The providers are not doing a good job in responding to the demands. For-profit organizations will be able to respond faster, but the concern is real that the quality of education may diminish. This aspect will create controversy in the future.

The university governance system is rooted in the faculty. As the individual institutions convert to dual mode teaching, the institutions will discover how cost effective it is to use courses from other institutions. They will move in a direction to reduce faculty counts drastically, and to hire adjunct faculty to do specific teaching tasks. This will force institutions to consider changes in the governance system. The consortia that will serve engineering education in the future will not be governed like existing institutions, but will be much more oriented toward participatory management.

The faculty reward system has been under pressure to provide more rewards for teaching. This trend will continue. There is a need to establish better techniques to measure the quality of teaching as well as the return on investment for education. Research universities will provide cutting edge courses which will be used at other institutions as well as in industry. Most U.S. institutions will see major shifts in the reward system for faculty.

Since distance learning will become very important in dual mode institutions, the organizational infrastructure of the institutions must respond. There must be changes so the distance learning infrastructure is not an appendage of the continuing education office. The educational technology must become a major factor in decisions affecting both on-campus and distance learning teaching. The infrastructure will become important to both research and teaching efforts. The research effort will become cooperative between institutions using the educational technology.

The goals and missions of institutions should be reconsidered in view of the impact of educational technology. Distance learning will become more important, and should be recognized in the mission. In addition, institutions should seek the best education for on-campus students, and sharing courses is an realistic method to improve course coverage and course quality. Faculty productivity can also be improved.

C. ENGINEERING EDUCATION

Conclusions include:

1. Engineers have created the devices and software which are basic to the information age.
2. Lifelong learning for engineers is necessary.
3. Practicing engineers are now using the information highway to create the devices and systems of the future.
4. Private for-profit organizations, as well as a few dual mode institutions, are providing engineering education of high quality.
5. Engineering education should encourage group collaborative courses with real problems, and reduce the competitive aspects of the experience.
6. Engineers have created the technology which has led to massive changes in how industry functions, how people live, and how experts view the future. The information revolution is the result of the expertise of engineers. A reasonable conclusion is that engineering education should be the leader in creating the new educational environments of the future.

Discussion:

Few engineering educators would argue with conclusions 1 and 2. Conclusion 3 is significant because it shows how distance learning can be used to improve all of engineering education. Since the practitioners are developing the uses of new technologies, engineering educators should work with them to develop the new methods for distance learning.

Conclusion 4 is apparent in the continuing education field. Large numbers of experts, including engineering faculty, are selling their knowledge in short courses. Many of the experts are connected with small companies created for the purpose of making profit from presentations, short courses and workshops. This is significant because it shows that some faculty already are operating as independent experts in engineering education. They and their presentations could easily become academic courses at interested institutions.

Conclusion 5 requires institutional action at the college level to implement. The desired course should be interdisciplinary requiring faculty cooperation to create. The NSF funded coalitions have started interesting efforts aimed at implementing this conclusion. A means of sharing these developments in a more effective way than publishing is needed.

Engineering education should lead the way in creating the educational technology of the future for the reasons outlined. In order to be the leader, an organizational structure is needed. The recommendations section of this chapter provides some ideas for an appropriate structure.

D. ENGINEERING FACULTY

Conclusions regarding faculty include:

1. Engineering faculty must create innovative educational programs to serve the needs of students which take advantage of distance learning methods and educational technology systems.
2. Faculty members must reassess the role they play in teaching and learn to use distance learning methods and materials to improve their productivity and course quality.
3. Engineering faculty with research interests should use educational technology to perform research with colleagues, and distance learning methods to increase their teaching productivity by sharing their materials with colleagues at other institutions.
4. Faculty must consider the possibility that they will become individual consultants operating as adjunct teaching faculty as distance learning materials are used for engineering education.
5. Engineering faculty who chose to change their role from disseminator of knowledge to a facilitator/planner of education must develop the methods for designing and managing educational programs for their students.
6. Engineering faculty with the cooperation of other professionals must invent new learning assessment techniques to properly measure learning outcomes.

Educational technology, distance learning, undergraduate, and graduate engineering education are all merging to create the engineering education of the future. Presently, educational technology relies on television technology because of the lack of computer-based or multimedia courses. Conclusion 1 will change this situation if action is taken to measure the quality of the courses and share the results nationwide.

Distance learning methods of developing self-study materials can be applied to undergraduate as well as graduate engineering education. Engineering faculty should use course materials created at other universities to allow them more time to interact with students.

Cutting edge teaching materials should be developed by researchers, and researchers should develop cooperative research and teaching efforts. Thus a number of institutions can work cooperatively in specific research areas with the results shared via educational technologies.

As institutions reassess their roles and options in the face of stringent financial constraints, some of them will choose to change the role of faculty. The institutions may choose to purchase course materials from vendors, other institutions or private for-profit sources. These institutions will retain a few faculty who are prepared to become managers and facilitators for engineering education. Other faculty, especially outstanding teachers, will become producers of educational courses in their fields of competence. They will do this as independent consultants or as groups of consultants banding together to get the best efficiency of production.

In order for the changes to proceed efficiently, professionals competent in assessing learning accomplishments must work with engineering faculty to create metrics. The measurements must be available to implement the changes needed. Without measurements, there is no basis on which to make decisions between the many options now available to engineering educators.

E. ENGINEERING STUDENTS

Conclusions include:

1. Engineering students must recognize that they are the customer that engineering education must serve.
2. They will realize that they have options which they will be able to exercise to get the best engineering education they can afford, and that they should shop around when seeking that education.
3. Present evaluations of engineering colleges rely on perceptions, not measured performance. Metrics to measure the learning accomplishments of engineering students will provide prospective students with the ability to choose education options intelligently.
4. Practicing engineers pursuing continuing and academic engineering education will be the best readily available source of evaluations, and they provide a good source of course quality evaluations because of their experiences.
5. The changing work culture in industry forces practicing engineers to assume responsibility for their technical competence, and therefore their lifelong learning efforts.

Prospective engineering students will benefit from the impact of distance learning on engineering education if the appropriate systems are established. All levels of education will also benefit from what is learned by using engineering education as an evaluation tool for distance learning methods as applied to face-to-face education in dual mode institutions.

IX. RECOMMENDATIONS

A. SPECIFIC RECOMMENDATIONS

The role of distance learning in engineering education is the title of this chapter. It is clear that on-campus learning and distance learning will merge in the future for dual mode universities. Engineering educational institutions will share educational resources in the future. Thus, distance learning will become integrated into engineering education, and will become a means for improving on campus education. A major challenge is to identify the best way to encourage and implement the best results. A Baldrige Award would provide guidance for implementing best practices.

Improved faculty productivity and course quality are the desired results of the recommendations. The recommendations must be practical and attainable. A means for measuring the impact of recommended actions must be identified and accepted.

A measure of faculty productivity is the number of semester credit hours taught by each person. This is a gross quantity measure obviously, but it is a start. A more precise measure would include the quality of the learning materials associated with the course. A measure of the course quality can be based on student evaluations. Both of these metrics could be improved with the help of professionals who have studied these questions. These observations lead to the first specific recommendations.

1. NSF should establish metrics that measure engineering faculty productivity and engineering course quality, and NSF should establish a data base for the collection of this data with professional access.

   Engineering colleges should be asked to provide data to NSF. The data base would provide engineering administrators with a means of comparing information and making decisions. The gross metrics mentioned above provide a suggested starting point, but better metrics may be available.

2. NSF should require all funded engineering teaching improvement programs to include the same metrics as a part of the final report.

   It is recognized that the metrics may not be appropriate for educational research efforts, but they should be required to raise the awareness of applicants to the need for metrics. In those cases where metrics are inappropriate, the proposers should be required to propose appropriate metrics for measuring the productivity and quality aspects of the proposed program.

   The proposed productivity/quality data base would be an important tool for establishing an educational technology test bed. This test bed would provide NSF with an opportunity to evaluate the effectiveness of funded proposals against data.

   Virtually every major engineering educational institution has an instructor evaluation system in place. NSF could easily initiate the data base recommendation for quality by collecting all of the existing data and creating a master data bank for engineering education. The NTU data is available to help initiate the project.

3. NSF should fund the development of user-friendly means for engineering educators to produce educational materials which can be delivered via technologies, and in all such programs require that the results be used in another institution.

   These programs would force cooperation among researchers, and would result in data which can be compared to the data base to measure outcomes.

   Additional problems which face dual mode institutions relate to accreditation and regulation. The problems associated with accreditation are being addressed by the appropriate organizations, and solutions are emerging. The problems associated with state regulations are not being adequately addressed. It appears that a national organization should consider this question to ensure protection of the students from poor education, while not constraining the flow of education across state and national boundaries. The development of a quality data base will provide a means for protecting the participants.

4. NSF should encourage the development of an educational infrastructure that can provide a test bed for new technologies and delivery methods by providing funds and some planning efforts.

   This recommendation could provide funds to bring the television-based system and the computer based-systems together. It should lead to development of tested educational technologies which can be used in other areas of education. The use of engineering education as a test bed is desirable, as noted elsewhere.

B. GENERAL RECOMMENDATIONS

These recommendations provide general guidance to readers.

1. A national distance learning for engineering education agenda should be created.

   The National Science Foundation has the responsibility to support the improvement of engineering education specifically as well as education in general. NSF should fund an existing organization such as the Institute for Distance Learning, or create a new organization, that will provide the coordinating leadership that will result in the rapid development of distance learning methods and the application of these methods to engineering education. Policy issue development should be part of the charter of this organization which should include representatives of all major stockholders in distance learning.

   The Department of Defense has established a Distance Learning Committee which may provide some leadership for the required organization.

   The organization should have national representation with states, consortia of states and educational institutions, higher education institutions, the business community, and appropriate government agencies. Goals include formal cooperative agreements for reciprocity on distance education with global implications.

   There is a need to develop measurements based on learning outcomes. Student evaluations seem adequate, but the measurements should be standardized so accurate comparisons can be easily implemented.

   The organization should create standards for distance learning organizations which are process based with the ability to evolve as knowledge is gained.

   The organization should address the questions of accreditation and licensing. This work has been started with the ALLTELL Project, and it should involve the National University Continuing Education Association Independent Study Division, and the American Council on Education. Coordination between regulatory licensing agencies and accreditation organizations is necessary for academic programs resulting in credentialing. Credentials are necessary to protect the public, since the new technology will allow non-competent suppliers to invade the market. Most of the corporate customers will police programs, but individuals will have difficulties identifying quality. The organization must define quality distance education and clear the way for its implementation, not only nationwide, but globally.

2. The NSF should fund research and small scale implementation projects which will allow innovative concepts to be explored with meaningful outcome based evaluations. Large scale implementation opportunities should be available through a structure for disseminating successful results.

   The NSF research agenda should rely on input from the national coordinating organization. Research in self-guided learning, assessment of learning accomplishments, best methods for learning, etc. should be included. Small scale implementation projects to evaluate new distance learning methods should be funded. Successful projects, as measured by evaluation results, should be identified, and dissemination of these results and the specific course materials to consortia for expanded use should be in place.

3. NSF should focus these efforts initially on engineering education, and should involve professional societies, especially the American Association for Engineering Education, for developing a common data base and reporting system among agencies and the engineering community.

   Data regarding the availability of distance learning opportunities are uncoordinated, and few are available. Engineering educators need a common, accurate source of information about available course materials. The data base should include quality evaluation data, and information about the cost and availability of the materials. A similar data base about teaching methods and technologies which are available should be developed. This data will serve the needs of engineering educators, and best practices can be implemented for other areas of education.

   In addition to the data base for educators, a data base for students is needed. Academic courses and programs should be listed with evaluation data. Interested students can access the data base for detailed course information. This data base will most likely be developed by distance learning suppliers. The focus on engineering education will provide the most competent large group of students to work with when developing new distance learning methods. The present distance learning programs are targeted at practicing engineers, and they are substantial. These students are using modern technologies in their everyday work environment. Therefore they can use the technologies and can evaluate their effectiveness.

   A major problem in realizing-computer based distance learning is the lack of course modules. Funding small implementation projects with large scale dissemination of high quality results will identify and bring the best of the modules to the student body. Training programs for faculty will be needed to reorient them to the new roles of the managing engineering education. The concept is to build on an existing base of distance learning activities to expand the opportunities.

   As the new technologies are developed, undergraduate engineering educators will need to accept the distance learning modules for use in the undergraduate engineering programs. As simulation technology and programs develop, again engineering education is the best environment in which to test the methods.

C. SUMMARY

The role of distance learning in engineering education should be one of providing a test bed environment for evaluating new methods of learning based on educational technologies. Total quality management concepts should be applied to the process. A leadership organization should be formed with broad representation. It should drive the development efforts. Dissemination of results to policy makers in education will spread the best practices to other educational arenas.

Engineering educators have studied and reported on engineering education for years in response to pressures generated by educators. Few of the reports have had any measurable impact, with the exception of the ASEE Grinter Report. The 1987 ASEE report, A National Action Agenda for Engineering Education, is a recent example of good recommendations having little impact. It is apparent to the author of this report that the forces pressuring for change are coming from outside of engineering education, and will force fundamental changes in course quality and faculty productivity. Competent faculty will be forced to respond to the external pressures, and should give serious thought as to how to respond. It is hoped that this report will help.

REFERENCES

1. **Denny, Brewster, C.**, Letters Section; *Science* Vol. 261, 24 September, 1993, pg. 1661.
2. Engineering Education Issues: Report on Surveys of Opinions by Engineering Deans and Employers of Engineering Graduates on the First Professional Degree; NSPE-PEE Sustaining University Program - NSPE Publication No. 3059, November, 1992.
3. **Lionel V. Baldwin and Kenneth S. Down;** Educational Technology in Engineering, prepared for the National Academy of Engineering; National Academy Press, Washington, DC; 1981 (out of print).
4. The Electronic University: A Guide to Distance Learning Programs; published by Peterson's Guides Inc., Princeton, N.J. in cooperation with the National University Continuing Education Association, 1993.
5. **Dr. Arnold E. Seigel and Cynthia Davis,** The Delivery of Televised Undergraduate Engineering Courses: A Survey of Engineering Schools with ABET Accredited Programs; An unpublished report, College of Engineering, University of Maryland, College Park, 1992.
6. **Maureen P. Gibbins,** The Effectiveness of Technology Applied to Instruction: A Summary of Research Literature; Center for Communications, San Diego State University; January, 1989.
7. **Gibbons, James F., Kincheloe, W. R., and Down, K. S.**; Tutored Video tape Instruction: A New Use of Electronics Media in Education; *Science*, Vol. 195, March, 1977.
8. **Stone, Harvey R.**; A Comparative Analysis of Interactive and Non-Interactive Video Delivery of Off-Campus Graduate Engineering Education, available from CPD Publications Office IOIEE Bldg., Haber Road, Iowa State University, Ames, IA 50011-3070.
9. **Greenberg, Milton**; Accounting for Faculty Members' Time; Point of View section of The Chronicle of Higher Education, Vol. XL, Number 9.
10. **Otala, Leenamaija**; Studying for the Future: Lifelong Learning in Europe, the U.S. and Japan; ASEE Prism, October, 1993.
11. **Stahmer, Anna with Jacqueline Bourdeau and Arlene Zuckernick**; Technologies and Lifelong Learning; Report for the Prosperity Secretariat, Vol. 1, Canada.
12. PARC Builds a World Saturated with Computations; Computer Science; Research News; *Science*, Vol. 261, September 17, 1993.
13. Upside-Down Thinking about the 21st Century Academic Workplace; AAHE Bulletin, September, 1993, Vol. 46, No. 1.
14. **Perlman, Lewis J.**; *School's Out*; publisher William Morrow and Company, Inc., New York, N.Y., 1992.
15. **Hewitt and Seymour**: They'd Rather Switch Than Fight; *Science*, October 18, 1991.
16. **Rumble, Greville**; The Competitive Vulnerability of Distance Teaching Universities; OPEN LEARNING an Open University publication, June, 1992.

17. **Walton, Mary**; *The Deming Management Method*; Peridgee Books published by The Putnam Publishing Co., 200 Madison Ave., New York, N.Y.

Acknowledgment: Thanks to Dr. Charles Elliott and Dr. William Cooper, pioneers in distance learning for engineers, for their review and comments.

Chapter 8

THE IMPACT OF TECHNOLOGY ON THE ENGINEERING LABORATORY

James P. Avery
Department of Electrical and Computer Engineering and Integrated Teaching Laboratory
College of Engineering and Applied Science
University of Colorado, Boulder

I. OVERVIEW

In order to assess the impact of technological change on the engineering laboratory, a number of assumptions about the role of the laboratory in the undergraduate curriculum must be made. However, the technological advances discussed here have the potential to change the underlying assumptions, which complicates the discussion considerably. I will use the University of Colorado, with which I am most familiar, as a basis for describing the kinds of changes I believe will occur in engineering laboratory practice. Where particular examples are needed, I will take them from electrical engineering. In addition, a new program at the University of Colorado, the Integrated Teaching Laboratory, provides a way of describing the implementation of the ideas presented here.

II. TECHNOLOGICAL IMPACTS

The technological changes that I believe are important to the future of laboratories in undergraduate engineering education are: the reduced cost of sophisticated computer equipment; the improvement of data acquisition and sensor technology; the widespread availability of simulation and computer-aided design software; the expanding use of multi-media based systems for delivering lessons; and the ability, not yet realized, of integrating all these advances.

A. LOW COST COMPUTER SYSTEMS

It is safe to assert that never before have advances followed advances so rapidly for so long as in the computer industry. Computational power continues to increase dramatically each year, with costs that have remained mostly constant over the last decade. Unfortunately, it seems that manufacturers are determined to keep prices constant and increase performance, rather than lowering the prices for systems of constant capability. Thus an Apple II of the 1970s costs roughly what a new Macintosh costs now. The implication for laboratories is that the power available for computation in the laboratory is much larger than it was, but the computers will still cost about the same as they would have a few years ago. The decision to implement computer-based laboratories therefore must be based upon the increased capabilities of the new machines, or other factors, such as industrial pressure to provide graduates with computer experience.

The simple data acquisition function often associated with the laboratory computer can still be adequately performed by machines of two generations ago. It is clear that this capability alone is not sufficient justification for the capital expenditure. However, now that the systems have much more power than is needed for the data acquisition function alone, their purchase may be justified if additional tasks are available that will enhance the quality of the laboratory. These tasks include better data acquisition programming systems, simulators and computer-aided design software, multi-media instruction, and the potential of integrated laboratory computation.

B. IMPROVEMENTS IN DATA ACQUISITION SYSTEMS

As recently as a few years ago, the implementation of all but the simplest data acquisition task required considerable programming and dealing with the computer at a very low level. The advent of LabView, VEE-Test, and other similar packages in recent years has simplified the process to one based upon visual programming concepts that are nearly intuitive and extensible in a logical fashion. While there is still a learning curve associated with the process, it is much less steep, and information acquired in the learning process is much more reusable. Our experience with HP VEE-test is that students quickly grasp the concepts and generally outstrip their instructors in its use after the second or third session. Similar reports have come from our mechanical engineering department, which has been using LabView.

In addition, the proliferation of computers has spawned a proliferation of data acquisition hardware, so that the necessity of using purpose-built hardware for educational systems has decreased. From IEEE-488 bus equipped instruments to "bare-bones" data-acquisition systems the performance, versatility, robustness, and economy of systems has improved. Still, the costs are higher than non-computer based alternatives. The ability to easily perform a multitude of experiments should make these systems cost effective.

The sensors used to convert physical properties to electrical signals are also less expensive than ever, more robust, easier to understand, and more accurate with less calibration required. All these improvements make the measurements easier for the students to make. The conclusion is that it will be economically possible for computer-based data acquisition systems to dominate the laboratories of the future. In order for these systems to be advantageously used it will be necessary for faculty to rethink the laboratory experiments and to decide how students should deal with the much greater amount of data that can be quickly gathered.

C. AVAILABILITY OF SIMULATION AND COMPUTER-AIDED DESIGN SOFTWARE

Modern, professional simulation and computer-aided design tools have become increasingly available to the undergraduate laboratory, and are compatible with the class of computer now typically installed in the laboratory. Indeed, it is this second use of the computers that provides a great deal of the added impetus for the installation of next generation computer systems.

In the area of drafting, computers have almost completely supplanted the drafting board. Traditional drafting courses have given way to courses in CAD tools, although in many cases an important component in free-hand sketching is retained. Hand-drawn circuit diagrams have almost entirely disappeared from formal laboratory reports, although notebooks still contain them; at the moment, the procedure for entering schematics is still not time-effective for all but the final finished design.

The use of simulation tools is also growing, but not without a great deal of soul searching and controversy. The reason is that simulation tools are often used to by-pass analysis, and analysis is very important in most undergraduate curricula. Students (like their counterparts in industry) find it easier to joggle schematics and press the simulate button than to perform the analysis traditionally taught in their classes. The important result is an understanding what is going on, and it seems clear that neither pure analysis nor pure simulation will yield that understanding. The complex circuits and structures currently being built defy closed form analysis in all but the least interesting of cases; while reliance on simulation tools whose provenance is uncertain is at least foolhardy, if not ethically wrong. Most, but not all, faculty in our department agree that the use of simulation tools as an adjunct to more traditional methods is useful. What appears to be evolving is the use of simulation to demonstrate the effect of changes and permutations in the circuits, and to check the results of analysis.

I believe that this method is fundamentally flawed, in that the simulation tools are not really taken seriously in the curriculum. The result is a student population who are increasingly adept with, and reliant upon the simulators, but who have no fundamental grounding in simulator technology or its strengths and pitfalls. As industry begins to increase its reliance on simulation, it would seem important not only to use it in the undergraduate curriculum, but to make it an issue, on a par with the other analytical tools used to design systems.

The extent to which simulation can replace experimentation is an area of active research. It is an important issue as the cost of laboratories continues to increase in the face of declining budgets. Innovative solutions, such as the laboratory simulators (for instance, the ELF system at Vanderbilt University) come into existence. In these systems, not only is the circuit simulated, but also the test equipment and the laboratory bench. Wires are placed on the virtual breadboard, components selected by color code, etc. As these systems mature, and allow the class of errors and partial faults so prevalent in the real world, it may be possible to do a number of experiments at home or in the dorm room, and then attempt in "real life" the one or two most promising. The systems do not yet replace the smell of burning resistors or the singed fingers that result from inserting an IC backwards: but perhaps the next generation of electrical engineer won't do these things, because the quality of the CAD and simulation tools will forestall such errors.

The utilization of full-scale CAD systems for the complete design of systems is currently not viable for any but the most dedicated of students working on sophisticated designs. A possible exception to this situation is the MOSIS project in the VLSI area. However, as the perceived complexity of the systems is reduced, it should be possible to use such systems in capstone design classes.

D. THE USE OF MULTI-MEDIA INSTRUCTION IN THE LABORATORY

The primary problem with laboratory manuals and equipment manuals is that they are static: you can see snapshots of the state of the equipment, but never the transitions. The use of an oscilloscope, and particularly a modern computer-based digital one, is a very dynamic and action-oriented operation. The use of probes, attachment of accessories, etc. is difficult to demonstrate in snapshots. At the next level, the overall operation of experimental setups is better shown in movies than words. All of these notions lead to the suggestion that multi-media based instruction has a significant role in laboratory education.

Additionally, students complain that the times allotted to laboratories is often too short; even those students who do perform the experiments in the allocated time complain that there is no time for unstructured exploration. Leaving the laboratory open longer hours would alleviate this problem, but having teaching assistants competent in all the experiments all the time may be difficult or too costly. Here again, the multi-media presentations of the experiment may be invaluable.

This idea has been tried before, using video tape technology. At CU, the results from the video tape era have not been uniformly positive. Fundamental to most complaints is the lack of professional production values in the tapes, and the inability to find what one wants to know quickly. The issue of production values is not automatically solved by multi-media based instruction; on the contrary, professional quality multi-media lessons may be even more time-consuming. However, whereas there was never a strong market for video tape production tools, there is a very strong market for excellent multi-media production tools. The NSF funded engineering consortia are already creating, gathering, and distributing multi-media productions, and in so doing raising the production standards. The ability to produce CD-ROM versions of lessons is beginning to be within the reach of larger departments or smaller colleges; as the technology progresses, costs will decline even further.

Properly scripted, multi-media systems do answer the needs of the student for access to particular portions of the experiment. This ability may be the strongest point in favor of multi-media based systems. Experiments now in progress at CU with multi-media based modules in fluid dynamics are being favorably received by the students; it is too early to tell whether the cost to produce such modules can be reduced to manageable levels as experience is gained.

E. THE INTEGRATION OF TECHNOLOGIES

For all of these advanced technologies to be worthwhile, they must be less trouble to use than the alternatives they replace. More than that, they need to offer significant advantages at relatively low learning cost, or they will not happen. The key to viable technologies is seamless integration between the parts. Students should not have to learn four different operation systems and ten different file conversion programs in order to perform a simple experiment. At present, such integration is becoming available, but is a goal, not a reality. Transitory proprietary "standards" abound, and conversion program writers discover they aimed a moving target: the "standard" changes faster than the conversion programs can be written.

The future of such integration looks bright, however, with demands by users in industry for real, open standards becoming greater all the time. In the meantime, systems have to be carefully constructed so that all the pieces are at least aware of each other.

But beyond this low level integration, the technologies must integrate into the curriculum and the engineering practice in a meaningful way. The laboratory must work with the coursework, and both must work to the goal of educating the student in engineering theory and practice. Thus the role of technology transcends simply learning how to use the computers, and instead becomes one of enabling educational opportunities that were not previously viable.

Properly executed, for example, a computer-controlled data acquisition system in an analog filters laboratory would allow a much larger number of filters to be built and characterized than if all the data for each had to be taken by hand. Comparisons between different circuits and the theoretically predicted response are fairly easy to manage in well integrated system. We are working to provide on-line laboratory manuals for students so that results can be cut-and-pasted directly into the computer-based manual. The cost of this reliance on computers is high, however, because what we are asking is that students use the computer as their primary tool for gathering, describing, and presenting information. Thus the demands on the computer network grow dramatically. In addition, it makes sense that the students have a single computer "home" regardless where they happen to be. Integrating all the computers in to a homogenous environment that nonetheless allows for the particular strengths of the individual computers to be exploited is something of a nightmare for system administrators. It is, however, how things must evolve.

Another level of integration is particularly important for systems design courses, including multidisciplinary capstone design experiences. The tools must be able to integrate designs and simulation results from all the disciplines involved in the design of the system. In industry, such systems are being brought together, particularly in the aerospace design groups, perhaps most notably at Boeing on the 777 project. The ability to deal with the mechanical ramifications of adding an additional signal line, or the electronic or thermal consequences of changing materials in an electronics bay in a systematic way is increasingly important in getting things done correctly the first time. Tools capable of doing this easily are years away from educational use, but planning for them should begin now.

III. THE FUNCTION OF THE LABORATORY

For technological change to have an effect on the laboratory, the laboratory must use equipment and systems which change meaningfully with technological advances. Some attempts to modernize laboratories have run afoul of inappropriate technology complicating already adequate experiments. Some of these problems could have been avoided by examining the function of the laboratory before implementing the changes. In general, there are three functions that are commonly performed by the laboratory. These are reinforcing theoretical concepts, developing engineering skills, and providing opportunities for design. The impact of technology depends strongly on which of these functions the laboratory is attempting to perform.

In the area of reinforcing theoretical concepts, simple experiments are often as good or better than their modern equivalents, and some thought must go into modern experiments that teach the same things. For example, moving coil meters have much physics and electronics to teach. Their less-than-perfect impedance can teach important lessons about loading that are difficult to duplicate with the near ideal behavior of FET input operational amplifiers. Similarly, understanding analog oscilloscopes can be invaluable in getting a "feel" for signals, and the lack of an "Autoscale" button requires the same common sense approach to a measurement that the "three digit but no decimal point" result of a slide rule required in calculations. The hair-trigger trigger circuits of the (really) old scopes required significant dexterity to acquire and hold a solid lock on a signal. Many other examples could be cited. I am sure they abound in other disciplines as well: for example, the use of double pan balances vs. single pan digital readout units in chemical laboratories. Those lamenting the loss of these principles are not unprincipled reactionaries; it is incumbent on the modernizer to provide adequate replacements, or else to move the debate to a different forum. For example, the digital scopes require a grounding in sampling theory and aliasing: we trade concepts.

In the area of engineering practice and design, the use of the modern technology is less controversial, because almost everyone realizes that modern engineering is conducted, even in the smallest consulting shops, with a daunting array of modern, technologically advanced tools. However, even in this arena, the implementation of the technology cannot be allowed to obscure the process of practice and design, unless the purpose of the exercise is to introduce the tools. Slide rule and drafting courses were once common; while the former has in many places been converted to some type of programming course, the latter has often simply disappeared. It might be argued that an "access to tools" class should be reinstated. Training high powered tools on the problem of picking the resistor and capacitor in a single-pole RC filter is not a satisfactory use of technology; however, such things can occur if the learning curve is too steep. By the time tool is mastered, the course is over.

The maturation of the tools may very well eliminate many of these concerns. The current generation of software is much easier to master than its equivalent of five years ago. It is also important to watch the development of cross disciplinary tools, which will facilitate the total design of systems rather than parts of components.

IV. THE INTEGRATED TEACHING LABORATORY

The Integrated Teaching Laboratory (ITL), to be constructed at the University of Colorado, Boulder, is designed to be a high technology test-bed for a number of the concepts presented here. The thirteen million dollar, 30,000 sq. ft. building is due to be completed in the fall of 1996. Every attempt is being made to allow the implementation of the technologies discussed here. In particular, considerable effort is being expended on the design of the computer infrastructure which will support the laboratory, and on making the laboratory hardware modules work with the simulation and design tools.

ITL will be a laboratory shared by all departments, separated into experimental areas (fluids and heat transfer, manufacturing, electronics and measurements, etc.) connected by commonality in experimental modules and computer networks. By providing a solid computer infrastructure, it is believed that the modern measurement, data acquisition, simulation and design tools can be implemented and integrated across disciplinary lines. Where possible, design decisions are being deferred until the last moment, so that the technology will be as advanced as possible. The entire laboratory is based on flexibility. We recognize that we know we can see into the future only well enough to be sure we do not know exactly what the future will bring. ITL will be designed to change.

We are already experimenting with modules in fluid flow and statics that will take advantage of the instrumentation and computer systems. Comparing flow in a water tunnel with the theoretical prediction using video capture technology is one example.

By sharing equipment widely between departments, we can achieve much better utilization, and take advantage of economies of scale in procurement. Possibly most important is that the laboratory will be large enough to justify college level staffing, as opposed to the partial staffing usually possible with department level resources.

V. CONCLUSIONS

In this paper I have attempted to show how the laboratory of the future will be changed by the technologies that are currently evolving. It is likely that I have underestimated the changes in technology, and correspondingly underestimated the changes to be wrought upon the laboratories. I believe that the changes will allow better laboratory courses to be brought to more students, and that those students will be better prepared as engineers upon graduation. The transition will involve considerable time and effort, especially until the tools mature. Currently little known technologies, like simulated laboratories, may refocus laboratory effort onto areas not easily simulated, but it is unlikely that the true "hands-on" laboratory will be allowed to disappear, unless equipment costs badly outstrip funds.

Some words may be appropriate on the entire issue of laboratories in undergraduate curricula. While most engineering faculty would state that laboratories are important to the educational enterprise, far fewer are willing to actually design and teach laboratories. In many cases, were it not for accreditation agencies, the press of new theoretical material would force the laboratories out of the curriculum. Indeed, at CU, one department had managed to get down to only two laboratory courses in its discipline. An additional complication is that modern laboratories—the high technology ones we are discussing here—require skills that the younger faculty are more likely to have than the older. But at most research institutions, it would be foolhardy for a young faculty member to devote too much time to instruction of any flavor, laboratories least of all. Perhaps the greatest technological achievement would be to make laboratories easy enough to run so that even faculty whose interests are largely theoretical would get involved.

Chapter 9

THE IMPACT OF COMPUTER AND COMMUNICATIONS TECHNOLOGY ON ENGINEERING EDUCATION

John R. Bourne and Arthur J. Brodersen, Vanderbilt University

ABSTRACT

This paper discusses how changes in computer and communications technology may influence engineering education over the next decades. Forecasts about the nature of universities in the early twenty-first century are given first, followed by an examination of how engineering education fits into projections of overall change. Pressures on engineering education that may be relieved by advances in technology are examined, along with the opportunities that may be available due to changing technology. Trends in current hardware and software are examined and forecasts for how this technology will change in the next decade are given. Scenarios of how projected change may affect engineering education are presented. Finally, recommendations are given for what engineering colleges should do now to take advantage of rapid change in computer and communications technology.

I. INTRODUCTION

There is no doubt that computer and communications technology is changing very rapidly. Spiraling decreases in cost, coupled with similar increases in computer speed, now provide affordable computer systems that are becoming as ubiquitous as the calculator was during the 70s and 80s. Change in recent years in communications technology has not been nearly as fast as in the computer industry; however, many believe that we are now poised for massive change in widely available communications capabilities during the upcoming years.[1] What do these changes mean for higher education and specifically for engineering education? How should educators react to this "new" technology, both in use in instruction and in teaching about the technology? Should we keep doing what we are doing now, or should we embrace new technologies as they appear? The purpose of this paper is to try to provide fodder to supply a context for thinking.

[1] Changes in communications may well be accelerated due to federal initiatives such as the high-performance computer and communications initiative (HPCCI).

There are economic and demographic drivers for change in higher education that will directly impact engineering education. Economic drivers include dealing with capping the continual increases in cost of education at both public and private engineering colleges that, if continued at the current rate, threaten to put higher education out of reach of all but the most wealthy. Demographic problems deal with delivery of education more uniformly to students who are unable to attend centrally located universities or, who by virtue of changing economic pressures, are forced to study part time. Various opportunities for improvement for technology-facilitated engineering education can be clearly identified; for example, sharing of curriculum among non-collocated institutions, collaboration, simulation, visualization, and sharing research, knowledge and data, are among the areas that can be assisted by the use of computer and communications technology. This paper will examine what changes may come about in the university pedagogy due to the availability of new computer and communications technology. Toward this end, forecasts of technological change are made, present and predicted characteristics of the populations of universities discussed, and potential for change analyzed. The approach taken is to first examine broad changes that may occur in institutions of higher education at the outset of the upcoming century. This broad brush overview is then followed by constricting our view to the study of the impact of technological change on engineering education. We will examine how technology can modify our pedagogical style and how our teaching about the technology itself may need to change. We have attempted to look beyond the "next" systems and to seek to understand the impact of "systems after next."

II. UNIVERSITIES IN THE EARLY TWENTY-FIRST CENTURY

A. INTRODUCTION

There is no real reason to think that universities in the twenty-first century will be very different from the universities of this century. After all, universities have not changed tremendously in character in many hundreds of years. Always an enclave for knowledge seekers, many universities retain, to some extent, the cloistered feel of earlier times. Although the "ivory tower" remains with us and the role of the university in society will likely retain its current character for some time to come, there is still the potential for profound change within the upcoming decades. Indeed, the roles of the university as the educator of our young and the seeker, sifter, definer and promulgator of new knowledge are likely to remain major elements in university charters. These characteristics of universities will not likely change, but the way that knowledge seekers and assimilators work will probably slowly change due to the nature of the technology that will exist. Indeed, one position that this paper takes is that the supporting infrastructure of the university of the future will change the way we work, but not what we do!

B. TECHNOLOGY DRIVERS

There are two main drivers for technological change in universities: computer and communications technology. Computer technology was discovered in the twentieth century, and has matured somewhat. Yet, there is no apparent end to the maturational process in sight. The same lack of maturity is essentially true for communications technology. As we near the end of the twentieth century, it is becoming clear that computer technology is invading every aspect of our lives, including education. Telecommunications technology has not yet experienced the same dramatic change as computer technology in recent years. However, it is straightforward to predict that there will be *very* significant change in the *very* near term. During this last decade of the century, America and the world has been wired with fiber optic communications capabilities, which will permit anyone to talk, view, work with, and exchange information with *anyone, anywhere, anytime*. Activation of this communications infrastructure is just now coming about. How this capability will impact society is unknown; in general, however, it is predicted that it will have a tremendous impact! The upcoming decade will witness very significant changes in the communications infrastructure of this country. In the university, we can only speculate about what the impact of these changes will be. To some extent, our forecasting is conservative; change is so rapid that even near-term forecasts are difficult to make. Due to the acceleration of change, many more changes will likely occur than are predicted below.

C. WHO ARE WE AND WHAT DO WE DO?

When we speak of universities, we are talking about the collection of individuals that make up a university and what they do. Bricks and mortar are much less important and potentially, due to technology, may become even less important in the future. For our purposes, we will first identify the more-or-less permanent denizens of the university. They are: the students, faculty, staff, and administration. Each perceives the structure of the university differently.

The role of undergraduate students in the university is to learn and mature, perhaps not in that order. Student life centers around social engagement—dormitory life, games, and interactions with others. Academic life is secondary in most undergraduates' lives. In their academic activities, students attend lectures and laboratories, utilize the library, read, write, meet deadlines, and complete homework assignments, among other things. These activities have gone relatively unchanged as the centuries have passed. Technology facilitation for undergraduates is limited currently to library searching, the use of computers for word processing, electronic mail (e-mail) and other tasks.

The role of graduate students is, in general, more intense than that of undergraduates at most universities. Technological engagement tends to be quite heavy, particularly in science and engineering education. Graduate students routinely use significant technology for conducting experiments in addition to extensive use of technology for word processing, library usage and e-mail.

The role of faculty is to teach, conduct research, and provide service to the community. Teaching is normally accomplished through lecture and discussion, with technology limited to use of overhead projectors. Research is much more technology facilitated, particularly in the sciences where instrumentation and computers are heavily used.

The administration of the university plans and organizes. Common technologies employed include databases, management information systems, and spreadsheets.

The staff in the university assists the faculty and administration in the day-to-day operation of the enterprise. Use of technology within the university is approximately in proportion to the use of technology by the faculty. Word processing and other tools are in heavy use.

Analyzing current utilization of technology by the entire group, we may observe that most use appears to be in using tools on common personal computers. Expensive computers and instrumentation are employed in research. Telecommunications, beyond the telephone and fax, is used by a significant segment of the permanent population. In general, undergraduate students tend to use e-mail less than faculty. The scenarios we present below offer the possibility that the use of computer and communications technology among these groups may shift then eventually stabilize.

D. PROSPECTIVE IMPACT AREAS

There are many possible areas in which technology could impact the university community of the future. Some of the more prominent ones are examined next.

1. The Classroom

Technology-facilitated classrooms are found in many, if not most, universities today. These classrooms typically consist of workstations at students' desks, projection equipment, and network connections among computers. Some institutions include multimedia workstations in these types of classrooms. While computer-facilitated classrooms have proved useful and popular, their universal use has not come about, probably because of the high cost of implementing the systems in many classrooms. A possible alternative to this type of classroom is the personal digital assistant (PDA)-facilitated classroom, in which students bring PDAs to class and plug into Ethernet (or similar) high-bandwidth communications facilities. If radio-Ethernet becomes widespread, a physical communication-backbone might not need to be employed because connections to library resources worldwide, to illustrative animations and explications, intelligent tutoring systems and personal knowbots would become commonplace in the classroom. These scenarios are almost doable now; hence, one might predict widespread implementation in twenty to thirty years.

2. The Student Laboratory

Student laboratories have changed little in engineering and the sciences during the last half century. Prospective changes include creation of virtual laboratories in which students can experience the use of equipment and preparations that do not exist except in a computer simulation. The extension of this idea to virtual realities appears plausible, as well. In the latter case, students would enter complex worlds that are generated in real-time by high-speed computers. In these worlds, students might be faced with simulated real-life scenarios in which they are required to solve problems. Anchored instruction[2] might also become commonplace in laboratory education.

3. The Research Laboratory

Research laboratories frequently require use of expensive equipment. In the university of the twenty-first century, it is possible that researchers may routinely use shared research resources that are located at different institutions but which are available virtually anywhere in the world due to the existence of high-speed communications networks. High speed would permit the connection of an instrument to a remote site; however, the use of physical materials would have to be handled by over-night shipment or matter transmitters.

[2]See papers by Kinzer and Dally, this volume.

4. Group Work/Teaming Experiences

Educational and research experiences have been typically individual or very small group activities. The very complexity of the world appears to mandate that people work together much more closely simply because individuals are incapable of solving the large problems that society faces today. Hence, universities must change to provide more group/teaming experiences. It will be commonplace for research activities to span multiple universities. Sharing of undergraduate educational experiences across universities will likely trail shared research experiences.

5. Distance Learning

High-bandwidth communications will permit currently expensive distance education to be delivered easily and cheaply to the home. How can universities take advantage of this prospective capability? Is it plausible to provide education to the home over the same cabling that provides home entertainment? The strong likelihood is that it will indeed be possible, given several fairly radical changes. First, universities would need to address the needs of the at-home market, consisting of learners that may want more focused education to help them in learning particular sets of skills. Second, universities would have to develop delivery mechanisms for interactive distance education. This activity would require building and reorganizing curriculum for electronic dissemination. Third, universities should probably not engage in these activities in a way that would erode the on-campus residential education that is now central to the survival of many, if not most universities. Fourth, universities may wish to reorganize across institutions to share electronic materials.

6. Knowledge Access

Libraries currently provide excellent knowledge retrieval support to the university community and, to a much lesser extent, to the surrounding urban/suburban non-academic community. The capabilities of libraries, augmented even more by technology, could become on-line digital resources for the non-academic community. For example, if materials could be secured on-line, over cable for example, there might well be a market for university library materials outside the university. In this scenario, universities could provide a "pay-per-view" access mechanism for their library resources.

7. Social Environment and Activities

In institutions in which undergraduate education is primarily residential, it may be anticipated that there will be some change due to technology. For example, computers in dormitories are currently linked to Ethernet in many institutions. During the upcoming five years, such networking capabilities will become more widespread. By providing electronic linking of students, group work can be facilitated, communications with instructors and teaching assistants can be implemented, access to digital library facilities can be increased, and access to the world over the Internet[3] will be provided. Also, computer-based learning modules, including multimedia-facilitated learning environments, will probably be employed more widely. It seems likely that learning modules of various kinds will become widely available over Internet or similar networks.[4]

---

[3] We use the term "Internet" loosely to designate all network services that might ultimately appear under the umbrella of the Internet.

[4] e.g., the National Engineering Education Delivery System is a prototype curriculum delivery system that became available over Internet in 1993.

E. IMPACT OF TECHNOLOGY ON THE INHABITANTS OF ACADEMIA

1. The Student Perspective

In the university of the twenty-first century, education of residential students will continue. There are likely to be additional learners at home and in industry who access lectures, discussions, electronic forums, and materials via high-bandwidth information pipes. Information will probably be delivered by cable, telephone access, or on EtherRadio.[5] As we proceed into the next century, it is possible that the use of hard-wire cable will give way to cellular radio for communication of all needs. Arguments against this latter scenario are that huge bandwidths are needed to utilize multimedia, for example. However, since this is a barrier, it can be anticipated that solutions will emerge to permit portable radio-based computing to become a ubiquitous unifying medium.

Classrooms may well become multimedia shows, or perhaps even virtual reality settings in which students engage in life-like simulations. Laboratories may well have the same character.

2. The Faculty Perspective

Faculty, as teachers and researchers, will have to adjust to a university of the future. Various roles may change. For example, the roles of curriculum creators and facilitators of student interactions with computer technology may increase for faculty. This change may be driven by a market in which many of the paying customers are at home or in industry and demand materials that are available over communications media. Faculty, as researchers, may interact with research teams that are distributed among institutions anywhere in the world. Further, expensive research equipment may exist only at remote sites, so that use of this equipment will require use of virtual instrument interfaces which permit faculty to use these facilities remotely.

3. The Administrative Perspective

As mentioned above, the administration plans and organizes. One would expect current planning and organization tools to develop during the next decades so that administrators could view, manipulate and understand the details of how their organizations work. Intelligent tools that permit "what-if-ing" and connection with pools of information worldwide will become commonplace, as will virtual meetings facilitated with group decision-making tools, supported by computers. The most significant advances will come from tools that permit more rapid decision-making by organizing vast amounts of information about an institution.

4. The Staff Perspective

The work of the staff will change, too. Staff will utilize computer and communications technology to facilitate every phase of their work. Just as it has taken decades for staff to begin to use computer tools throughout universities, routine work will become more automated over the upcoming decades.

F. IMPACT ON PERSONS OUTSIDE THE UNIVERSITY

During the twenty-first century, universities have the potential to serve a broader audiences in both local and global communities. Telecommunications will provide

[5] i.e., Ethernet delivered over radio.

instantaneous connections to and from universities to anyone, anywhere, anytime. If developed properly, the impact that higher education can have will be enormous. The after-college market is likely to grow due to ever increasing rates of speed at which new knowledge is generated. However, rapid change will place significant burdens on faculty, as the distillers and purveyors of new knowledge, to work ever harder to put knowledge in forms that mass consumers can use for improving skills.

G. THE ROLE OF CULTURE IN CHANGE

A predicted change in the twenty-first century is in the faculty culture. During the middle years of the current century many faculty sought knowledge for knowledge's sake, supported by the institutions for which they worked. Knowledge seeking in the form of research became a central focus in the faculty culture during this time. This situation is likely to change to make faculty work even more customer-driven. Research, already strongly customer-driven will become even more so. Education will also turn toward a customer orientation since more and more people will need to meet specific education targets simply to survive in the world of the future. Our student customers will have educational requirements that must be met just to enable them to take part in the technology-intense work world of the twenty-first century. Those universities that provide for these needs will survive. Research will survive, as well. However, it is unlikely that the current models for measuring faculty achievement in research in universities will continue. Tenure is likely to disappear and faculty will become much less protected against unemployment than they are presently. Thus, the culture of the university of the future may well be much like American industry today: insecure, goal-driven, fearful, and customer-focused. This scenario is not a pretty picture from the perspective of the late twentieth century; yet, it is a picture that may become reality because of economic and social pressures.

H. THE VISION

Will universities exist in the twenty-first century? Yes. Will universities be just like they are today? No. Will technology be the driver of change? In part. First, universities are likely to retain much of their current character from a student's perspective. Computer and communications facilitation is almost certain, and most work habits will be altered even though the mission of education and research remains unchanged. Second, universities will become more distributed. Still geographically concentrated to serve resident students, universities will nevertheless spend ever increasing time on serving the needs of the growing continuing education community and the needs of students who do not participate in residential education. Technology will facilitate this change through implementation of teaching tools, curriculum and group meeting technologies designed to bring the faculty to their customers. Third, universities will become more customer-driven and less secure havens in which to work. Faculty, administration and staff will have to work harder and will likely be not rewarded as well. In summary, the future appears as a mixed forecast: there will be improvements in learning and communication facilitated by technology; however, technology-created societal drivers will probably make the university of the twenty-first century a very different place in which to live and work. To the external observer, the university of the twenty-first century will not be very different from universities of the past. Internal to the university, however, the way we work and live may well be quite different.

Only time will tell! Next, we will examine the impact of these prospective changes on engineering education.

III. ENGINEERING EDUCATION OF THE TWENTY-FIRST CENTURY

In the early part of the 20th century, engineering education was oriented toward design. Following World War II, the Grintner report [Grintner, 1955] and other reforms stimulated a redirection in engineering education toward a more scientific base. The curriculum delivery model that evolved during the 1950s remains with us in the 1990s. In most institutions, information delivery is still typically via lecture and laboratory experiences. While our curriculum has remained largely stagnant in terms of delivery and excitement, the world has changed and so have the customers of engineering education — the students. No longer content to tinker in a basement with vacuum tubes, transistors, or their present-day equivalents, students of today are captured by glitzy video-birthed entertainment and rarely by the fine points of circuit analysis, spherical coordinates, differential equations or finite element methods. Some would argue that massive change is needed in engineering education in order to reach this "new" generation of students [Tobias 1990]. In fact, there are many experiments apace in institutions of higher learning today. For example, examination of just-in-time strategies [Winkel and Froyd, 1992], conservation-based learning [Glover et al., 1992], mechanical dissection [Agogino, 1992], and anchored instruction [Bourne and Brodersen, 1992; Dally and Zang, 1993] are among the novel ideas being investigated. Although there are many experiments currently being conducted, the lecture/laboratory paradigm that has served us well for decades will likely remain in place. Therefore, it is essential to provide evolutionary change to this paradigm and not try to replace it with revolutionary pedagogical approaches.

The ability to change quickly has not been a hallmark of engineering education. In fact, change in engineering education appears to require decades. As noted above, the last major change in this field was just after World War II when the Grintner report [Grintner, 1955] recommended that engineering education become more scientific and less "trade" oriented. It is clear some fifty years later that indeed that engineering education has indeed become more scientific. It is equally clear that much of the curriculum and delivery methodology has not changed during this fifty year period. It now appears that change should occur once again. Why? First, the requisite time for change has elapsed — engineering educators are beginning to be pressed out of their comfort zones and are forced to recognize that the world is changing. Forecasts for the availability of engineering jobs in the future show clear decreases in the need for the types of engineers that we have educated during the last several decades[6] [Bell, 1993]. Second, the demographics of the world are changing, and more students of different backgrounds and abilities need to be educated. Elitist engineering education cannot continue in the current mold — that is, educating only those who are like those that came before. Shiela Tobias [Tobias, 1990] points out in her seminal work on educational paradigms, that "they're not dumb, they're different". It is clear that engineering educators do not currently recognize the need to educate the "different" ones. Typical engineering students of the last decades have conformed to the mold of "technogeek"[7] rather well, as did their predecessors. Typically in the past, mathematically inclined, introverted high school students who built radios or took apart cars chose to become engineers. These types of students have virtually disappeared — and have been replaced with a more renaissance-type

[6] e.g., there is decreased demand for engineers in the aerospace industry.

[7] i.e., a technology-centered, nonsocial individual.

high school student who is video-nurtured and technology-comforted. These students aren't dumb—they're different! There is no evidence that students will ever return to the mold of the "technogeek." Nor, is there any evidence that they should. It is more reasonable that engineering education should change to fit the needs and society, and serve the student populations in ways that are more relevant to their futures as technologists and scientists of tomorrow's society.

Engineers of the future may play different roles in society from their current roles. For example, engineering students for the last decades have been prepared for design and manufacturing roles in industry. Technology has changed industry to such an extent that there is a much stronger need for problem solvers and thinkers than for "grunt" designers and manufacturers. Engineering education must prepare students to assume these modified roles that society will demand.

There are many pressures and opportunities that can be envisioned that are associated with the current rapid technological changes in computers and communications technology. For example, there are pressures exerted on the engineering education system by students who are accustomed to video and computer-delivered information and who are easily bored in the traditional classroom. There are cost pressures; providing new technology obviously has a cost and a benefit. How can currently financially strapped educational institutions afford to adopt the newest technologies for delivery of educational materials? On the other hand, there are tremendous opportunities. For example, we now have an opening to learning methods in which technology can truly augment how much and how fast students learn. Connection to knowledge sources such as remote libraries, multimedia resource materials, supercomputer simulations, and animation, and the capability of tapping into the best knowledge sources nationally and internationally on-demand, are all tremendous opportunities that argue for taking risks in shifting engineering education to a more technological base. If we do not do it, others will. Given the government investment in networking [McClure et al., 1991] and the current atmosphere of willingness to "change the paradigm," the mid to late 1990s may well be the time to make the move to change engineering education. There is an apparent possibility of moving our nation to preeminence through embracing technology as a vehicle for education.

IV. TRENDS IN HARDWARE: COMPUTER AND TELECOMMUNICATIONS HARDWARE

Without question, one can predict continued decreases in cost for given computer capabilities, increases in compute power for a given cost, increases in speed of communications and a decrease in cost for communications. It appears close to certain that the trends of the last decade will be mirrored in the upcoming decade. Assuming this scenario, what innovations might we expect by the beginning of the twenty-first century? Table 1 shows some of the possibilities:

INNOVATION	POSSIBLE IMPACTS
Computer miniaturization	Computer assistants in every aspect of life
Decreased cost of computers	Computers available to everyone
Increased compute power	Complex tasks possible
Fiber optics into the home/school	Home learning/inter-institution-home networked learning
Personal digital assistants	Replace notebooks, desktop computing
World-wide digital communications	Group work, virtual corporations, virtual universities
Ubiquitous computing	Computing integrated with life/education
High speed graphics engines	Visualization capabilities
Ubiquitous networking	Network used much like current telephone services

Table 1. Innovations and Impact of Change in Hardware

A. CHARACTERISTICS OF LATE TWENTIETH AND EARLY TWENTY-FIRST CENTURY COMPUTERS

It is always difficult to forecast the future; it is especially difficult when dealing with rapidly changing technology. Hence, the predictions made below are just that: predictions with little underpinning except what is known about past trends and current technology.

First, it is likely that trends toward miniaturization will continue. In 1993, Personal Digital Assistant (PDA) systems began to appear (e.g., Newton, General Magic, etc.) which, within several years, should have the compute power of desktop systems in use in the early 90s. Probably PDAs will be remotely linked with high bandwidth networks (e.g., perhaps Radio PDAs or EtherRadio will be used) which will permit instant linkages to large data and knowledge bases on the Fibersphere.[8] Hence, compute power in a PDA could be used for tasks at hand without having to dedicate large local storage. Local storage probably won't be a problem, however, since mass storage will probably be large, light, and inexpensive. Second, computing is becoming a consumer product just like televisions, dishwashers, or mixers. In fact, the appliances, computers, phones, etc. will likely merge in technology to have common interfaces and all will be networked (probably without your knowing it). We may anticipate a doubling in computer speed coupled with a 50 percent reduction in cost every two to three years. As speed limits for single processor systems are reached due to physical connectivity (e.g., at somewhat over 100 Mhz), multiple processors will become much more widely used[9]. Hence, as an example, by 1998, one may expect to find, at a minimum, the most common processor to have a processor speed of at least 100 Mhz and cost in today's dollars less than $1,000. The physical nature of computer housings, power supplies, keyboard, pointer, etc. will limit the reduction in cost along a linear scale. It is probably reasonable to project that the cost of minimal computer systems will not decrease much below five hundred of today's dollars due to hardware costs[10]. However, delivery of computer services over TV and in other appliances may provide computer services at even

[8] The term "fibersphere" comes from the recent book by George Gilder entitled *Telecosm* (1993). The fibersphere is the interlaced global network of dark fiber which could ultimately support ubiquitous light-speed communications. Education may well be a major beneficiary of this capability by providing virtually instantaneous inter-institutional sharing of curriculum and connection of homes to universities. See below for a more complete discussion of the fibersphere.

[9] However, software issues for multiprocessing will have to be solved for this to occur. (See below.)

[10] This effect can be already observed. Low-end computers simply dissapear from the market; they do not continue to get cheaper.

lower costs. Merging network and computer services into high definition TV would permit new services to be delivered to the home very inexpensively.

The physical nature of future computers is open to considerable question. For example, typing will probably continue, due to the easy eye-hand coordination and acquired skills for producing documents. Continued growth in features for facilitating document composition will occur; however, it is unlikely that pen computing using writing styluses will replace keyboards. Speed of information entry is vastly better using keyboards. Also, it is interesting to speculate whether voice recognition will enable replacement of keyboards. At this time it seems unlikely. Even assuming that voice recognition becomes excellent in the upcoming years and that voice can be easily transcribed to text, it is problematic whether or not people will change to this entry modality. There is some comfort in being able to quietly type and organize one's thoughts on a keyboard. It is well understood that dictation and keyboarding skills are not identical.

B. TELECOMMUNICATION POWER AND COSTS

America has been wired with fiber by the regional bell operating companies (RBOCs) and others. In fact, many thousands of miles of fiber exist in the US in 1994. Considerable fiber is in use; however, much remains dark (i.e., unused). The RBOCs support point-to-point communications using complex switches. For increased bandwidth, some choices include: ISDN, SONET, T1, T3, and ATM. ISDN is an integrated digital services network already in commercial and home use. T1 is 1.544 Mbps service, and T3, a 45 Mbps service. Asynchronous Transfer Mode (ATM) is a service expected to begin operation in 1994 at the 155 Mbps rate or higher. SONET is a family of standards for multiplexing synchronous channels on fiber optic cables at rates up to 2.4 Gbps.

In a recent book entitled *Telecosm*, George Gilder introduces the concept of switchless telecommunications that trades bandwidth for switching capacity. In effect, Gilder argues, a telecosm can be created over fiber that is analogous to radio broadcasting. The fiber infrastructure of America and the world would lie in quiet repose simply broadcasting signals over the fiber optic implemented systems without switching. If this type of infrastructure was ultimately in place, there would be virtually no need for the RBOCs and associated carriers as we know today. Furthermore, the existence of a passive, instantaneous and world-wide communication network would permit tremendously different ways of thinking about sharing and use of educational material, and pedagogy in general.

The first steps toward merging of communications and computer power were taken in the early 90s with the introduction of the National Research and Education Network [McClure et al., 1991] and the High Performance Computer and Communications Initiative. The first legislation on NREN was introduced in legislation by Senator Albert Gore in August of 1986. Senator Gore introduced Senate Bill S. 1067 on May 18, 1989 (identical House Bill H.R. 3131 was introduced on August 3, 1989) [S.1067, 1989], the National High-Performance computer Technology Act of 1989. NREN operates in the US as the NSFnet in 1994, supporting around 20,000 different computer networks. Ultimately, the network will be completely supported on fiber-optic media and will enable such activities as close interactions between individuals who are not collocated and are sharing audio, video, animation, etc. (that is, any shared data, information or knowledge which have high bandwidth requirements).

C. VIRTUAL REALITIES

Virtual realities (VRs) (see article by Dede, this volume; [Reingold, 1991; Helsel and Roth, 1991]) are not yet a mature science, although work began in the 80's. In VRs, computer simulations attempt to duplicate experiences in the real world. By using computer methods to deceive the visual and auditory senses, one can experience simulations of realities that do not exist or are similar to situations that do exist in the real world. VRs are computer-generated by using high speed presentation of scenes fed to binocular views in a helmet that is also equipped with binaural sound. Typically hand, head, and body position can also be sensed and presentations coordinated with information secured from the helmet. Are there situations in which VRs could provide engineering students with realistic industrial experiences that they could not get otherwise? The answer is a guarded yes, with the caveat that there may well be a decade or more before we see this type of innovation in the teaching arena. The ultimate success of VRs is tied both to the creation of low cost hardware, both for sensing and for computing, and to software required to create the virtual experiences.

Even though there is significant research being conducted on VRs, there is only minor evidence, as yet, that educational VRs created in the same way would be really useful. One example is the creation of Virtual Laboratories (VLs) which appear to have the potential for significant impact in engineering education. VLs are computer simulated laboratories that look, operate, and produce the same results as actual laboratories. Several reasons for using VLs in engineering education can be cited. First, VLs can provide students with experiences that often cannot be achieved in a real laboratory, because of a lack of expensive equipment. Second, a VL can provide a precursor experience to a physical laboratory. If students conduct VLs prior to working in a physical laboratory, they can gain valuable experience and understanding in a self-paced format. Further, the cost of operating physical laboratories can be decreased due to a decrease in the amount of physical laboratory time required.[11] Third, VLs can provide on-line help and guidance for students while they are learning about laboratories.

V. TRENDS IN SOFTWARE

While cost/power trends in computer hardware are relatively easy to observe, and even predict to some degree, it is much more challenging to predict changes in software over the upcoming decades. Although the cost/power trends of hardware appear phenomenal, it is likely that software changes will equal and probably exceed hardware changes.

Programming tools have evolved over the last several decades. Early extensive use of assembler coding has been significantly reduced as efficient compilers have made it unnecessary to code programs in assembler. Only a decade ago, engineering educators felt that it was necessary to teach assembly level language programming to students. Now it is clear that only a small minority of students will ever need these skills in their careers. Early high-level languages such as Basic and Fortran have continuously improved over the years. For example, the Basic language has recently experienced a Renaissance in the form of Visual Basic [Visual Basic, 1993], a visually oriented programming language pioneered by Microsoft Corporation. Likewise, other stalwart languages including Lisp, Smalltalk, and C have undergone transformations during the decades. Even C has evolved into C++ and more

[11] e.g., see [42]

recently, Visual C++. What is happening with all these languages? Probably they demonstrate a trend toward empowering users of computer languages by making it easier and easier to program. Once the province of computer scientists or engineers alone, the capability to create computer applications is being delivered to the masses by products such as Visual Basic which permits programming almost without knowing how to program! A significant paradigm is the more universal acceptance of object-oriented programming, which reduces programming complexity by reducing all constructs in an application to the concept of object. The only communication with an object is the sending and receiving of messages. Objects hide their internal structure, thus reducing complexity. Another idea is the use of inheritance, a technique in which abstract objects can be specialized. This technique reduces programmer time and effort by making objects reusable. Smalltalk was the earliest example of object-oriented programming [Goldberg and Robson, 1983]. It has taken over twenty years for the marketplace to respond to the excellent ideas originated at Xerox Parc which have driven the creation of contemporary object-oriented programming tools.

During the mid-90s, it is highly likely that there will be an upsurge in the use of object-oriented programming languages. Simple languages such as Basic/Visual Basic will acquire object-oriented features and more complex languages such as Visual C++ will acquire additional features such as incremental compilation and inheritance visualization. However, most importantly, the trend toward automatic programming will continue with languages being produced containing more and more features that permit the programmer/application builder to even more easily create applications. The obvious trend is to empower users to build applications more quickly and easily than is presently possible. This capability has obvious implications for engineering education/teaching or whatever.

As computer hardware increases in speed and requirements for additional power become evident, parallel hardware will become more prevalent. Hence, software for programming parallel computers also will be required. While currently relatively primitive, parallel software will need to be developed to the same level as software application programming indicated in the previous paragraph. It is more likely that parallelism may be handled by the operating systems created in the late 90s in such ways as to make programmers understanding of programming in parallel unnecessary. Given the trend toward permitting application builders to program visually with little code, it is unlikely that complex parallel programming paradigms will become widely used. It is much more likely that the programming of parallel processors will be completely transparent to the user. One can easily imagine simply adding more processors to the desk computer and having the operating system manage the allocation of the processors to the processes running on the machine.

Given the scenarios presented above, we hypothesize that the teaching of engineering students about programming must change. First, it has already become clear during recent years that using and understanding application packages is a more important skill for engineering students than learning specific programming languages.[12] Traditionally, engineering students learned programming languages in order to be able to create numerical solutions for problems. As increased numbers of tools have appeared (e.g., matrix manipulation programs, spreadsheets, simulation systems of various kinds, etc.), the need to program "from scratch" an algorithmic implementation of a solution to a problem has become less and less needed. The percentage of engineering problems that can be solved using computer tools has inched ever higher as time progresses. The natural extension of this trend is that within a few years, tools will exist that can solve most engineering problems. What of languages? Should we continue to study languages?—Yes, because there will always be special cases and there must be people who write the application programs that

[12] A result of the survey conducted by the study sponsoring this paper. (See results compiled in this volume).

people use. However, it may be quite unnecessary, and, in fact, undesirable for all engineering students to become expert programmers!

Another factor in this discussion is the changing nature of programming itself. As programming becomes more and more object-oriented, the conceptual nature of programming changes. Students will have to learn to deal with organizing and understanding collections of objects and their interaction—a quite different skill from learning to code sequential lines of assembly code.[13] In many ways, however, the learning of object-organization as a paradigm that mirrors engineering problem solving is highly desirable, since the methodology parallels skills that are taught in engineering.

What should engineering students learn in computer languages? First, it is important to recognize that some facets of computer languages will remain an essential element in the engineers' "bag-of-tricks" during the upcoming decades. Perhaps the type of language learned will need to change and the extent and depth of exposure will vary greatly depending on the areas of interest of the student. Second, it is equally important to recognize that tool use, integration, and synthesis will be essential in engineering problem solving. Hence, we believe that language understanding and exposure must continue, reorganized to reflect changes in the character of languages that are currently evolving. However, at the same time, we also believe that learning about tools that support the problem solving nature of engineering must begin to take a central role in engineering education. All engineers will (or do) use computers for problem solving; engineering educators must embrace this idea and change the curriculum to reflect the reality of the present day.

Engineering students of the future will need more grounding in knowledge integration and utilization. For example, the ability to use tools to sift and sort knowledge in global data and knowledge bases to secure information for benchmarking, the collection of causal factors and reasoning paradigms, visualization of physical phenomena and modeling and simulation are likely to become central to our engineering curricula. This assertion is not well recognized today, since little emphasis in our engineering education curricula is placed on informatics-related topics. With increased use of Internet [Krol, 1993; Lynch and Rose, 1993] and similar networks, it is likely that some facets of engineering education will migrate to utilize inter-university and inter-human communications capabilities afforded by networking. The information, libraries, up-to-date discussions and interactions provided through this technology-based advance should become part of the armaments of engineering education paradigms.

VI. AREAS OF TECHNOLOGY IMPACT IN ENGINEERING EDUCATION

A. THE CLASSROOM

The classroom has remained for many years the standard mode for instruction in engineering and other disciplines. Why? Because single lecturer/multiperson classes is an efficient way of imparting information, most likely. Although efficient, it is by no means clear in engineering instruction that this method is the best! Hence, engineering education and the sciences have typically utilized laboratories to supplement/augment lectures (see below). Due to economic pressures, it is likely that classroom instruction will remain a major

[13]It is not the intention to suggest that assembly code is obsolete; indeed, for small controllers, highly specialized systems, time critical applications, etc., there is likely to be a need for understanding assembly code until compilers produce code as good as assembly code produced by expert programmers. But the question remains: If people no longer learn assembly code, what will happen when knowledge of assembly coding technology is lost?

element in engineering education for years to come. Better models of instruction might be found in one-on-one instruction, individual teaching and laboratory/discovery and apprentice systems, all of which are much too expensive to consider, given current economic constraints. Hence, what can technology do to facilitate the current instructional model?

First, various incremental improvements in technology facilitated classroom instruction can be made. For example, technology that permits demonstration and student interaction can be added to most classrooms. Already, many universities teach in classrooms with computers at each desk. However, this mode has proven less effective than originally thought, since software paradigms are not yet present for smooth presentation and seamless integration of lectures/demonstrations and student interaction. Software for facilitating the classroom needs to be created. Second, most engineering education classrooms in America are not computer-wired. One can easily envision how classrooms might be computer facilitated. For example, all classrooms could be equipped with computers at every desk. This solution has several drawbacks. First, it would be outrageously expensive. Second, management of computers in every (or most) classrooms would prove quite unwieldy. Third, students would have to adapt to each computer in every room, unless standards were imposed. A better solution might be to follow the trend of requiring students to purchase portable notebook computers or enhanced Personal Digital Assistants (PDAs). In 1994, it seems possible to predict that within five years, engineering students will be able to use PDAs as easily and cheaply as calculators. PDAs in the upcoming engineering education environment could be equipped with Ethernet (or similar) connections and each seat in a classroom could be equipped with plug-in Ethernet connectors so that students could attach their PDAs to connect to the current lecture or to access materials outside the classroom. Alternatively, radio-PDAs (r-PDAs) can be envisioned in which wireless Ethernet (or equivalent) could be used so that students would have access to knowledge, data, lectures, etc. wherever they are. This solution would permit classroom learning to be easily augmented without dedicated computer facilitated classrooms.

Given the potential plausibility of computer-facilitated classrooms using radio PDAs, what changes can be envisioned in the classroom environment? Could classes become more Socratic? Could demonstrations and on-line exercises become the norm? Could virtual classes be held in which participants are not collocated? What is the value of collocation anyway? How can we determine the optimum learning mix?

A current strong suspicion is that the traditional lecture is almost dead (at least from the multiple-decade perspective!). There is abundant evidence that our video-nurtured generation of students is "bored to tears" by our traditional lectures. Having grown up with exciting media-facilitated information input, it is no wonder that lectures delivered in a traditional manner leave current day students "cold." What can we do? We can shift the paradigm! Meet students in their own milieu by providing educational experiences that match what they expect.[14] Technology can easily provide the shift and, if properly managed, at a cost that all can afford. The major impediment to change is probably faculty attitude, which is typically "we should do what we have always done."

[14]This position is "customer-driven" and is tremendously unpopular with engineering faculty (as expected). However, in these times when the customer walks if needs are not satisfied, only those engineering schools that satisfy the customer will survive.

B. THE LABORATORIES

The laboratory experience in engineering education is supposed to supply students with a degree of "hands-on" realism that mimics the real world. In truth, however, laboratory experiences in engineering education are often dull enactments of rote procedures. That is, students are given a series of experiments to perform using laboratory instrumentation and then required to complete a laboratory write-up that details their experiences. Innovative project laboratories involve students in creating physical entities. Capstone design courses create ending-year design experiences for many students. Typically, labs are conducted using shared and often out-of-date equipment. The high cost of sophisticated laboratory equipment makes it impossible for college laboratories to always be equipped with the most recent equipment. Ideal laboratories would provide learning experiences that go well beyond simple measurements and rote exercises. Ideal laboratories would also provide instruments for each student and would have students engage in interesting and useful learning experiences that reflect both the lectures and the knowledge they need to acquire to work in the real world. Sadly, few engineering education laboratories meet these ideals.

What can we do in the upcoming years? One possible concept is to create virtual laboratories (as mentioned above) in which laboratory learning is mimicked using computer programs. For example, laboratory instruments and experiments can be simulated [Mosterman et al., 1993] and students can learn outside of a laboratory the concepts and measurement techniques that they would learn in a physical laboratory. This technique may prove useful in situations in which either (1) laboratory equipment is unavailable or antiquated or (2) is too expensive to be acquired in a typical college laboratory. With advances in computer power and graphics, virtual laboratories could become a major element in tomorrow's engineering education laboratories.

Another laboratory transformation that is possible is the intertwining of classroom and laboratory experiences. A prospective pedagogical style is "anchored instruction" in which learning is tied to concrete experiences whereby students learn by doing. In this paradigm, laboratory experiences and classroom experiences could all be tied together using networked systems. This possibility becomes more likely if the rPDA[15] technology postulated above becomes a reality. Then, students could evoke laboratory simulations in the classroom and in project-oriented learning experiences. Indeed, given sufficient advances in technology, students might be able to engage in learning experiences in computer-generated virtual realities (see article by Dede, this volume).

C. SIMULATION

Simulation has been a mainstay of engineering education for decades. The sophistication of modeling and simulation has evolved over the years from simple computer implementations of mathematical models to rather complex interactive simulation packages that permit simulation of complex engineering systems such as aircraft, automobiles, and electronic systems. As computer power increases while cost decreases, it is likely that the current trend will continue — production of simulation packages that will permit visualization of physical phenomena based on mathematical modeling at every students rPDA. This prediction should be easily fulfilled in the near term (e.g., less than five or ten years).

---

[15] rPDA = radio personal digital assistant

Another aspect of simulation/modeling is the creating of pedagogical models for assisting learning. Textbooks today are rather dry, non-interactive materials that students often find uninteresting and difficult to understand. Textbooks might well be augmented with interactive simulations that would serve to illustrate and explicate concepts in the text. A simple example can be given as follows. It is difficult to teach the concepts of resonance in electric circuits since the student has to relate circuit element values to the effect of passing different bands of frequencies in a passband. It is fairly straightforward to explain the concept in a text but difficult to see, unless the student is actually using an oscilloscope and signal generator to test various filter components. However, if a filter simulation is created, students can better understand how a filter works without the use of physical equipment. One can envision the textbook of the future with interactive learning materials (even multimedia) embedded. These examples might ultimately replace current figures that are used for explaining the material.

In many ways, simulation is the heart and sole of engineering education. It is highly likely that advances in providing simulation to students will be made in the upcoming years.

D. SCIENTIFIC VISUALIZATION

Scientific visualization may become a cornerstone of engineering education. For decades we have required students to relate mathematical equations to n-dimensional graphical representation. As computer power has increased, so has our ability to animate multidimensional representations of mathematics. Packages such as Mathematica (see article by Black and Getz, this volume) have provided a new generation of students with the capability of visualizing mathematics. Further, high performance computing has permitted advances in visualization and understanding in topics as diverse as weather forecasting and dynamic modeling of the heart. There is the clear potential for providing increased high-performance computing over high speed networks, including the national research and education network [Campbell, 1993]. However, at the current rate of progression, in terms of increases in computer power coupled with concomitant decreases in computer prices, what was once thought of as high performance computing will be available on the desk top in the near future. Computer speeds well exceeding 100MIPs should be available in PCs at the retail $1K range within only a few years. These speeds could be dramatically higher if problems with parallel processing are solved such that multiple low-cost CPU can be easily parallelized.

E. TEAMING/GROUP WORK

One area that can be predicted to be very important during the upcoming years is the impact of team learning and group work on engineering education. Traditionally, engineering education has been focused on the individual; students have been encouraged to learn individually and, most assuredly, to take tests individually. This tradition has come about in our society and educational system due to the need to measure individuals. For example, industries come to universities to recruit the "best" students as measured by individual testing, rather than those who work best in teams! Engineering education overall has yet to conclude that students must be trained to work as part of teams. Industry requires team effort (the Japanese have recognized this for many years); yet, so far American educational institutions have not embraced the idea. Some mix of individual and team learning is probably the proper direction to take for engineering education.

Cooperative learning [Johnson et al., 1991] produces higher achievement, more fruitful relationships, and healthier psychological adjustment for students than do competitive or individualistic experiences. Although the use of teams in undergraduate engineering laboratories is a common practice, these effects do not automatically appear. The cooperative learning experience must be carefully structured. In cooperative groups, students can form communities of inquiry that allow them to make public their reasoning, thinking, and explanation about proposed actions. Such activities can help them reflectively learn with understanding [Cognition and Technology Group at Vanderbilt, 1992; Campbell and Andrus, 1991; NCTM, 1989; Palinscar and Brown, 1984; Palinscar and Brown, 1989; Vgotsky, 1978]. Moreover, public performance can help students check monitor one anothers' work so the group stays on track and individuals' misperceptions are minimized [Campbell, 1993].

Data from Adams et al. [Adams et al., 1988] and Lockhart et al. [Lockhart et al., 1988] support the notion that knowledge is more likely to be used when it is acquired in a problem solving mode than in a passive factual knowledge mode. In a passive situation, students may memorize solutions but have little ability to link, interpret, explain, and use it to create new solutions. Whitehead [Whitehead, 1929] termed the results of such passive learning "inert knowledge." Two studies [Cognition and Technology Group at Vanderbilt, 1990; Goldman and Robson, 1989] indicate that students in the upper stanines of mathematical achievement can solve basic word problems, but they do very poorly when they must generate the relevant subproblems and determine what information is required to satisfy the subgoals that they generate on their own. Self-generated information is better remembered than passively-received information; cooperative learning greatly assists in problem and solution generation [Bransford et al., 1989; Perfetto et al., 1987; Campbell, 1993].

Computer-supported cooperative work (CSCW) is a rapidly growing field dedicated to the proposition that computers can assist human to human collaboration. While significant bodies of research on CSCW have been generated during the last five years [Baecker, 1993], there has been less than widespread use of CSCW in higher education [Landow, 1990], and virtually no work dedicated to understanding how CSCW can facilitate engineering education. Studies in CSCW and groupware have been conducted in various areas. For example, group authoring [Ellis et al., 1991]; calendar management [Grudin, 1988]; coordination [Action Technologies, 1989]; and group decision support [Robinson, 1990] are among the types of applications in the field. CSCW supports both synchronous and asynchronous interactions among people. An example of synchronous interactions are video on-line interactions in which people can see each other. Asynchronous interactions are best illustrated by electronic mail. Improvements in the latter category are more useful to consider for facilitating students in university settings, since synchronous communications can be achieved simply by scheduling face-to-face meetings. Students work together on groupware-facilitated systems [Landow, 1990] to comment, critique, and discuss aspects of various liberal arts and sciences courses. Early experiments conducted using the hypertext framework of the Intermedia system developed at Brown [Landow, 1990] support the idea that this is also a potentially fruitful line of inquiry for engineering education.

What will be the impact of technology on student learning? How will a broadly fiber-optic-connected world affect the way we team and the way our students learn? In a word: dramatically! First, in terms of the way faculty work together, we may anticipate that a richly interconnected fibersphere will permit us to utilize teaching materials and information drawn from many sources. A practical example would be the ability to coalesce electronically captured slides for lectures, a capability that exists only in a minor way at the present time. One can envision network browsers (for example, similar to wide-area information services (WAIS, on Internet) that permit browsing of materials on different class topics from universities across the nation. Services of this type might be offered by publishing companies so that faculty could receive royalties analogous to the royalties received for textbooks. For students, the impact of densely connected networks is apt to become a central learning vehicle. By providing learning experiences that can be captured off-line in the dormitory or a home, we will provide students a mechanism for securing knowledge when they want it. Critics among us will argue that students will never want the knowledge! However, structured learning similar to classroom centered experiences can be constructed in which specific requirements must be met in a timely fashion to participate in a learning experience. It is important for us to realize now that the classroom mode of instruction may not be the only mode in the future. Cultural change will be required if we move to a networked learning environment. Faculty will face high levels of discomfort in learning to work with networked communications.

F. DISTRIBUTED EDUCATION

This volume contains a paper (Mattson) on distance education. One might envision technological enhancements to this mode of education: such as interactive distance education in which students are directly linked to the instructor. Linking might occur in different ways. for example, sets of "talking heads" on an instructor's screen, or perhaps summaries of polling from hundreds of students interacting with a lecture/discussion. This latter example might even be expanded to include the idea of distant students using learning tools that continually assess what the students know and don't know. This information would be fed to an instructor and analyzed on-line to permit the instructor to tailor remarks, examples and information presented to the needs of the class of listeners/interactors. Ultimately, distance education might become distributed education.

G. THE INFLUENCE OF MULTIMEDIA

Multimedia in the mid-1990s is a combination of on-screen animation, sound, text, and video. Multimedia refers to many types of media, and, hence, might well incorporate more than these most simple types to implement the media. A sense of touch and smell might convey other information and hence could be considered to be part of a multimedia experience. The utility of multimedia for engineering education is probably significant but not overwhelming. Let us examine how multimedia can impact engineering education and then discuss areas in which multimedia may be more or less useful. One of the very clear ways that multimedia can impact learning is in the explication of abstract and physical constructs that are difficult to explain using words or pictures alone. A simple example is in the understanding of how gears work together to facilitate work, or how a mechanical system with many linkages actually operates. Pictures and word explanations will help a student learn this information. However, if textual information and diagrams are coupled with an animation, a video of a real-world situation and perhaps audio for explanation, understanding can come much more quickly to students because of the ability of the medium to show information easily. The abstract can become concrete quickly. Reading about a set of gears meshing is not nearly as instructive as seeing gears mesh or changing gear ratios in a simulation and viewing the result. Interactive multimedia should provide even more enhancements to learning. In the above example, permitting a student to change gear ratios and immediately see the effect should powerfully influence learning. Also in this volume, Thomas and Katz provide a detailed look at the use of multimedia in engineering education.

Another application for multimedia is in the creation of anchored instruction paradigms for enhancing learning. In anchored instruction [Bourne and Brodersen, 1992; Dally and Zang, 1993; McLarty et al., 1990; Bransford et al., 1990], a scenario is posed perhaps using video. Students then interactively solve problems. For example, Bourne et al. [Bourne et al., 1992] created an anchored learning experience in which students were cast in the role of a consulting engineer who is given the task of determining how to modify the amount of water in a reservoir needed to keep a wildlife population alive while maintaining enough water flow to permit white water river rafting to continue on a river emanating from the reservoir. To solve the problem, students were required to make assumptions and solve equations. The basic notion was to cast students in a realistic problem-solving environment and to provide the computer tools needed for students to solve the problems posed. This type of application of multimedia appears to be quite promising for future directions in engineering education.

One may also consider virtual realities to be a kind of multimedia in which spatial location and tactile sensing are included in the set of senses that make up the supporting media. Virtual realities are typically thought of as fooling the senses by presenting visual, auditory, and tactile sensations that give the user the sensation of being in another environment. Visual scenes are deceived by binocular presentation that subtend a person's entire view. Likewise, headphones can shut out all sound except that presented binaurally. Spatial location of a person in a virtual reality can be sensed by head-mounted sensors, and hand and/or leg location can be sensed by exoskeletal sensors and position sensors. Taken in combination, coordinating these sensory stimuli can effectively place an individual in almost any simulated environment desired. More than likely, environments will be created for entertainment, but there also appears to be vast opportunity for education, as well. For example, much like individuals can explore simulated worlds in a virtual reality [Rheingold, 1991; Helsel and Roth, 1991], engineering students of the upcoming decades may be able to explore the interior of semiconductor devices or the working of complex chemical interactions to make new materials or fabricate nanorobots in a virtual learning environment.

H. LEARNING FACILITATION: KNOWBOTS

Concepts from the field of Artificial Intelligence [Barr and Feigenbaum, 1981] may well enable the teaching and learning methods of tomorrow. Knowbots, for example, appear to have the potential for facilitating teaching and learning. Knowbots are autonomous agents that conduct tasks in intelligent or semi-intelligent ways to assist humans in their informatics work. In terms of teaching, knowbots may help instructors organize materials, find information, and organize and present lectures. In terms of learning, knowbots may act like friendly coaches that assist students in understanding materials, watching for misconceptions, and providing remediation, as needed. Coupled with intelligent tutoring systems [Wenger, 1987] and with on-line distributed education capabilities to contact human instructors, the combination could provide an alluring instruction entity for the future.

VI. TEACHING ENGINEERING STUDENTS ABOUT ENGINEERING AND TECHNOLOGY

It is clear that technology will affect the modalities of teaching. What do we do when teaching? In engineering, most of us teach what we know, a situation not much different from the sciences or mathematics [Cross, 1993]. Due to current faculty focus on research, teaching what we know is only logical, given that we have spent our lives learning facts and methods to conduct engineering research. However, is this the right thing to teach our students? We argue that in an information intensive environment of the future, we may want to modify our current curricula rigidity that binds learning in colleges of engineering to a dramatically monolithic model. Fostered and bred throughout the last half of the twentieth century, the current model places a premium on learning what faculty learned when they were taught. One thesis of this paper is that this model of teaching students is not thorough enough; students should learn to think, solve problems and use technology, rather than focusing exclusively on the "what is" of technology. The teaching of laws and principles and their application is appropriate; however, students must learn to work in teams; make presentations; write correctly and convincingly; be tested in intense intellectual experiences in groups; and be required to invent, learn, search and research. The rote regurgitation of facts has little place in the engineering curriculum of the future.

Engineering is a model-driven discipline. Casting aspects of the world into models that can be manipulated in order to understand how to change the world is central to engineering. Yet, in our experience, few undergraduate students understand this concept. They are more concerned with how to solve the next homework set! We feel that educational experiences in engineering should somehow reflect the ideal of understanding, abstracting, problem solving, discovery, teaming, and knowledge manipulation.

Must we teach students about hardware and software? Begrudgingly, the answer is probably yes! Purists in the engineering education community would probably say that students should be taught how to learn, how to solve problems, and apply and evaluate knowledge. In truth, engineering education accomplishes such lofty goals only sporadically. We do teach constantly about things. In terms of hardware, we teach about new types of microprocessors or new type of chemical processes. Engineering provides an ever changing panorama of "things" that we feel duty bound to teach something about. After all, our students will need to know what microprocessors (or similar) things are when they graduate and go to work for an industrial concern. We have to teach about hardware — both computer hardware and other engineering hardware. We cannot teach how to think and solve problems solely; there must be an interweaving of literature and knowledge about the current world with higher level cognitive skills that will persist and enable a lifetime of learning.

What about teaching programming skills? During the mid twentieth century, programming became a very popular skill for engineering students to acquire. However, as the century comes to a close, it is more apparent that programming will not be a skill acquired by all engineering students. The reason, of course, is that more and more engineering tasks can be accomplished without detailed programming skills. Visual programming makes it possible for many tasks to be accomplished without having to ever write a line of code. Hence, the conclusion is that the teaching of detailed programming skills will probably disappear from the engineering curriculum in the upcoming decades. Only those who intend to pursue careers of developing tools for programming will continue to learn these skills.

VII. SUMMARY AND RECOMMENDATIONS

It is patently clear that computer and communications technology will impact the teaching and learning of engineering education in the decades to come. There is also little doubt that change will be slow because the culture of colleges of engineering will resist change. Nevertheless, there is the hope for real change in the relatively near term due to (1) the impetus of National Science Foundation funding for assisting colleges of engineering in moving in innovative directions, including the engineering education coalitions, (2) the dramatic price drops in computers of recent years and (3) the all-but-certain inexpensive high-bandwidth communication media that will be in-place by the end of the 1990s. National acceptance of changes in teaching and learning in engineering education, coupled with affordability, will bring about changes. In part, change will occur simply because of the market; students will be attracted to institutions in which change occurs. Hence, institutions that do not change will become obsolete.

Where will the biggest changes be? A guess is that the most significant early changes will occur in the classroom, where small investments in technology can reap the appearance of significant change. The same is true of multimedia; the attention-grabbing capability of multimedia offers early returns to institutions that implement it. The glamour of "high-tech" education presented via multimedia has the potential of attracting more students than "dry" lectures presented in the traditional fashion.

In terms of substantive change, a likely area of change is in the education of post-college learners, including continuing education for industrial employees. Strictly speaking, mechanisms for distance education have been in place for many years. However, penetration of this market has not been exceptional due to the lack of a ubiquitous communication medium. As discussed above, the fibering of America should remove the barriers to widespread distributed education, thus permitting institutions of higher learning and K-12 to become involved in education of students anywhere, anytime, and in different modalities. There is the potential for this style education to eclipse residential learning during the upcoming century.

Another substantive change will likely focus around computer-based learning. Current computer-based learning systems are not very good; programs do not adapt well to the learner, computer-based learning modules are very limited and very significant amounts of time are required to build learning modules. All of these disadvantages to computer-based learning are likely to disappear during the upcoming ten or more years. Intelligent tutoring systems that contain student models that permit programs to adapt to the learner will become more prevalent, as more people build modules and distribute modules. More acceptance and better modules will result and authoring tools will improve so that creation of modules will not take enormous amounts of time.

A third technology-facilitated change will likely be in the area of group learning. Already, learners in our colleges are using remote libraries, news groups, and e-mail to acquire knowledge outside their constrained geographical area. This trend will grow to permit access of information of any type, from any location, and at any time. Working with others and acquiring information from others will become much more commonplace than it is today.

The changes postulated above will drive universities to assume new roles in society. Continuous education of the population is a role that our universities can fill. In engineering, knowledge changes at such a rate that continuous education will almost be essential within the foreseeable future. For survival, it may be required that the culture of the residential engineering college change such that faculty knowledge workers serve a much broader population than has ever been served before. The media for providing these services will be in place within only a few short years. Hence, the colleges of engineering of America must begin to plan now for this evolution.

The short term recommendation for engineering colleges is to begin planning for how to take advantage of the coming evolution in engineering education due to the availability of computer and communications technology. Long term strategies should begin to orient educational activities toward prototype studies including investigations of how to bring about cultural change. Beginnings such as the requiring of e-mail communication between students and faculty; demonstration classrooms; creation of multimedia and simulation software that fits into current classes; exploration of ways of delivering education at a distance; and the instituting of courses that require group work by computer and other topics will be the first steps in the journey toward this new era in engineering education.

REFERENCES

1. Action Technologies, The CO-ORDINATOR II, 1989.
2. **Adams, L., Kasserman, J., Yearwood, A., Perfetto, G., Bransford, J., and Franks, J.**, The effects of facts versus problem-oriented acquisition, *Memory & Cognition,* 16, 167-175, 1988.
3. **Agogino, A., Sheppard, S., and Oladipupo, A.**, Making connections to engineering during the first two years, *Proceeding of Frontiers in Education*, Lawrence P Grayson (Editor), 563-568, November, 1992.
4. **Baecker, R.M.**, *Readings in Groupware and Computer-Supported Cooperative Work*, Morgan Kaufmann, San Mateo, CA, 1993.
5. **Barr, A., and Feigenbaum, E.**, *The Handbook of Artificial Intelligence,* Heuistech Press, Stanford, CA, 1981.
6. **Bell, T.**, Jobs at risk, *IEEE Spectrum*, 30(8), 18-36, August, 1993.
7. **Bourne, J.R., and Brodersen, A.J.**, The virtual engineering reality: An introductory freshman engineering course, *Proceedings of the Undergraduate Engineering Education IV Conference*, Santa Barbara, CA, July 26-31, 1992.
8. **Bourne, J., Brodersen, A., Cantwell, J., Debelak, K., Kinzer, C., Koussis, A., Haden, G., van der Molen, H., and Droes, S.**, Intelligent Hypertutoring: Investigation of Intelligent Tutoring Systems for Use in Undergraduate Engineering Education, *Proceedings of New Approaches to Undergraduate Engineering Education III Conference*, Banff, Alberta, Canada, pp. 301-312, July 28-August 2, 1991, United Engineering Trustees, Inc., copyright, 1992.
9. **Brandford, J.D., Sherwood, R.D., Hasselbring, T.S., Kinzer, C.K., and Williams, S.M.**, Anchored instruction: Why we need it and how technology can help, in *Cognition, Education, and Multimedia*, D. Nix and R. Spiro (editors), 115-141, Hillsdale, NJ, Lawrence Erlbaum and Associates, 1990.
10. **Bransford, J.D., Franks, J.J., Vye, N.J. and Sherwood, R.D.**, New approaches to instruction: Because wisdom can't be told, in *Similarity and Analogical Reasoning*, S. Vosniadou and A. Ortony (Editors), 470-497, Cambridge University Press, NY, 1989.
11. **Campbell, J.O.**, Discussions on cooperative learning, Vanderiblt University, 1993.
12. **Campbell, J.O., and Andrus, G.R.**, Cutting Development Time for Computer-Based Instruction, *Proceedings of the Ninth Annual Conference on Interactive Instruction Delivery*, Warrenton, VA, SALT, 52-53, 1991.
13. Cognition and Technology Group at Vanderbilt, Anchored instruction and its relationship to situated cognition, *Educational Researcher,* 19(6), 2-10, 1990.
14. Cognition and Technology Group at Vanderbilt, The Jasper experiment: An exploration of issues in learning and instructional design, *Educational Technology Research and Development,* 40(1), 65-80, 1992.
15. **Cross, P.**, On college teaching, *Journal of Engineering Education*, 9-14, January, 1993.
16. **Dally, J.W., and Zang, G.M.**, A freshman engineering design course, *Journal of Engineering Education,* 82(2), 92-100, April, 1993.
17. **Ellis, C., Bibbs, S., and Rein, G.**, Groupware: Some issues and experiences, *Communications of the ACM* 43(1), 38-58, January, 1991.
18. **Fromm, E.**, GATEWAY: Opening doors to the future, *Proceedings of Frontiers in Education*, Lawrence Grayson (Editor), 354-358, November, 1992.
19. **Glover, C.J., Lunsford, K.M., and Fleming, J.A.**, TAMU/NSF engineering core curriculum. 1. Conservation principles in engineering, *Proceedings of Frontiers in Education*, Lawrence P. Grayson (Editor), 603-608, November, 1992.
20. **Goldberg, A., and Robson, D.**, *Smalltalk-80: The Language and Its Implementation*, Addison-Wesley, Menlo Park, 1983.
21. **Goldman, A. and Robson, D.**, *SmallTalk-80: The Language,* Addison-Wesley Publishing Co., 1989.
22. **Grinter, L.E.**, Report on evaluation of engineering education, *Journal of Engineering Education,* 46(1), 26-60, September, 1955.
23. **Grudin, J.**, Why CSCW applications file: Problems in the design and evaluation of organization interfaces, *Proceedings of the Conference on Computer-Supported Cooperative Work*, ACM, New York, 1988.
24. **Helsel, S.K., and Roth, J.P.** (Editors), *Virtual Reality: Theory, Practice, Promise*, Meckler, Westport, 1991.
25. **Ingraffea, A.R.**, National engineering coalition, *Proceedings of Frontiers in Education*, Lawrence Grayson (Editor), 359-362, November, 1992.
26. **Johnson, D.W., Johnson, R.T., and Smith, K.A.**, Cooperative Learning: Increasing College Instructional Productivity, ASHE-ERIC Higher Education Report No. 4, 1991.
27. **Krol, E.**, *The Whole Internet*, O'Reilly and Associates, Sebastopol, CA, 1993.
28. **Landow, G.P.**, Hypertext and collaborative work: The example of intermedia, in *Intellectual Teamwork: Social and Technological Foundations of Cooperative Work*, 407-415, 423-428, Galegher et al. (Editors), Lawrence Erlbaum, Hillsdale, NJ, 1990.
29. **Lockhart, R.S., Lamon, M., and Gick, M.L.**, Conceptual transfer in simple insight problems, *Memory and Cognition,* 16, 36-44, 1988.
30. **Lynch, D.C., and Rose, M.T.**, *Internet System Handbook*, Addison-Wesley, Reading, MA, 1993.

31. **McClure, C.R., Bishop, A.P., Doty, P., and Rosenbaum, H.**, *The National Research and Education Network (NREN): Research and Policy Perspectives*, ABLEX Publishing Co, Norwood, NJ, 1991.
32. **McLarty, K., Goodman, J., Risko, V., Kinzer, C., Vye, N., Rowe, D., and Carson, J.**, Implementing anchored instruction: Guiding principles for curriculum development, in *Literacy Theory and Research: Analysis from Multiple Perspectives*, J. Zutell and S. McCormick (editors), 109-120, 39th NRC Yearbook, Chicago, IL, National Reading Conference, 1990.
33. **Mosterman, P., Dorlandt, M., Campbell, J., Burow, C., Bouw, R., Brodersen, A., and Bourne, J.**, Virtual engineering laboratories: Design and experiments, *Journal of Engineering Education*, in press.
34. National Council of Teachers of Mathematics, *Curriculum and Evaluation Standards for School Mathematics,* March, 1989.
35. **Palinscar, A.S., and Brown, A.L.**, Instruction for self-regulated reading, in L.B. Resnick & L.E. Klopfer (Eds.), *Toward the Thinking Curriculum: Current Cognitive Research*, Alexandria, VA, American Society for Curriculum Development, 19-39, 1989.
36. **Palinscar, A.S., and Brown, A.L.**, Reciprocal teaching of comprehension-fostering and comprehension monitoring activities, *Cognition and Instruction,* 1, 117-175, 1984.
37. **Perfetto, G., Yearwood, A., Franks, J., and Bransford, J.**, The Effects of Generation on Memory Access, *Bulletin of the Psychonomics Society,* 25, 151-154, 1987.
38. **Rheingold, H.**, *Virtual Reality*, Howard Rheingold, Summit Books, New York, 1991.
39. **Robinson, M.**, Pay bargaining in workers' cooperatives, in *Self Referencing in Social Systems*, F. Geyer and J. van der Zouwen, (Editors), Self Referencing in Social Systems, Intersystems Publications, 1990.
40. National High-Performance Computing Technology Act of 1989, U.S. Congress, Senate, 101st Congress, May 18, 1989.
41. **Tobias, S.**, *They're Not Dumb, They're Different*, Research Corporation, Tucson, Arizona, 1990.
42. Visual Basic, Versions 1.0 - 3.0, 1989 - 1993, Microsoft Corporation, Redmond, WA.
43. **Vygotsky, L.S.**, *Mind in Society*, Cambridge, MA, Harvard University Press, 1978.
44. **Walker, M.L., Bucciarelli, L.L., Sommer III, H.J., and Regan, T.M.**, ECSEL: A strategy for educational reform, *Proceeding of Frontiers in Education*, Lawrence P Grayson (Editor), 350-353, November 1992.
45. **Wenger, E.**, *Artificial Intelligence and Tutoring Systems,* Morgan Kaufman Publishers, Los Altos, CA, 1987.
46. **Whitehead, A.N.**, *The Aims of Education,* Macmillan, New York, 1929.
47. **Winkel, Brian J., and Froyd, Jeffrey**, An Integrated First-Year Curriculum in Science, Engineering and Mathematics: Progress, Pitfalls and Promise, *Proceeding of Frontiers in Education*, Lawrence P Grayson (Editor), pp. 557-562, November, 1992.

Chapter 10

NEW PARADIGMS FOR ENGINEERING EDUCATION USING THE NATIONAL INFORMATION INFRASTRUCTURE AND ELECTRONICALLY MEDIATED INSTRUCTION

Randy H. Katz
ARPA/CSTO
and
EECS Department
U.C. Berkeley
Berkeley CA

Robert J. Thomas
School of Elect. Eng.
Cornell University
Ithaca, NY

Joseph G. Tront
Dept of EE
Virginia Tech
Blacksberg, VA

I. INTRODUCTION

Our ability to create information more rapidly and in different forms by using new computing and communication technologies is changing the world in which we live. Among the first to feel the change have been the institutions of higher learning. Their need to communicate knowledge is understood by all. Students and faculty alike have been influenced by the range of computing that has become available. The ability to easily publish with a desktop environment and still use that same environment for sophisticated simulation and visualization to analyze vast amounts of data has had a profound effect on all who teach and learn. Yet it is the current evolution in digital networks that has the potential to link people and resources over vast distances that is likely to change forever how humans interact, communicate, and learn. Representing information using diverse media will become paramount as new human links are created for the purpose of exchanging, understanding, and acting on the material. Thus, a serious re-examination of how we teach and learn, how we perform research, and how we interact and communicate among ourselves and the world will need to occur.

To change engineering education in the United States to make it more relevant to the new scheme of things will require the development of innovative technological support for instruction and learning. Such support will demand radically new settings such as high-technology classrooms/laboratories that go beyond traditional "computer labs." And the evolution of electronically mediated instruction will require pedagogical changes involving new roles for teachers. The use of technology will spawn new pedagogies, such as teaching-by-walking-around in a simulation center. It will foment the creation of interdisciplinary gateways and encourage a new collegiality among both professors, staff, and students. Modern information technology (interactive computer graphics and simulation, video recording, laserdisk storage, and hypermedia, among others combined with networked computers) has a tremendous potential for helping students understand, correlate, and access information. This potential remains largely untapped; huge informational resources are effectively unavailable to a large segment of our faculty and students. A major activity should be the continued development, testing, and dissemination of hardware and software tools for accessing and using electronic information. What will have revolutionary impact is a national system that makes it easy to access and explore modifications of existing pedagogical options, and to embrace new ones as they are developed.

The National Information Infrastructure (NII) is a vision for a pervasive web of communications capabilities and information services supporting critical applications of national importance. To be effective, an infrastructure must provide a high degree of leverage and reuse of its elements across a broad range of applications. Of current interest are the so-called "National Challenge Applications," including health care information systems, crisis management, environmental monitoring, design and manufacturing, and most importantly for this paper, education and training. These National Challenge Applications will exploit an emerging information infrastructure to achieve new economies of scale for information enterprises that are of national (and international) scale. In so doing, these applications will yield new efficiencies, productivities, and capabilities. In this paper, we introduce a conceptual framework and architecture for the NII, while focusing on its specific application in the education and training arena. An example of an implementation of a challenge application that could provide the setting we seek for engineering education is the Synthesis coalition's National Engineering Education Delivery System (NEEDS) [15,16]. Before discussing the NII and a specific implementation aimed at engineering education, we discuss the current situation in engineering instruction and some impediments that need to be addressed before large-scale acceptance can occur.

II. TOWARDS A NEW PARADIGM FOR ENGINEERING INSTRUCTION

An underlying philosophy of many teachers and many institutions is that "students acquire knowledge from being told" [1]. While the traditional lecture format is the most common method of instruction, it is not necessarily the most effective. In this format, students often learn out of context and consequently are able to repeat facts and formulae, but are not able to recognize situations to which the facts and formulae apply. However, learning can be kept in context by several means. Olsen and Bruner's theory of education is that students learn best through direct experience. Papert is more general, and allows different media to vicariously substitute for direct experience. All would agree, however, that a diverse instructional environment that provides multiple representations of a situation is the most effective way for students to learn. Learning styles vary, and information in multiple forms allows students to select the method in which they learn best. Seeing a video tape of a car going up an inclined road under icy conditions, along with an animated graphic showing the effects of gravity, friction, and acceleration, along with the ability to reference a text on inclined planes, all provide a student with more opportunities to learn than does a lecture. Simulations allow students to scrutinize situations that are costly, dangerous, or impractical. Simulations can provide a great deal of insight into the inter-relationships of a complicated system. Furthermore, if the student is in control of the simulation, she/he can adjust the pace and direction of the presentation, something which is not possible in a lecture.

Instead of a standard classroom organized for lectures, imagine instead a classroom where students obtain individual instruction using computer displays. Imagine the rich graphic imagery that is possible, and the reinforcement through repetition using different words and images that can occur. Electronically mediated materials makes this possible by increasing the accessibility of information. While students are one category of user, instructors may also use the material to prepare for class, or may even display it directly in class. The ability to call up elements from a database to a computer screen makes it practical to use various media for teaching. Practically speaking, the early users of electronically mediated material would probably be the instructors. As networks develop and network speed increases, students could be given access to the same material for creating reports or for study and review. But since instructors will initially create and use some of the database elements, it is important to examine the constraints they will face in these endeavors.

A. GENERAL PROBLEMS FACING THE ADOPTION OF NEW TEACHING METHODS

The diffusion of innovations is an incremental process. Most of a population that has an opportunity to change will watch the efforts of the few early adopters, and then seek the opinions of people they respect before deciding whether to adopt a new approach. If a program to encourage faculty to use technology in their teaching is to succeed, it must be ready at the outset to meet the needs of the early adopters, and to provide accurate information to those seeking it.

The use of technology in teaching has several problems associated with it [1], two of which are applicable in a university setting, and which need to be addressed:

1. Teachers are often unwilling to undertake those things necessary to institute a role change

In order for instructors to consider using electronically mediated material, the value of that material must be apparent to them. For example, they need to recognize that the material is:

- superior to, or at least, equal to, the traditional methods
- compatible with current educational content, purpose, and institutional priorities
- cost-effective, and
- not too complex [2].

Instructors need school-based support for their efforts to change [3] so that there is:

- access to resources, such as hardware, software, and training
- available technical expertise and access to credible information
- rewards for innovation, and
- indications that the organization's values are conducive to change [2].

If an institution provides the impetus for change, and the courseware is valuable, it still seems unreasonable to assume that these two factors alone will cause the instructor to use electronically mediated material; an interim level between the institution and the courseware is needed. “When novice computer users have someone to whom they can turn for knowledge about computers as well as emotional support and reassurance, they are more likely to begin integrating technology into the curriculum” [3]. Others feel that hands-on workshops, access to technical expertise, and personalized training are necessary to encourage adoption [2]. While a change-agent makes important contributions to awareness and information, “the institution has a far greater influence on actual adoption of technology”[2]. Instructors need school-based support for their efforts to change [3].

Last, instructors need to be able to use the material in class by either having the classrooms equipped to present electronically mediated information, or by having portable equipment available. When instructors want to illustrate a point in a typical classroom, they draw on the tools that are available to them: chalk, overheads, possibly slides. If the classroom is equipped with a computer and video projection, they might consider using other media such as simulations, graphics, animation, and video tape. The existence of an electronic classroom alone will not necessarily encourage the creation of new material consistent with its potential, but its absence will certainly discourage the creation of such material.

To summarize, an instructor interested in adopting newer methods of teaching has four concerns: that the material meets class needs and can be practically used; that the institution supports and recognizes efforts to improve teaching using these means; that there is technical support available for his or her efforts; and that there are opportunities to use the material in a class or lab setting. Of these four, the role of the institution is perhaps the greatest contributor to change.

2. High Quality Courseware is not Available in Suffucient Quantity and Variety

Another concern that is often raised is the lack of availability of suitable content material. In many engineering colleges this is not as big a problem as it would be in other departments or schools. For example, many schools have computer-aided design instructional facilities that support several graphical analysis programs that might find immediate use. However, due to the rate of change in technological fields such as engineering, this concern cannot be entirely dismissed. A constant effort is needed to keep materials in step with technological developments. Also, it is well known that simulation material exists and is available for instruction in far larger numbers than is actually used. It is our thesis that it is an inability to embed these simulations in linear "chalktalk" lectures that hampers their use in engineering education. The integration of simulations in the highly visual material we seek to create should proliferate its use beyond what exists today.

The new teaching materials we are interested in should not have to be created by each teaching individual; just as an instructor can draw on several published texts to teach a course without having to author one, so should s/he be able to draw on electronically mediated material without having to create all of it. It is critical to turn the creation of electronic materials for instruction into a truly national-scale activity. At the outset of an effort to encourage the use of electronically mediated material, limited amounts will be available. One way to start would be to transform traditional forms of material such as text into a digital format. If there is a core amount of information, instructors could recombine pieces from a variety of sources into new assemblages, perhaps even adding one or two pieces of their own creation. These assemblages could be stored as units for use by others. With institutional encouragement, with production resources available, and with technical assistance, it is quite possible to begin increasing the number of database elements that are available. Networking the database through the Internet (and the future NII) would further increase the pool of authors.

II. THE NATIONAL INFORMATION INFRASTRUCTURE

We believe that the NII will emerge only through a process of developing new information architectures that achieve broad commonalties across a wide spectrum of critical network-capable, information-centered applications. The NII will provide widely available, easy-to-use, and easy-to-interface computing and communications services, including the underlying communication channels (which we call bitways) and new kinds of information technology capabilities (which we call services).

The NII will be created in large measure by the private sector computer and communications industry. Nevertheless, government plays the leadership role in catalyzing the NII's development while progressing on solutions to key National Challenge Applications. The federal government's research and development strategy should stimulate growth in the critical areas of technology that underlie the NII, while collaborating with key stakeholders in NII and the National Challenge (NC) areas to create effective prototype information architectures that can scale-up to national-scale enterprises. Government can also seed and accelerate the development of the NII by becoming an early and friendly user of the NII itself, using network-based information technology to re-engineer and reinvent government work processes.

NC Applications are fundamental information-processing intensive applications with broad impact on the nation's economy and its future well-being. These include extensively inter-networked systems to provide health care information, digital libraries of environmental information and satellite imagery, planning support for crisis management, integrated design and manufacturing systems, and computer-based education and training systems. Information infrastructure creates new opportunities to achieve significant economies of scale by creating a diverse collection of network connectivity and information processing capabilities. Among the NC Applications, education and training is one in which the public sector must take a lead in the development.

What is a NII? A national infrastructure is one that makes shared national-scale resources widely available to industry, government, and individuals, while enabling more productive and efficient activities. Electric power, water distribution, and telephone systems are examples of "utility" infrastructures that have their roots in the early twentieth century. A national-scale air traffic control system or nationwide highway system are examples of "transportation" infrastructures that evolved in the mid-twentieth century. For the twenty-first century, the information century, whole new dimensions of infrastructure resources will be needed, linking together diverse enterprises through low-cost, widely-available, and easy to use information services [4], [5]. There is little doubt that the new dimensions in information are driving a re-engineering of mature infrastructures such as the electric power system and the telephone system.

The Clinton administration, under the leadership of Vice President Al Gore, Jr., has made the NII a high priority initiative [6], [7]. U. S. industry was at first concerned that the infrastructure would be created and run by the federal government, counter to the long tradition in this country of private sector ownership of telecommunications systems. Over time, the government's role has become clarified to focus on the crucial issues of information and telecommunications policy, to simultaneously encourage private sector development of new information and communications technologies, and to protect citizens' rights of access and privacy in the electronic age, balanced with law enforcement and national security concerns. The emerging federal research and development technical program is building on existing federal programs in information technology as well as the efforts of the federal High Performance Computing and Communications (HPCC) program, which is developing base technologies of advanced computing and communications. In addition, there are significant government-funded programs in specific applications areas, such as education, and these will be refocused to exploit the NII as it becomes available and cost-effective.

The federal government also has a critical role to play in reducing the risk in creating a NII, while ensuring that the technology will be cost effective and appropriate for developing implementations to the National Challenge Applications. Information infrastructure technologies will play a critical role in the federal government's own plans to re-engineer its work processes. In an outstanding articulation of the government's role in information infrastructure, Vice President Gore has drawn an analogy between the NII and the first use of telegraphy in a limited government-sponsored demonstration (between Washington and Baltimore), subsequently followed by widespread private sector deployment on a national scale.

Thus, the NII will be implemented as a cooperative effort among private- and public-sector organizations. The role of the federal government will be to (1) seed the long term underlying technologies through proven strategies of federal research and development across the academic, National Laboratory, and industrial technology systems, (2) verify the new technologies and their application to National Challenge applications through early testbed demonstrations, and (3) work closely with industry as an informed and active "early user" of the technology. In more specific details, the Federal government's role will be to:

- Foster creation of the NII through close collaboration with industry and researchers. Through the Federal government's research and development strategy, progress in key technologies can be accelerated. Government can stimulate commonality in critical interfaces and components to facilitate the scale up of NII technologies to a national scale. And the government can support rapid and continual transition of new NII capabilities into commercialization and widespread use.

- Collaborate with involved parties to address NC Applications. This would lead to the development of specific, applications-driven NII capabilities. In addition, a close coupling to applications would enable fundamental change in the business practices of the NC industry sectors. In particular, exploiting commonalties within the NII should eliminate barriers to economies of scale while creating a foundation for a competitive supply of services.

- Lead in the development of early systems. Government-sponsored testbeds can demonstrate scalability in the underlying technologies by supporting multiple independent experiments unified by a common architecture. These testbeds should span the range from basic technologies to actual application usage in a pilot mode. This will make it possible to demonstrate how application users can gain immediate benefit from the availability of widespread information infrastructure capabilities.

If the National Information Infrastructure works as envisioned, the benefits would be (1) whole new ways of doing business in critical national applications, (2) a stronger economic future providing the basis for long term national security, and (3) an infrastructure providing a sustainable base for the future.

A. INFORMATION SUPERHIGHWAYS

Due in part to Vice President Gore's father's crucial role in drafting the legislation that led to the interstate highway system, it has become popular to draw the analogy between transportation highways and information highways. These infrastructures share many common characteristics, even though transportation deals with the movement of physical objects while information transmission is concerned with virtual objects, encoded in bits. Of course, the physical objects can also be information objects such as books or paper mail. Yet just as the interstate highway system is much more than roadways, a national information infrastructure is significantly more than computer networks.

As technology improves, significant advantages accrue by having specialized "highways" for information objects that are capable of dealing with their unique properties. For example, traditional information objects like this document can be converted into "bits" which can be moved much more efficiently than its paper version. It can also be "digitally signed" to insure that it contains the same information as intended by its author, and its content has not been modified during that transmission. Finally, it can be encoded in such a form that only selected recipients can view it, perhaps for a small fee that is collected for the author and automatically transferred to his bank account. Furthermore, much more information is being generated natively as "bits" rather than paper, and this representation can be directly stored and processed in this form.

Table 1 illustrates some of the concepts that are common to infrastructures in general and transportation and information infrastructures in particular. For example, to permit leverage, an infrastructure must adhere to certain engineering procedures and standards. In the transportation system, these include the choice of materials for road beds, the curvature of the road, and the structural engineering methods for constructing bridges and overpasses. In the information highway, these procedures and standards deal with network protocols, issues of interconnection and interoperation, and the provision of standard interfaces to applications.

Infrastructure Elements	Transportation Highways	Information Highways
Engineering Procedures and Standards	Road beds and materials, road curvature, bridge and overpass construction	Network protocols, interconnection and interoperation, application interfaces
Services	Gas stations, restaurants, lodgings	Electronic mail, news, information services, database search and retrieval, remote access
Regulations	Tolls, highway patrol, speed limits; usage metering and charging; trust and authentication; network operation and oversight	Amenities
Amenities	Scenic overlooks	Information browsing/grazing
Access	On/Off-Ramps, ubiquitous road net	Ubiquitous access by phone or direct connection
Enablers	Interstate trucking	Electronic commerce, distance learning, electronic town hall

Table 1: An Information Infrastructure is More Than a Communications Network

The transportation system also provides services for its users, such as gas stations, restaurants, and hotels. In the information domain, the notion of services maps onto the infrastructure's capabilities for mail, news, information search and retrieval, and the remote access of resources anyway in the infrastructure. Just as the transportation infrastructure regulates usage, in the form of tolls, speed/weight limits, and enforces its regulations through the highway patrol, the information infrastructure must charge for service, authenticate access to the infrastructure, and provide mechanisms for network operation and management.

Amenities along the transportation infrastructure include scenic overlooks and stopovers, while the information infrastructure analog is its capability for supporting the browsing of information. Access to the transportation infrastructure is through on/off ramps and easy interconnection with local road networks. This same ease of access to the information infrastructure is provided through ubiquitous connectivity, either by telephone dial-in or connections through local networks. Infrastructures enable new kinds of industries and activities. In the case of transportation, it is the interstate trucking system. For information infrastructure, new forms of commerce, learning, and interaction will be made possible by the infrastructure.

Table 2 takes another cut on the common effects of transportation and information infrastructures. The fundamental utility of the interstate highway system is that it makes possible the movement of goods with a time and cost that is much less than the cost of local production. This enables new economies of scale and resource sharing: apples grown in Washington can be sold in Florida and clothing manufactured in Georgia can be sold in California, as long as the cost of transportation is low enough to advantageously outweigh the cost of local production. In this way, the transportation infrastructure can enable whole new national-scale enterprises.

Overall utility is further increased through the introduction of commonality that provided increased efficiency of the transportation systems. For example, the modern highway system is engineered to support large tractor trailers traveling at sixty-five miles per hour and below, with a commonly understood signage system, and gasoline, food, and lodging services appearing frequently along the routes. A lack of common engineering practices and services would reduce the utility of an infrastructure. For example, during the American Civil War, the South suffered because of an inadequate railroad network compounded by the lack of a standard railroad gauge. Careful exploitation of such commonality can enable continuous improvement; new innovations and new capabilities can be inserted into the framework of the infrastructure as long as they meet the restricted interfaces imposed by the common architecture.

Infrastructure Elements	**Transportation Highways**	**Information Highways**
The ability to move objects rapidly and cost effectively	Physical: may include information	Information: may be transformed to physical
Movement (time, cost)	<< Local Production	<< Local Production
Access (time, cost)	= Local	= Local
Fundamental limits	Moving mass and Logistics	Speed of light, computing, and conceptual understanding
Rate of change	Longer than a human generation	Shorter than a human generation

Table 2: Infrastructure Enables

An information infrastructure provides the ability to move information rapidly and cost effectively so that local production of information can be enhanced by information sharing. The rate of change of information technology has increased dramatically as the result of electricity-based communication over the last several generations, and more recently by transistor-based computation. Prior to these recent developments, a given information technology spanned many human generations to advance even a single technology generation. As an example, consider the centuries-long use of smoke signals as a means of communications among nomadic tribes versus the decades-long transition from analog to digital telephony. Since the new electricity-based (since the mid-nineteenth century) and transistor-based (since the mid-twentieth century) technologies have emerged, many generations of information technology have evolved within the time frame of a single human life. It is a key attribute of the information age that computing and communications technologies are both evolving and converging rapidly. Information technology has become the fundamental enabling technology underlying all sectors of the economy.

B. APPLICATIONS, SERVICES, AND BITWAYS

1. National Challenge Applications

Numerous lists of NC Applications have appeared. For example, the Computer Systems Policy Project has identified Design and Manufacturing, Education and Training, and Health Care as the National Challenges [4]. Here, we have selected as the NC Applications those with a pervasive impact on American society coupled with an opportunity to leverage extensive communications and information processing capabilities. The NC applications include (but of course are not limited to) the following:

- Crisis Management. These are systems that exploit information technology to insure the United State's national and economic security. This is accomplished by providing timely data collection and intelligence fusion, advanced planning tools, rapid communications with our defense forces spread around the globe, and a command and control ability to respond quickly to crises. The same basic capabilities can be deployed on a smaller scale to respond to local emergencies, such as devastating hurricanes, earthquakes, or fires.

- Design and Manufacture. These are systems that integrate engineering design with product manufacture, to reduce the time to create new products, to lower production costs, and to increase product quality. In a wider sense, a pervasive design and manufacturing system should couple suppliers to their customers throughout the production chain. Customers provide their desires as inputs to the system, leading to more responsive product design, manufacture, and just-in-time warehousing and product delivery.

- Education and Training. These systems provide access to on-line instructional and research materials, anywhere and anytime. Once created and made accessible, instructional materials may be reused and evolved by instructors around the country. Educational use of the information infrastructure will enable distance learning, where remote students can gain access to specialized instruction, while training use will exploit simulation coupled with remote access to actual apparatus.

- Environmental Monitoring. These systems integrate ground, airborne, and space-based sensors to monitor (and potentially respond to) environmental changes. They may be used to discover a nuclear accident in progress, and oncoming climatic effects such as smog conditions, or they can be exploited for longer term studies such as climate change.

- Government Information Delivery. Citizens have a right to ready, low-cost access to government information for which they have already paid to be collected. These include economic statistics, trade information, environmental and land use information, and uniform "one stop shopping" for government services such as veterans and social security benefits.

- Health Care. These systems use information technologies to improve the delivery of health care, by providing ready access to patient records; remote access to medical expertise; support for collaborative consultations among health care providers, and rapid, paperless claims adjustment that can help reduce health care costs.

Success in scaling up solutions to these applications, in terms of larger numbers of users spread across wider distances and exploiting more diverse communications and service technologies, depends critically on achieving underlying commonalties. A plethora of interfaces and incompatible modules is unlikely to yield an architecture that can scale up to a nationwide system. In the absence of common interfaces, methods, and modules, it may be possible to demonstrate prototype solutions to specific applications through monolithic "stovepipes," but these solutions are unlikely to provide leverage across other applications. Systems constructed in this fashion are neither able to gracefully evolve over time nor can they sustain the kind of incremental improvements made possible through continual competition among the underlying technology alternatives.

2. Three Layer National Information Infrastructure Architecture

One possible architecture for the National Information Infrastructure has three major interconnected layers as shown in Figure 1. These layers represent NC Applications which are supported by diverse NII communication and computation Services, which in turn are

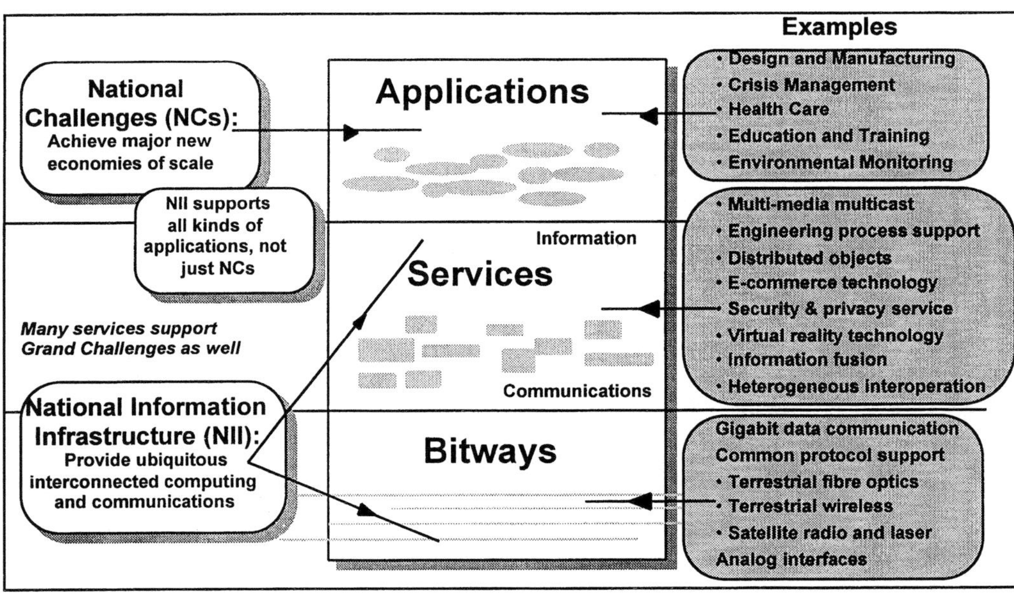

Figure 1. Three-Layer Architecture for the NII. One possible model architecture for the National Information Infrastructure is shown above. Bitways provide the communications substrate, the applications layer supports the implementation of the National Challenges, and the services layer provides the bridge between communications and information.

supported by high performance NII Bitways. Each layer sustains a diverse base of technologies, supports a broad base of suppliers of that technology, and maintains a continuously increasing capability for users over time.

To support a rapid rate of growth in a competitive environment for many suppliers of technology, developments in each of these layers focus not only on stimulating the new technologies, but also on determining common mechanisms, the "commonality," by which that capability can be delivered. For example:

- The diverse Bitways technologies can deliver communications capability in a uniform manner through use of standard protocols, such as SONET/ATM. This has the effect of insulating developers of NII services from the details of the rapidly evolving communications technologies used to deliver information capabilities to end users and applications.

- Similarly, services such as multimedia multicast can be delivered to developers of Application capabilities through proper adherence to common protocols. With well-designed protocols and interfaces, rapid growth in multimedia capability and capacity can be delivered to end users and applications developers without requiring major re-engineering of the whole applications-level systems.

- Finally, the keys to scaling up in NC Applications are often in the choice of common application-specific protocols. For example, manufacturing applications require shared representations for product and process descriptions to support widespread interoperability among design systems and tools.

This architecture addresses directly the challenge of scale-up within each of the three layers. The ongoing validation of concepts is achieved, in each layer, through large-scale testbed experimentation and demonstration conducted jointly with industry, users, and suppliers of new technologies and information capabilities. If the evolution of the NII architecture proceeds as hoped, the result will be the integration of new capabilities and increased affordability in the NC Applications. Each layer supports a wide range of uses beyond those identified for the specific NC Applications. For example, generalized NII Service and Bitway technologies can also support smaller-scale applications, extensions of existing services, ad hoc distributed computing, and so on.

3. Elements of an Information Infrastructure

As already stated, an information infrastructure must be more than bandwidth, switching, and ubiquitous access. It must also yield an environment within which to create new kinds of pervasive, network-based NC Applications. This would be an enormous task if the development of each application were undertaken from scratch and in isolation. Every large-scale application contains elements of interfaces (e.g., speech recognition, data visualization), application building blocks (e.g., planning subsystem, imaging subsystem), data/process management (e.g., search and retrieval, action sequencing), and communications (e.g., IPC). A critical element of an information infrastructure (as opposed to a transportation infrastructure or a communications infrastructure) will be the creation of development environments, toolkits, operational protocols, and data exchange and action invocation standards from which a diverse set of service solutions can be combined, integrated, and reused across the variety of applications efforts to be undertaken.

A key goal of an information infrastructure should be to create reusable pieces of applications from which large-scale applications can be constructed more quickly and more effectively. The evolution of common tools, environments, and procedures are a critical byproduct of the information infrastructure. In somewhat more detail, the critical technical components of the information infrastructure that support the development of reusable portions of applications are the following:

- Tools, Libraries, Databases, Data Sets: Major, complex software systems that provide implementations for significant portions of the NC Applications already exist. For example, large collections of computer-aided design (CAD) software are already used extensively in engineering design domains. Similarly, relational and object-oriented database management systems provide extensive capabilities for structured data storage, indexing, and management, The best of diverse sets of software tools and subsystems, as well as the creation of whole new tools, must be integrated into coherent applications development environments to form the development base with which to implement the NC Applications. Similarly, diverse libraries of program components, databases of data elements, and less structured data sets must be composed and integrated to the extent possible into the development environment.

- Composition and Integration Frameworks: Special toolkits already exist in certain specific domains to assist in the composition and integration of tools, libraries, databases, and data sets. For example, the CAD Framework Initiative (CFI) accomplishes this by providing interface specifications for tool-to-tool communications and tool-to-database communications. In addition, the Initiative has developed prototype implementations of these capabilities. These can form the basis of value-aided and commercially supported packages and software toolsets.

- Building Block Object Sets: Due to the commonality that characterizes many of the service needs of the NC Applications, we would expect a shared base of software objects (that is, actions, operations, and protocols as well as data structures) to emerge which can be reused across multiple application development efforts. For example, a "schedule" object, which provides operations for allocating limited resources to critical tasks, could be used as a component of several different applications.

- Application Customized Objects: Leveraging the evolving building block object sets, we expect the objects from which the applications are implemented to be customized and extended for the application at hand. For example, while there is much in common in terms of command, control, communications, and intelligence (C3I) for an intensive care unit and an environmental spill, we would expect the details of sensor integration, strategies for alerting, and demands for real-time response to be somewhat different. The elements of the underlying object base will need some customization for their use in specific NC Applications. The degree of commonality across applications, which one hopes is large, remains to be discovered.

A high degree of commonality among the elements of the services layer of the NII will be critical for its success. The complex components that constitute the NC Applications consist of many interacting subsystems. Without a highly leveraged common service layer, each application is likely to result in the development of point or "stovepipe" solutions to its critical internal components. The result would be little commonality of effort across the various applications. Each would be built upon its own inherently unstable structure of services and subsystems and likely to be subject to collapse and failure. The individual implementation efforts would be expensive and would take too long to yield effective systems. Therefore, for the best chances for success, the diverse underlying technologies must evolve into common, shareable, and reusable services that can be leveraged across multiple NC Applications. Alternative implementations of a frequently used service, such as display window management, eventually will lead to the identification of "best practices" that can be embodied into a common services layer, for example, for interfaces.

One possible service taxonomy classifies generic services into the following categories: interfaces, applications building blocks, data and process management, and communications. Interface services include window managers (e.g., Motif, Nextstep), tools for speech handling and integration (generation as well as recognition), handwriting recognition, data visualization packages, toolkits for audio and video integration, etc. Applications building blocks include planning packages, scheduling packages, data fusion, collaboration support, virtual reality support, and image processing and analysis. Data and process management services consist of capabilities for configuration management, shared spaces, process flows, data integration, data exchange and translation, and data search and retrieve. Communications services include ubiquitous access through a variety of communications mechanisms (e.g., wireless as well as wired connections into the bitways); mobility services to support users as they move through the various points of connection into the network; interprocess communications and remote process call mechanisms to support distributed processing; and trust mechanisms such as authentication, authorization, encryption, password, and usage metering.

The service layers themselves will evolve, as new underlying technologies come along that provide new functionalities or dramatically better ways of doing things. However, the key point is that commonality will emerge from diversity: that generic data structures and operational protocols will reveal themselves by examining the alternative realizations of a given application service.

A critical aspect of this model is the need to develop a "construction kit" with which to build the layers of commonality. Elements of this kit include software environments for developing applications; evolution of standard operational and data exchange protocols, software toolkits and software generators for building or generating well-defined portions of applications; and frameworks for integrating tools and data into coherent, interoperable ensembles.

Such an architecture for the NII should yield lower costs for application development by leveraging commonality across the services needed to implement such applications. In addition, development times will be reduced, thereby bringing cost-effective systems on line more rapidly, leading to lowered costs. The value of a common services layer is conceptually indicated by Figure 2. In (a), the lack of a common services infrastructure leads to stovepipe implementations, with little commonality among the service capabilities of the various NC's. In (b), a common set of services is leveraged among the NC's, aided by a collection of toolkits, integration frameworks, and applications generators.

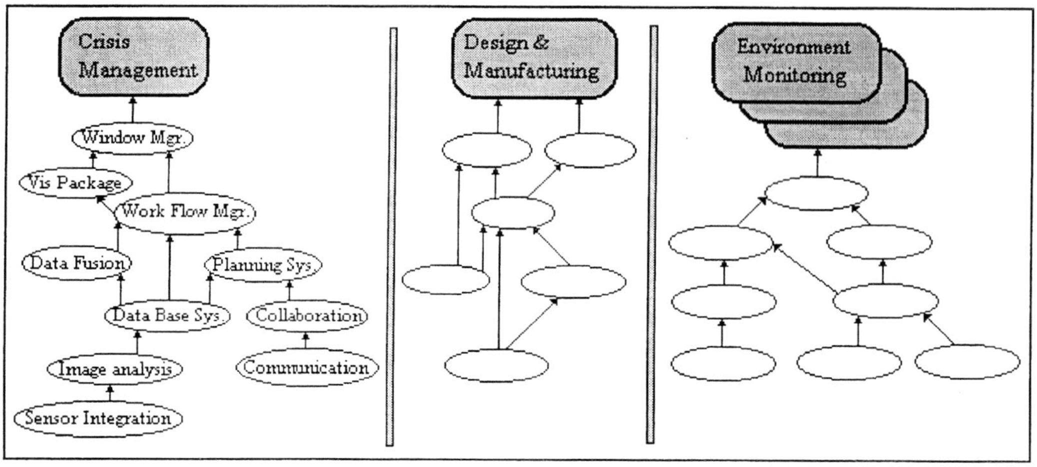

(a) Stovepipe implementations without a common services layer.

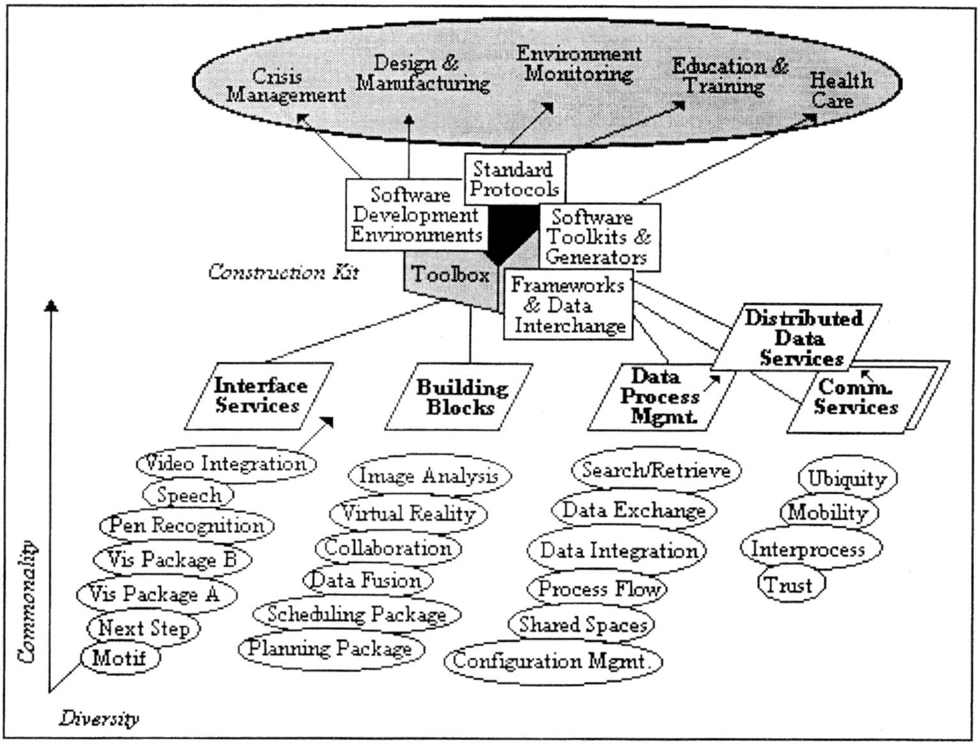

(b) A common services layer coupled to toolkits for building applications.

Figure 2. Technical Challenges in Building a National Information Infrastructure

4. Information Enterprise Elements

The essence of the successful evolution of the information enterprise is utility through scalability, scalability through commonality, and cost effectiveness and growth through diversity of supply. By this we mean that a capability adds usefulness to the information infrastructure if it can work in small yet scale up to the very large. This is most easily achieved by exploiting commonality in protocols, operations, and data structures. But before that commonality can be agreed upon, there must be a diverse collection of implementations, to reveal what is in common, and to provide large numbers of alternative functionalities and approaches. In a sense, diversity means "let a thousand flowers bloom," yet progress is only possible when some common framework arises from the diverse alternatives.

Figure 3 shows examples of the critical elements for each of the three layers of the NII Architecture (that is, applications, services, and bitways).

	Diversity **What technologies require rapid growth, diverse supply**	**Commonality** **How capabilities are presented to client systems**	**Utility** **What value is provided to the end user**
Applications Capability to users	• Universal computing and communication • Decision support and heterogeneous databases • Modeling and simulation • Sensors, teleoperation, advanced displays	• Manufacturing–product and process representations • Education–electronic public library protocol • Health care–patient record model	• New ways to do business in critical national applications, leading to fundamental improvements
Services Information service foundations for applications	• Image repn/compression • Security technologies • Virtual reality, simulation • Systems engineering tools	• Multi-media protocols • Security protocols • Engineering frameworks • Heterogeneous databases	• Information processing and management to enable broad range of applications
Bitways Communications channels among services	• Fiber Optic • Laser Space • Wireless • ISDN	• Transport protocols • Transmission standards • Wireless to wireline interface protocols	• Ubiquitous access to information

Figure 3. Illustration of Diversity, Commonality, Utility Concepts

In Figure 3, the three columns indicate:

- Diversity. These are the key areas where innovation, rapid technological growth, and diversity of supply are essential to cost-effective delivery of increasing levels of capability. For example, a competitive supply of fiber optic technology is needed in order to provide ubiquitous access to high performance bitways. Also, continued improvements in laser and wireless communication improve affordability of high performance mobile communication. Thus, diversity arises from rapid change over time, and from a rich and competitive array of supporting technologies.

- Commonality. Scaling up (in capacity, number of users, capability, etc.) and delivery of continual improvements in capability can only occur when there is an identification of areas where commonality is to be maintained. This commonality creates a framework within which the growth and diversification of technology occurs, enabling the technologies to be made accessible to clients (in the next layer up, for example) with acceptable risk and cost-effectiveness, lowering entry barriers for new users of the technologies and sources of new capabilities. This is the essence of the principle of "open architecture." For example, transport protocols for bitways provide users of communications services with a means to access the service independently of the particular choices of underlying component technologies.

- Utility. Each layer, of course, provides specific value to users or to clients in the next layer. For example, the bitways are intended to provide ubiquitous data communications, and in a manner such that designers of applications need not know whether the communications links are through fiber, wireless, or some combination of links.

This chart shows examples for each of the three major layers. In the charts that follow, each layer will be examined in detail. This organization of concepts focuses attention immediately on two critical issues that must be resolved before the architecture can gain acceptance:

- Scalability. We believe that it is crucial to identify the needed commonalties in order to avoid inappropriate dependencies among capabilities. Such dependencies introduce risk as we try to scale-up the system. For scalability to succeed, there must continue to be an orderly competition for sources of supply of new capabilities and their implementations.

- Legacy. In addition, by identifying the commonalties of new capabilities, we provide a basis for interoperability with existing systems. Scale-up is the major driver of infrastructure design. Scale-up can occur in many dimensions (such as performance, number of users, and extent of capability).

A high diversity among infrastructure capabilities and their implementations is the essence of innovation. In the case of an information infrastructure, the pace of change is rapid, and there is a wide range of approaches to achieving a given capability (e.g., in the communications bitways, physical media may consist of optical fiber, land mobile wireless radios, or laser communications). The challenge for both end users and applications developers is how to exploit the continuing innovation while remaining insulated from the continual change. That is, the consumer seeks to "ride the curves" of growth while avoiding continual re-engineering. In other words, with high diversity, the choice of commonalties is critical to whether investment can be made in upper layers with acceptable risk.

C. EDUCATION AND TRAINING APPLICATIONS OF THE NII

1. High Leverage Areas

Education and training is clearly a major national-scale enterprise that could benefit significant investments in a shared information infrastructure. Other applications like entertainment or health care delivery will be driven by private sector investments and development; education and training will need to be driven by the public sector. While there are many aspects of education, in this section we focus on those high leverage aspects that can take advantage of a national-scale infrastructure. These areas include distributed simulation, remote presence (sometimes called "telepresence"), and educational software. Distributed simulation is based on a model that represents a target system in software. The description focuses on the system's functional behavior and makes possible the study of how the system behaves as a function of time as well as the study of relationships between input causes and output effects.

A major focus must be on more realistic user interfaces. The tools to manipulate the system under study and the displays that represent the status and output of the system must be as similar to the real system as possible. The fidelity of the simulation should be tailored to the scenario. For example, for basic training, a simple model of behavior can be used; as the student becomes more adept at controlling the system being studied, a more complex and more sophisticated model should be used. Simulation is a more advanced technology than more traditional computer-aided instruction in that participants can navigate the model and exert their control at will.

Simulation provides some significant and unique advantages for training, For example, it is possible to control time in dynamic systems, by pausing, going slow, speeding up, or even backing up to review previous behavior. It promotes "what if" trials. On-line simulations can also provide on-line help and explanations.

A key future direction for simulation is to incorporate increased reality, including real equipment to be controlled as part of a mixed real and virtual simulation. Real process data, instrument outputs, etc. will be used to drive the simulation and it will be possible to manipulate real objects in simulated scenarios. This will make it possible to increase a student's hands-on experience, while simultaneously reducing the risks associated with operating real equipment. Furthermore, the infrastructure makes it possible for students to interact with live remote participants.

This leads to another new dimension for education made possible by leveraging a shared communications infrastructure. Students can gain first hand knowledge of team planning and crisis management. Users experience real interactions with a remote environment. This means that audio, visual, and tactile sensing will need to be supported as integral input and output modalities. The system will enable the manipulation of real objects and systems. Examples include customer training and service and ongoing interaction with experts. In fact, telepresence may be the real presence of the future. The NII will support networked access to remote environments for multiple participants.

Software of many kinds will be needed in order to support educational applications. These include dedicated courseware, general applications (for example, word processors, spreadsheets); professional software tools (such as computer-aided design software, perhaps scaled down or simplified as appropriate for student use); simulations; "executable" descriptions of apparatus; and machinery with embedded documentation.

It is clear that the development tools used to build educational applications should be rooted in the mainstream of software development tools. Education is an important applications area in its own right. The usual mandate to support application-specific development by experts in the educational domain applies. What is needed is integrated frameworks for: construction of distributed, collaborative work environments, simulations, and the intelligent creation, population and retrieval of multimedia information. Furthermore, such environments should be usable by distributed groups of end users.

2. Educational Applications and the National Information Infrastructure

Educational applications meet the NII in terms of distributed digital libraries of multimedia, interactive course modules that are available anywhere, be it in the classroom, in the dorm room, on the job, or at the remote location. Furthermore, these materials should be made available anytime, either in real time during scheduled lectures, or time-shifted for access on demand. Where possible, collaboration opportunities can be exploited and enabled by a national communications infrastructure, through inter-campus team teaching, course projects, guest lectures, and tutorials.

There are several implications for NII services level. First, there should be generic support for diverse media, integrated into compound, interactive documents. Second, we will need cataloging services to register/publish materials and to assist in their discovery across the network community. Third, the infrastructure will need to provide network services for multimedia document distribution and delivery. Fourth, authentication services to identify and protect authorship and intellectual property will be needed to make possible the development of high quality, for-fee educational applications. Fifth, we will require electronic commerce services to support collection of royalties where appropriate. And lastly, the network infrastructure will need to support videoconferencing and video multicast.

The diverse internetworked, information-rich world envisioned by the NII will make it possible for learners to gain access to enormous educational resources, packaged as multimedia courseware modules stored in distributed digital libraries. The effects on publishing are potentially enormous, as textbook authors are replaced by courseware creators whose works are combined by individual teachers ("a unique textbook for every class, a unique textbook for every student"). While the instructor will be seen as a synthesizer and organizer of the available materials, the students will have direct access to them, pursuing them based on their own interests and at their own pace.

As described above, educational applications require access to media-rich, "active" documents that can be viewed on a diverse collection of hardware platforms, with adequate performance across a wide range of bandwidths and computational capabilities.
The technical capabilities to compose materials must be supported while simultaneously ensuring that the intellectual property of the courseware developers is protected. Developers have every right to expect to be paid for their efforts, and so the infrastructure must be able to measure usage and provide billing for services rendered.

A key technical challenge is providing rich enough cataloging services so new materials can become widely known throughout a very large community in an expeditious manner, while searchers of materials can find what they need equally quickly. The impact on the educational enterprise of the location-shifting and time-shifting of the learning and studying experience, made possible by technology, remains to be determined, and is worthy of intensive piloting activities.

The elements of the education and training delivery system that must be interconnected by the information infrastructure include electronic classrooms, instructors' offices, authoring studios, and student dorm rooms (or commons or libraries or workplaces or homes, etc.). These are summarized in Figure 4 and described in more detail below.

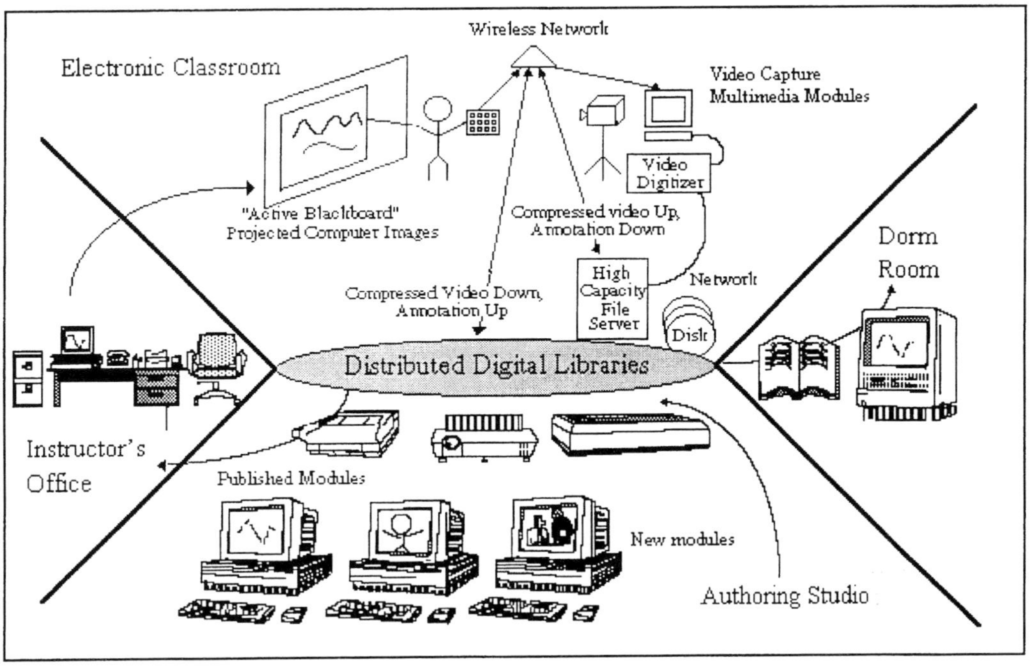

Figure 4. Linkages Among Elements of the Electronic Courseware Delivery System. The key elements are the electronic classroom, the instructor's office, the authoring studio, and the dorm room. These are interconnected by communications networks and digital libraries.

a. Electronic Classroom

The primary delivery vehicle for multimedia materials will be the high technology electronic classroom. The instructor becomes the "conductor" of multimedia presentations, previously assembled in his or her office from relatively short modules extracted from the courseware library. Lectures can be video taped and added to the library as video annotations. Students at wireless terminals can annotate the lecture with their own notes, which are uplinked and indexed with the lecture.

b. Instructor's Office

The instructor's office is likely to contain high bandwidth connections to the courseware library, although it may be distributed around the country. The primary use of the office computer is for the instructor to rapidly identify relevant materials for incorporation into his/her lecture presentations.

c. Dorm Room

Student rooms will be equipped with high-speed access to the local campus communications infrastructure, allowing access to locally developed course materials, including video taped lectures, as well as related materials out in the infrastructure. Eventually, as the communications infrastructure extends itself to become ubiquitously available, these capabilities will be generally available to the populace at large.

d. Authoring Studio

The authoring studio contains more specialized equipment than a typical office: high performance computer-based video editing systems, recorders and mixers, etc., and may be manned by specialized technicians and curriculum developers. Working closely with lecturers, who are the domain experts, the authoring studio staff work with faculty and teaching staff to develop course materials for local use and eventually for insertion into the national infrastructure. The authoring environment is thought to be pivotal at this point in time and is discussed as a specific example of an important component of the NII in Appendix A.

D. NII REQUIREMENTS

The NII must be able to support distributed simulations, multimedia productions, collaborative distributed projects, intelligent information retrieval, virtual field trips, remote mentoring, administrative applications, and person-to-person communications and more. This leads to the following collection of requirements for educational applications to be constructed on top of the NII:

- Low threshold for interconnection
- Support for scalability through open standards
- Affordability
- Ease of use
- Operation and management without "wizards" across a diversity of platforms
- Extensive, dense local networks,
- Easy coordination of distributed activities
- Good support for personal location and connectivity.

The requirements for NII technologies and the underlying service layer include the following:

- Diverse collections of simulation modules
- Plug-and-play operating environments
- Authoring tools for simulation and instructional modules
- Virtual reality simulation hardware and software
- Ubiquitous network access and bandwidth
- On-line multi-media libraries
- Directory services, and information navigation

- Easy-to-set-up wireless, portable local networks

In addition, the NII needs to provide flexible interfaces to the following:

- Light-weight, portable head-mounted displays
- Mosaic wall-size displays
- Glove interfaces with tactile feedback
- Low-cost remote camera and sensor systems
- Speaker-independent speech recognition and synthesis
- Ubiquitous, highly integrated multimedia workstations

Education is an excellent opportunity as a targeted NII application area. First of all, there is a national focus on educational reform. Education is attractive because it is an application that can be built upon networking, human-computer interaction, and simulation research. Furthermore, there exists a motivated research community familiar with education who is anxious to work on this application area. Technology developed as part of the educational application can provide infrastructure for other applications testbeds.

Testbeds and pilot program infrastructure can enhance the infrastructure for existing educational network testbeds and consortia. This will allow testing of new NII services across large and diverse learner populations and over a wide range of new types of educational applications. It should be possible to catalyze industrial partnerships for these educational testbeds.

In terms of the NII service requirements for these educational testbeds, the first priority must be low-threshold, open growth across all testbed sites and participants (from K-12 schools to community colleges, universities, informal learning "at home," and learning in the workplace settings). Increasingly the innovations will require heterogeneous participation across institutions.

Planned testbed opportunities include a thousand K-12 schools and districts "on-line," two hundred community colleges, continuing education, ten vocational schools, twenty Department of Defense Dependents Schools, universities (e.g., twenty engineering education departments), and homes and community centers.

Current testbed investments include some of the following: the Ameritech Super School, a vision of the future for education and technology; the State of the Art Home Learning Center; 24-hour hotlines; desktop conferencing; student workstations that integrate audio, video, data and text; an advanced resource center multimedia library; access to the Supercomputer Weather Lab for thunderstorm simulation; distance learning classes between U.S. cities; support for low-enrollment classes (e.g. calculus, Latin, etc.), and the Administrative Office of the Future.

In more detail, let us consider the Ameritech school of the future. The school supports home learning channels that support access to information libraries from home. The students make extensive use of electronic mail to create study teams and to share research results, Bulletin boards are extensively used. Personal computers equipped with video conferencing capabilities are used to enhance the learning process. The school also supports distance learning from remote locations.

In the short term (two-three year time frame), infrastructure developments should focus on: adaptive quality of service; authoring tools for multimedia libraries and collaborative projects; local school infrastructure, including the distributed management of NII services; infrastructure augmentation for ongoing testbed participants; educational applications of distributed networked simulations; and experimental testbeds to stress NII services.

The middle term priorities (three-seven year time frame) include low-cost ubiquitous access, NII services available to support the "Teaching Factory" (e.g., remote presence); plug and play across diverse platforms; services extended to widespread use of broadband applications; wireless, low-cost devices supported by all NII services; integrated frameworks for software services, and a well-defined layered infrastructure.

True measures of progress will be needed. With respect to the effective transition of technology for educational software to the market, the following questions will need to be addressed. How many educational products are companies actively commercializing? What is the quality and the measured effectiveness of these products? With respect to infrastructure for dissemination of education, how many more students (within and outside of institutions) are on-line, and how fast is this increasing? What are their demographic profiles? What fraction of federal government information is available on-line at nominal cost?

VI. SUMMARY AND CONCLUSIONS

It is well known that the most effective learning takes place through "learning by doing." The NII will make available a wide variety of simulation-based education and training services that can be accessed through home or industry telecommunication systems. The simulation-based training systems will be tailored to a specific domain and will allow a student to live in that domain. The system will present a variety of skill-based and cognitive-based problems, provide coaching and remedial instruction, and tailor follow-on assignments to address student weaknesses. Fundamental to this vision are authoring tools, digital libraries, associate systems, and the ability to collaborate across space and time.

Underlying this educational delivery system is an architecture for the NII, based on the three layers of applications, services, and bitways. Bitways provide the communications capabilities. The services layer provides a collection of capabilities that transform the data objects of the bitways layer to the information objects of the applications layer.

Undoubtedly, education and training will be a key application of the NII, and this is one application in which the government can rightfully seed the development. A critical opportunity for the NII will be the development of the underlying software technology for K-12, vocational and personal education, as well as university-level instruction. It is our proposition that the federal government should act as a catalyst between the software industry and the educational establishment. Furthermore, the federal government should provide a common infrastructure to the numerous ongoing testbeds and initiate selected new testbeds.

Finally, we have laid out some specific requirements for the NII to support educational applications. These include diverse collections of simulation modules, plug-and-play operating environments, authoring tools for simulation and instructional modules, virtual reality simulation hardware and software, ubiquitous network access and bandwidth, on-line multimedia libraries, directory services and information navigation, and easy to set up wireless, portable local networks.

REFERENCES

1. **David Hawkbridge**, *New Information Technology in Education*, Johns Hopkins University Press, Baltimore MD, 1983.
2. **D.L. Morehouse, S.H. Stockdill**, Understanding the Influence of a Change Agent on Technology Use by University Faculty and Staff, *Educational Technology* 31:61-2, July, 1991.
3. **Shira E. Persky**, What Contributes to Teacher Development in Technology, *Educational Technology* 30:34-8, April, 1990.
4. Computer Systems Policy Project, Perspectives on the National Information Infrastructure: CSPP's Vision and Recommendations for Action, January 12, 1993.
5. Institute for Information Studies, A National Information Network: Changing Our Lives in the 21st Century, Aspen Institute, Queenstown, MD, 1992.
6. **Al Gore, Jr.** , Infrastructure for the Global Village, *Scientific American*, V. 265, N. 3, pp. 150-153, September, 1991.
7. **William J. Clinton, Albert Gore, Jr.**, *Technology for America's Economic Growth, A New Direction to Build Economic Strength*, February 22, 1993.
8. Information Infrastructure Task Force, The National Information Infrastructure: Agenda for Action, September 15, 1993.
9. Information Technology Association of America, National Information Infrastructure: Industry and Government Roles, July 1993.
10. **Brian Kahin**, Information Technology and Information Infrastructure, in *Empowering Technology: Implementing a U. S. Strategy*, L. M. Branscomb, ed. , MIT Press, Cambridge, MA, 1993.
11. **Ed Krol**, *The Whole Internet User's Guide and Catalog*, O'Reilly & Associates, Inc., Sebastopol, CA, 1992.
12. **J. Motiwalla, M. Yap, L. H. Ngoh**, Building the Intelligent Island, *IEEE Communications Magazine*, V. 31, N. 10, (October 1993), pp. 28-34.
13. National Computer Board of Singapore, A Vision of an Intelligent Island: The IT2000 Report, March 1992.
14. Office of Science and Technology Policy, FCCSET Initiatives in the FY 1994 Budget, April 8, 1993.
15. Synthesis Standards Study Project Report, March 1993
16. **A.R. Eide and R.J.Thomas**, The Impact of Technology on Undergraduate Education, a paper presented at and published in the proceedings of the 1992 Frontiers in Education Conference, Nashville, TN, Nov 11-14, 1992, pp 625-627

APPENDIX A

FORMING A COURSEWARE DEVELOPMENT AND PLAYBACK ENVIRONMENT FOR ENGINEERING EDUCATION

In order for courseware to be developed and delivered by even the least computer literate faculty member, an environment must be created that is easy to use. Additionally, the environment must also be capable of serving the needs of the very sophisticated user who is attempting to get the maximum use and performance from the technology. Next, we will describe an environment that is achievable using today's technology, is mindful of cost limitations, and looks towards future advances in technology that will be easy to incorporate into the system. The technical details described pertain primarily to personal computer-based systems, though the general information and comments also can be used when configuring systems based on workstation class computers.

I. THE COURSEWARE DEVELOPMENT ENVIRONMENT

The development environment is distinguished from the delivery environment because of the more extensive and expensive needs of the former. Development facilities are generally a superset of those needed for delivery, except for the need to have projection equipment in classroom delivery systems. Some of the functionality that an instructor could expect to be available in a courseware development facility includes:

- Software for creating and displaying presentations which include: still images; digital video; 2D and 3D animation; simulation results; textual material, as well as software for controlling devices such as laser disks, video tape machines, and CD ROMs (this conglomeration of software is generally referred to as *authoring software*)

- Hardware development platforms with sufficient speed, interface devices, storage, and a network connection

- High quality still-image capture and digitization from a variety of source materials including: hardcopy, slides, video tape, laser disk, and directly from video cameras

- High quality video capture, and digitization from a variety of source materials including: video tape, laser disk, and directly from video cameras, as well as the facility to edit and compress digital video

- Output to printers, film printers, video tape, and other output devices;

- Basic video production equipment including: Hi-8 and/or SVHS production and editing, dubbing, format conversion, with some special effects generators

- Sound production facilities with: sound digitizing and editing, as well as MIDI sound generation and edit

- CD ROM and laser disk mastering equipment

In addition to the functionality items listed above, a development facility must also be staffed with personnel charged with scanning the technological and pedagogical horizons in order to make recommendations on the appropriateness and use of technologies based on instructors' needs. With the diversity of target environments, it is advantageous to have cross-platform development capabilities available. This would include hardware and software for developing under DOS/Windows, Apple, and UNIX. Finally, the development facility should provide the means for authors to access copyright information and obtain clearance for use of intellectual properties.

A. THE HARDWARE

At the core of the development environment is the computing system and its associated peripheral devices. Many of the perceived or real limitations of a system might come from the choice of a sufficiently powerful central processing unit (CPU). Systems that process information and perform operations so slowly as to underwhelm even the most patient user are generally perceived to be incapable of performing the required task. Although the task would eventually be completed, if the user is not willing to wait for completion, the system is deemed to be incapable of completing the task. What this points to is that the choice of a CPU is quite subjective as well as being dependent upon the types of operations that will be most repeatedly performed. For example, a 386-based 25 MHz CPU is quite capable of performing most present day word processing tasks at a reasonable speed. However, using that same CPU to do a shading operation on a 1024 by 1024 pixel by 64K color image would take an unreasonable amount of time and the CPU would be deemed unsuitable for this operation. This example also points out the difficulty that arises as technology advances. Ten years ago such a shading operation would not be attempted on desktop systems due to their relative lack of processor capacity. Today the operation is fairly routine for most high-end personal machines. As more and more complex operations migrate to desktop machines, performance demands on these systems will continue to grow. Therefore, in order to choose an appropriate CPU, it is necessary to have a set of desired operations in mind. The basis operations envisioned here are those that are commonly performed to develop and deliver from simple through complex technology-based presentations.

1. The Central Processing Unit

An effective way in which to choose CPUs for a development facility is to first divide the facility into workstation groups by function. Certain functions will require the highest performance (and most costly) CPUs while other functions may be accomplished on less costly units. Typical groupings might include three classes of machines: image and video capture and processing stations; image, video and sound edit stations; and presentation integration stations. Not all developers will use the full range of workstation groups. Some will require only the use of a platform that can assemble a presentation from already developed resource material. Other users will develop their own resource material on the higher end workstations and then proceed to assemble the material using the lower end stations. Most development facilities will require more assembly stations than those used for the other two classes of functions. However, given the reduced requirements of an assembly station, it becomes cost-effective to locate this type of station at each active developer's office site, reducing the number of low-end stations housed in a central facility.

The capture station is likely to be used in the most CPU intensive tasks. Typically, images are captured at the highest resolution allowed by the interface hardware and subsequently reduced to a resolution compatible with the target playback system. Captured images may be 1024 by 1024 pixels by 1024K colors which translates into as many as 3 Mbytes of information. Considering just the time it takes to write a file this size to disk requires a knowledge of the speed of the processor as well as that of the disk drive. A 25 MHz CPU with a reasonably fast hard drive might have a write cycle time of 10 microseconds per byte. This means that it will take about 40 seconds just to store a 3 Mbyte image. Actually the bottleneck in the image capture operation is not in the storage time but rather is in the time it takes to input the image through the interface device and to convert it to a digital format. This part of the capture operation is both dependent upon the speed of the CPU as well as the speed of the image collection hardware. In most cases, the image collection hardware takes as long to process the incoming data as it does to store the data, resulting in an overall image capture time of about twice the time it takes to move an image to a disk.

There are two ways to improve the situation described above. First, the required resolution of the collected image may be reduced such that a smaller amount of data is collected. This compromise is a less desirable solution, since the quality of any subsequent display of a collected image is dependent upon the quality of the original image. That is to say, it is always better to collect an image at the highest resolution that is practical, since subsequent image operations, including image compression, tend to reduce the quality of the likeness. The second means of improving the situation is the obvious one—increase the speed of the CPU, the disk, and the image collection hardware. The tradeoff here is cost. The workable solution is one that is a concession between resolution, cost, and operator discomfort.

Capturing video images in real-time is even more CPU intensive than capturing still images. Video images may be thought of as individual frames, each containing a still image, that are played at a rate varying from ten to thirty frames per second. Using the resolution quality described previously regarding the collection of still images, (i.e., 1024 by 1024 pixels by 1024K colors) will generate 90 Mbytes of data for every second of video collected at thirty frames per second. Displaying images at less than about twenty-two frames per second tends to cause the perception of a flickering movie for some people, though a frame rate of as low as fifteen frames per second is tolerable for other individuals. However, even if the flicker produced at the rate of ten frames per second could be tolerated, 30 Mbytes per second of data is still being processed. Today's CPUs are not capable of working with a data rate that high. Therefore, it is necessary to effect a compromise in resolution, capture rate, frame rate, or CPU speed in order to be able to operate in real-time.

Capturing a video image sequence at the resolution of 640 by 480 by 64K colors by 22 frames per second generates approximately 13 Mbytes of data per second. Upon playback, the digital video would be fairly pleasing to view with little perceivable flicker and an image quality that is not too grainy. This reduction in resolution over the previous example might be considered a reasonable first-step compromise. However, considering just data movement, the necessary data rate would require a byte to be moved approximately every seventy-five nanoseconds. This comes close to the rate at which individual single instructions are executed on today's high-end desktop computers. Another consideration is that most video is accompanied by sound and that sound must be digitized and captured along with the image information. A balance between video resolution and CPU speed is necessary in order to make video digitization and capture viable in a broad arena. A further discussion of how this balance is achieved can be found in the section on video image capture. This section describes the video capture hardware and software that can be used to produce a workable system, and looks toward what the future holds in this area.

The discussion contained in the previous paragraphs is meant to point out the fact that video capture is one of the most exigent tasks that the development system must undertake. Demands placed on the image capture workstation require that the platform chosen to implement this system must be of the highest performance affordable. It must have a CPU whose clock rate is at least 33 MHz, preferably higher. A better measure of the speed of a system would be a standardized benchmark rating such as millions of instructions per second (MIPS) when performing a specific task. Unfortunately, a widely accepted standard that characterizes the types of operations performed when doing still image and video capture has not yet been developed. In the absence of such a standard, the best specification that can be given for the image capture workstation is that it should be a machine with the highest combined clock rate and throughput that is affordable.

For many CPUs, processor cache memories can provide an additional measure of increased throughput. Processor cache memories vary in size from 8 Kbytes to 256 Kbytes. Preference should be given to CPUs containing larger processor caches, as they will provide a speed advantage over processors with the same clock rate that contain no cache.

Primary memory, usually referred to as RAM (random access memory), is an integral part of the CPU. The amount of memory included in a CPU can be a principal factor in determining the overall system throughput. As the size and complexity of applications programs have increased, so has the demand on the size of the RAM. Many software packages will compensate for small main memories by caching information on disk, but this tends to slow the operation of the software. Other software will simply not operate without a sufficient amount of RAM being present. For systems used to perform capture operations, 12-16 Mbytes of RAM memory is generally considered to be a minimum amount. Given that in the near future software demands on memory are sure to increase, a more forward-looking approach would incorporate at least 16-24 Mbytes of RAM into the CPU. It should also be pointed out that it would be prudent to choose a CPU that has the capability to expand onboard memory up to 64 Mbytes.

Video and sound editing stations are somewhat less demanding on CPU speed than the corresponding capture operation, allowing a slightly lower performance system to be used. However, due to the nature of the editing operations, it is necessary to have more disk space available in this second level system in order to retain all of the segments of data that are generated during edit operations. If money were no object, the video and sound edit workstation would be the same as the capture station. Since that is almost never the case, a reasonable tradeoff is to use a CPU that is no more than 10-15 percent slower than that of the capture station. This tradeoff will generally produce a corresponding 5-10 percent reduction in the cost of the system. The danger in trading hardware capital investment for speed is that the tradeoff is really between user discomfort and cost. But creating too much discomfort for the users could cause them to not make use of the technology.

The presentation integration workstation is least demanding on the performance of the CPU. In part, the demands this operation will place on the CPU depend upon the nature of the material being developed and on how well tasks were performed at the other high-level workstations. Typically this system may be as much as 20 percent slower than the highest performance station, but will still provide satisfactory performance. Again there is the danger that saving money by reducing performance may cause instructors to bypass the use of technology in their teaching/learning materials development. Better to have fewer, more costly stations (i.e., higher performance stations) with contention for their use than many low-performance stations that go unused because of user frustration.

2. Display Systems

A display system for a computer consists of the actual cathode ray tube (CRT) that shows the image, as well as the display generator subsystem that performs the internal manipulation and storage of the image. A myriad of configurations are possible, ranging from simple, inexpensive, low resolution displays, to complex, costly, very high resolution assemblages. Display systems are enhanced by improving the performance of the CRT, the image generating hardware, or by increasing the size of the image display memory contained in the display generator subsystem. In most computers, the display generator subsystem is contained on the system motherboard (the planar printed circuit board into which all components and other cards are plugged) or as an interface card that is added to a system to upgrade its capability.

It is preferable that the display system for all of three types of development stations be capable of showing an image that is a level above the resolution of the target presentation. The resolution of the target display system will be determined by the economics of purchasing projection systems or multiple individual displays, and is usually directed towards the least expensive displays that will produce reasonable results. Choosing development systems that have a higher resolution than the target system builds an over-design into the product courseware at a nominal cost and allows for accommodating future improvement in the target display cost/performance ratio. The minimum resolution for a target system should be 640 by 480 pixels by 256 colors. In order to be able to manipulate near-photo quality images, development system displays should be able to provide a rich assortment of colors. The next best compromise in display technology is to provide up to 64K colors. Thus using a display system with a resolution of 640 X 480 pixels by 64K colors should be considered a minimum for development systems. In order to accommodate very high-quality image capture for special purpose development, it is reasonable to consider using a display system capable of 1024 X 768 X 64K for the high-end image capture stations.

Another factor affecting the visual quality of a display is the size of the individual dots projected onto the CRT to form an image. The size of these dots is referred to as the dot pitch, with a smaller dot pitch being preferred because it will produce a less grainy picture. Today's displays have dot pitches that range from 0.20 mm through 0.40 mm. CRTs having a dot pitch of about 0.28 mm produce satisfactory results in development as well as playback systems.

The amount of memory contained in the display generator subsystem in part determines how fast an image can be drawn on the screen as well as the number of pixels that can be included in an image. Many off-the-shelf computer systems contain display memories of 512 K to 1 Mbyte with on-board expansion up to 2 to 4 Mbytes possible. All three types of development systems should contain at least 2 Mbytes of video memory which can usually be obtained at a fairly nominal cost.

3. Disk Storage

The amount of disk storage available in a development system can be critical to the usability of the system. Multiple distinct users must have sufficient space to store intermediate and final products and must have a convenient means of transporting this information from machine to machine. The primary disk storage should be large enough to be able to store applications software as well as user files. Most applications packages along with their supporting files consume from 5 to 25 Mbytes of disk space each. Users can be expected to generate files that are from 1 to 5 Mbytes for small applications and up to hundreds of Mbytes for applications that involve an extensive amount of video and audio. Anticipating a reasonable average user file size of 20 Mbytes and assuming about 20 active users requires about 400 Mbytes of disk space. A standard 500 Mbyte disk drive would accommodate these users and 100 Mbytes of applications software. This scheme is reasonable but requires a certain amount of discipline in requiring users to clear their unneeded files from the primary hard disk on a regular basis. Thus a 500 Mbyte primary hard disk is the recommended minimum sized drive on the development systems.

To augment the primary storage device, it is necessary to have a means of archiving user files as well as a means of transporting large user files from one type of workstation to another. Additionally, each workstation needs a means of conveniently reading software and courseware distributed by vendors and other development sites. The standard 3.5" diskette drive is a fundamental mechanism for moving relatively small files easily and fast, and as such, should be included in every type of development system. Removable storage, such as magneto-optical diskettes that hold on the order of 120 Mbytes, can be used as a cost-effective archive and transport device. Removable cartridge drives of larger capacities can also be built into a system. For example, 650 Mbyte removable cartridges have become quite cost-effective over the past few years and are regularly used in image processing configurations. The cost of a magneto-optical 120 Mbyte diskette is presently on the order of $75, while 650 Mbyte cartridges cost nearly $400 each. Though the cost per Mbyte of the larger cartridge is lower, the reduced initial buy-in makes the magneto-optical cartridge an option that more users will be able to afford. Another storage medium that should be available to each development station is the CD ROM reader. These devices allow workstation users to access up to 650 Mbytes on a read-only basis from CD ROMs, a medium that is becoming very popular for information distribution. In particular, more and more software developers are distributing their products via CD ROM, making CD readers a necessity.

The storage devices described above can be permanently attached to a station or can be added into a system using a SCSI interface device. Including a SCSI interface allows for the easy reconfiguration of systems as well as providing some standardization of the device interface, both of which are beneficial to those who maintain and upgrade the system. Further, the standardized interface will accommodate new types of storage devices as they are developed in the future. It is strongly recommended that the storage devices be configured permanently into the system, rather than be used as portable add-ons that are moved from one system to another repeatedly. Though this type of portability is perhaps possible, it will eventually take a toll in system upkeep, both from the perspective of connector breakdown and from the standpoint of ready availability.

Speed of operation of the various storage devices plays a part in setting the price of the device. Access speed for the primary storage device should be as high as is possible since it is the principal potential bottleneck in system throughput. Hard drive access speed is usually measured by two parameters: the time it takes to find a piece of information on a disk (seek time) and the rate at which the data may be moved once it is found. For the types of applications being considered here, the latter value is likely to be the most important. This is particularly true for productions that contain substantial amounts of video, sound, or animation. Sustaining high data rates when transferring these particular types of information is critical to maintaining high-quality playback.

Typically, hard disk drives used as primary storage devices should have a seek time of 12 ms or less and data throughputs of greater than 2.5 Mbytes per second. This speed is attainable with most medium- to high-performance hard disk drives on the market today. Magneto-optical diskette drives are typically two to three times slower than primary storage. This slower speed is reasonable given that the nature of the use of the diskette drive is mostly to archive data. CD ROM drive technology has recently made advances in the data delivery rate. Early drives delivered information at the rate of 150 Kbytes per second while recent high performance releases have a delivery rate of 600 Kbytes per second. Again, the nature of the task performed by the CD ROM reader on the development system is such that the medium-priced 300 Kbytes per second (generally referred to as a double-speed) drive will suffice.

Another storage option that is becoming popular is to use a networked interconnection of workstations in which one of the workstations acts as a storage host. The storage host contains one or two very large hard drives, some facility for removable storage, and one or more CD ROM readers. The advantage of this arrangement is the economy of consolidating most of the storage in one location and of having the information somewhat readily accessible from all workstations. The disadvantage is that as the load on the network increases and as the size of the files being stored increases, the performance of the network will deteriorate. Given the present state of networking technology, the approach to storage advocated at this time is the use of large local disk drives rather than a centralized storage server.

Though the use of networked primary storage may not be advantageous in a courseware development environment, the ability to access information from local and remote sites via a network, on a sporadic basis, is very important. Additionally, implementation of the network connectivity allows for the sharing of centralized resources like printers, tape backup units, and infrequently used (but necessary) software. Therefore each development workstation should have a network connection that allows for local information exchange as well as for access to remote information sites.

4. Networks

There are two major network protocols today: TCP/IP and OSI. The newer OSI protocol is still under development and not yet commercially available. Once available, OSI is expected to out perform TCP/IP, though it will not be compatible. A physical network may consist of any combination of electrical wire, fiber optic cable, microwave, satellite, or infrared transmission links. Various types of bridges, gateways, and interface cards are used to interconnect the different types of physical networks and the computers at the leaf nodes of the network. There are three distinct types of sub-networks, classified according to their size and capability: local area networks (LANs), internets, and wide area networks (WANs). LANs are applications-oriented and generally constrained to a limited geographical area—usually within a group of offices, a building, or a campus. LAN topologies can be Ethernet, Token Ring, FDDI, or CDDI. Internets are information transport-oriented, providing a vehicle for moving information over many different sized domains. WANs are physical interconnectivity-oriented, providing digital services from 1.5 Mbytes/sec T1 Internet transmission to very high-speed SONNET service.

The ability to share common physical resources is but one of the reasons for interconnecting development systems via a network. Access to remote databases, including newly developing visual databases, can be obtained through a strategically configured network. Collaboration between local and remote courseware developers is also supported and encouraged when network access is present. To facilitate these capabilities, a local area network environment that provides multi-user multimedia information sharing must be implemented. Further, the local area network must provide a pathway to access national and international wide area networks.

A variety of LAN operating system software exists including programs such as Novell Netware, Microsoft LAN Manager, Windows for Workgroups, Banyon Vines, PC TCP/IP to name but a few. This software can be used to enable local computers to communicate with one another, as well as enabling connection to the Internet. The decision as to which overall network package (hardware and software) is appropriate for a particular installation depends on the mission and needs of the specific facility, and the preference of the facility manager.

5. Still Image Capture

Still image digitization is one of the important operations that will take place on the image capture stations. Two basic still image acquisition subsystems are typically used: a flatbed scanner, and a 35 mm slide scanner. A hand-held scanner is essentially another form of flatbed scanner that produces less desirable, lower quality results at a considerably lower price. The resolution capabilities of a scanner are measured by counting the number of distinct positions at which a quantity of image data is collected. This measure is generally referred to as the number of dots per inch. Examples of an image scanned at two different resolutions are shown in Figs. 1 and 2. Flatbed scanners come in a variety of resolutions, with the 600 dots per inch model being quite satisfactory for most applications.

Figure A-1. Image scanned at 40 dpi.

Figure A-2. Image scanned at 300 dpi.

Scanners come in black and white or color, with color scanners being preferred for most courseware development. Another specification that should be examined when choosing a scanner is the speed at which the scan is performed. This speed is usually measured in inches per second, and obviously the faster the scan, the better.

Flatbed scanners are used to digitize printed images, photographic prints, or otherwise. Slide scanners are used to convert high-resolution 35 mm photographic slides to electronic form. The specifications on 35 mm slide scanners are similar to those of flatbed scanners and the choice of an appropriate scanner should be made based on the same criteria used to choose a flatbed scanner. A third type of subsystem, an electronic still camera, is becoming affordable and may eventually enjoy more widespread use. Using an electronic still camera, images are digitized at the time the picture is taken rather than being digitized by a scanner. These digitized images are then directly transferred to the development system through a standard digital interface such as a serial or parallel port. Electronic still camera resolution can be similar to or better than that of a scanner.

Many scanners have provisions for performing Optical Character Recognition (OCR) which means that a scanned page of text is stored as ASCII characters and format information rather than as a sequence of black or white dots. This means that the output generated by an OCR scan can be conveniently manipulated by a text editor. Perhaps more important is the fact that in this form the information will consume far less storage space as compared to a bitmap of the textual image. OCR allows an instructor to blend together text, not originally in electronic form, with other entities of a presentation to form an overall congruous computer-based presentation. The cost of OCR capability is fairly nominal and as such this capability should be considered a worthwhile addition to most scanners. However, OCR is ordinarily a software operation and as such can usually be added to a scanner configuration once it is needed.

Depending on the number of users that will access the facility and the nature of the courseware that will be developed, one or more still image scanning devices of each type will be needed. Typically, an image scan can be completed in about two minutes, after which a certain amount of cropping, color adjust, etc., takes place. The overall process, including transfer of the data file, may take as long as five to ten minutes per image. There should be a sufficient number of scanners in a facility such that there is a reasonable probability that a user will be able to access a scanner on an impromptu basis.

6. Video Capture

Video image digitization is the second significant operation that will be performed using the image capture stations. This capture operation entails the digitization of both moving picture images and the corresponding synchronized audio. Video is delivered to the digitizing subsystem from one of several possible sources: a Super VHS player, a Hi 8 mm player, or a laser disk player. The digitizing subsystem hardware that resides in the host computer is sometimes referred to as a *frame grabber*. Player units are controlled by the frame grabber which in turn is controlled by the capture software.

Depending upon the video capture hardware and upon the resolution of the images and the sound, video can be captured in real-time, that is, at the speed that the tape unit operates for normal viewing, or frame by frame. Typical capture resolutions are 160 x 120, 320 x 240, 640 x 480, and 1024 x 768 pixels, with the numbers of colors ranging from sixteen colors through millions of colors. The choice of the number of pixels and the number of colors depends on a broad-ranging tradeoff between the speed of capture and playback, the amount of memory needed to store a video, the complexity (and thus the price) of the frame grabber, and the amount of hardware necessary to play back the video.

Producing a digitized video capable of playback on a personal computing system is usually accomplished in two steps. First the sequence of frames is digitized along with the corresponding sound and stored on disk. The resultant file is usually very large and would require far too much storage to allow use of digital video to be practical. Over 844 Kbytes per second are collected for a video of size 160 x 120 pixels by 64K colors by twenty-two frames per second. If the video contains a single (monaural) sound track sampled at 22 KHz at 16-bit sound resolution, the total number of bytes collected rises to over 888 Kbytes per second.

To reduce the number of bytes stored and manipulated during playback, an image compression operation is generally performed on the picture portion of the video. Many different compression routines have been developed, each having a particular advantage as well as weakness. The type of compression algorithm used depends upon the target play back environment. One type of compression scheme is used to produce a data file that can be played back without any special purpose video card in the PC. This is generally referred to as software playback. Other algorithms are used to compress files that are targeted to play on specific video playback hardware, which is naturally referred to as hardware playback. Using the algorithms available in 1993, and targeting systems that have no special purpose hardware for playback, the compression operation can generally reduce the size of the video data file to about 175 Kbytes for every second of video (based on a 160 x 120 x 256 video at twenty frames/sec). Another way of stating this is that one Mbyte of data will produce six seconds of video at the previously mentioned resolution. Using hardware playback generally reduces the amount of storage necessary for a particular resolution allowing the user to choose to have larger higher resolution images at higher play back rates.

7. Sound Capture

Digital sound can be used to highlight portions of a presentation, providing implied authority through the use of a familiar voice, a demonstration of an actual situation, simple narration, or emphasis by the very fact that voice over or a sound effect is used. Sound can be captured during the capture of video as explained above, or it can be captured independently through the use of a sound digitizing board, also referred to as a sound card. Sound cards vary in price and complexity, from the simple single-channel (monaural) 8-bit card to high fidelity stereo 16-bit cards.

Sound is digitized by sampling an analog waveform, as shown in Fig. A-3, at discrete time points and converting the sampled data point into a byte of information. The sampling is done by an analog to digital (A/D) converter. To reproduce a sound waveform from a file of digital values, the values are fed into a digital to analog (D/A) converter where they produce a waveform as shown in Fig. A-4. The waveform is smoothed both electronically and by the natural filtering of the human ear to produce the sensation of a continuous sound waveform.

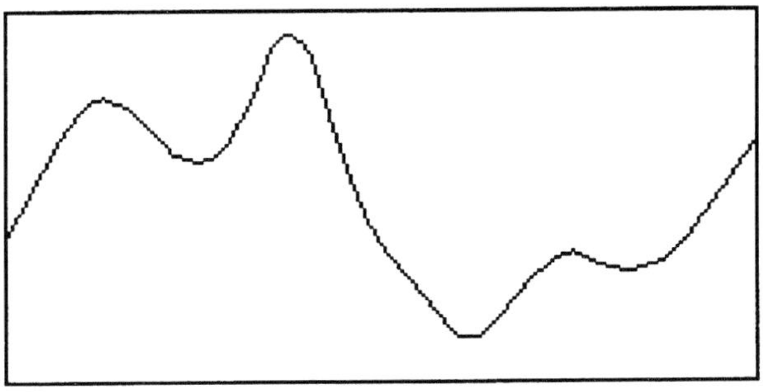

Figure A-3. Analog Sound Waveform

The size of the file generated by the capture of an interval of sound depends upon the quality of the capture board and the sample rate of the sound card. High sample rates produce larger amounts of data, but also provide higher quality sound reproduction. Typical sample rates vary from as low as 5 KHz to as high as 44 KHz. Sound sampled at 5 KHz is very low quality while sample rates of 44 KHz produce CD quality sound that is generally unnecessary for most pedagogical productions. A sample rate of 22 KHz is usually quite acceptable.

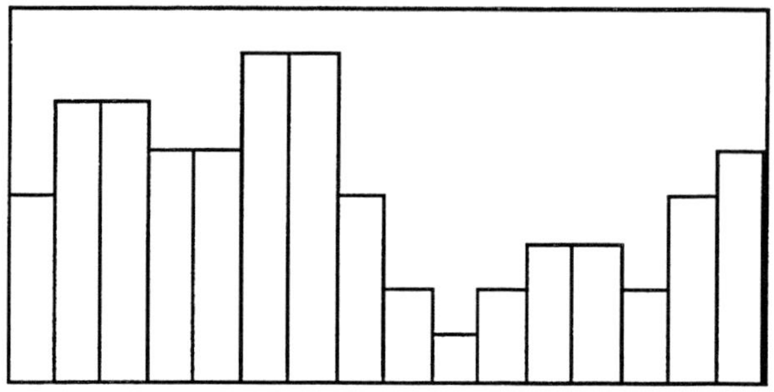

Figure A-4. Digitized Sound Waveform

A second factor determining the size of the file is the resolution of the analog to digital (A/D) and digital to analog (D/A) converter that is used on the sound card. Two types are generally found in PCs and workstations, those with 8-bit resolution and those with 16-bit resolution. The resolution of the A/D converter refers to the precision to which the digitizing process can approximate the actual value of the analog sound waveform. Higher resolution is obviously better, but a price is paid in both the increase in the size of the files generated as well as the increase in the cost of the sound card. Sampling a waveform at 22 KHz using an 8-bit sound card produces 22,000 bytes of data for each second of sound, while a 16-bit sound card produces twice as many bytes. It is not difficult to calculate that a few minutes of sound can accumulate to several Mbytes of data.

Sound cards are used to both capture as well as play back audio. Most newer sound cards contain an internal amplifier capable of producing up to a few watts of audio power. This is enough to easily drive a headset and is generally enough to drive small speakers. For most applications it is desirable to have a reasonably high quality set of amplified stereo speakers attached to the development machines. Headsets are also a necessity if the development environment is a shared facility and the sound of the playback would interfere with the work of other users.

Another means of generating sound is through the use of a MIDI (Musical Instrument Digital Interface) synthesizer. This type of sound is typically used to highlight a point in a presentation or to provide musical background sound for segments of a presentation. Some applications software packages that are capable of generating data based on various relations, for example, *Mathematica*, are capable of generating MIDI files corresponding to this data. The concept is to use sound to help the learner get a better feel for a somewhat physical representation of the data. This is similar to the way in which a 3D color plot of data lends a visual conception to abstract data.

Many of the upscale sound cards on the market today have the capability of accepting MIDI code as input and synthesizing tones using onboard circuitry. MIDI code is simply ASCII text that is used to command the hardware to generate a certain musical note as played by a specified musical instrument. MIDI code files can be generated using an uncomplicated text editor, though it is much more practical to use a keyboard or other instrument to produce MIDI data files. A second systematic means of generating original MIDI files is to use a MIDI editor that allows for the input of musical notes directly from sheet music. MIDI files are much smaller than a file containing a digitized form of a sound pattern. For example, ten seconds of a digitized waveform, sampled at 22 KHz, uses up 220 Kbytes of disk space while a MIDI file that produces a similar amount and quality of sound typically requires only about 5 Kbytes or less.

MIDI files are played by sending the data to a tone synthesizer(s), typically an oscillator that generates a waveform whose shape and frequency is based on the code received. There are several keys to good quality synthesized sound—one is to be able to simultaneously produce more than one tone. Usually a sound card capable of producing five or more simultaneous tones is sufficient. Synthesizer circuits based on frequency modulation (FM) tone generation are satisfactory though there are higher quality techniques available to audiophile developers.

8. CD ROM Writer

Earlier, CD ROMs were mentioned as a means of storing up to 650 Mbytes of data on a single removable storage medium. In the past, it was only possible to create a CD ROM by sending data files to a manufacturing site and having the CD pressed by the vendor. The economies of the situation required that the CD be produced and consumed in mass quantities in order for the process to be cost-effective. Today it has become cost-effective to generate single CD ROMs within the development environment by installing CD ROM writing equipment ($7,000 to $10,000) and blank writable CD ROMs priced at around $20 each. One restriction with today's CD ROMs is that they are WORM (Write Once, Read Many) devices, making them unsuitable for some mass storage applications. Advances in technology may soon overcome this limitation with the development of an affordable re-writable CD storage medium.

Two effectual applications of CD ROM technology include using the devices to distribute software and courseware, as well as to archive medium to large volumes of information. The recent rise in the popularity of CD ROMs has caused the prices of readers to dramatically decrease, further fueling the widespread incorporation in off-the-shelf computing systems. This trend can be expected to continue as advances in reader technology continues to improve performance. Based on this trend, inclusion of a CD ROM writer (burner) as a peripheral to one of the workstations in a courseware development facility is warranted.

Burning a 650 Mbyte CD ROM generally takes place in two steps. First, the information to be placed on the CD is transported to the host workstation and an image of the CD is created on a hard disk drive. This requires that the host workstation have sufficient available disk storage space to accommodate both the raw files and the CD image, or up to a Gigabyte (1,000 Mbytes) of storage available. The amount of time required to produce an image file depends upon the speed of the computer, generally, taking about thirty minutes to generate a full 650 Mbyte image. The required time decreases proportionally to any decrease in image file size. The second step in the process is to actually transfer the image file from the hard drive to the CD burner, and subsequently to the writable CD. Seventy-four minutes are required to write a full 650 Mbyte CD, though some CD burners offer double speed operation which halves the write time. The resultant CD can be read by a CD reader or can be sent to a reproduction facility to be used as a pre-master for mass duplication purposes.

Chapter 11

PARADIGM SHIFTS IN ENGINEERING EDUCATION: SUMMARY AND CONCLUSIONS

J. Bourne and A. Brodersen, Vanderbilt University

Engineering education grows and changes as each decade slips by. For those caught within the web of constant faculty meetings, curriculum deliberation, retention of old ideas and out-of-hand discrediting of new ideas, change appears to occur at a glacial pace. There are some that say, "Paradigm Shifts? Who are you kidding! Why change? We'll always do what we are doing now." Wrong. Things are changing, albeit at a pace that is accelerating strongly on some fronts and almost not at all on others. In this study, the opportunity to hold meetings and have many conversations has given us the chance to reflect and to make predictions about what the future may hold for engineering education. This chapter is the primary authors' attempt to extract meaning from the meetings, papers, comparative analysis and discussions that we have had over a two year period. The contents of this chapter represent only the opinions of the primary authors and not those of the collective study group.

We wish to examine a set of rather controversial questions in the light of the information that has been presented in this book. Questions that organize the topics that we plan to discuss in the remainder of this chapter are listed in Table 1 below. As a test, we ask that you read these questions and see what your answers are before reading our answers. Hint, if you try to mimic our answers, we err on the liberal side. Of course, there is no assurance that our answers are the correct ones, since no one can predict the future. In the materials that follow, we will do our best to attempt to be as unbiased as possible. Think back on what the authors of the chapters that you have read have said when you consider each question. Also, read the results from our summer meeting, because the finding in these materials also influence the answers we give to these questions.

Questions about the Future of Engineering Education
(Test yourself now!)

- Why change?
- Change in what direction?
- How to change?
- What is engineering anyway?

Table 1.

- Must engineering education be science based?
- Is the curriculum wedded to a four-year tenure?
- Will students always be predominantly campus based?
- What about tools for teaching?
- Is multimedia going to be an effective teaching and learning tool for engineering?
- How can we bear the high cost of creating multimedia presentations? How effective will such presentations be?
- Is simulation a viable paradigm in engineering education?
- What is meant by simulation anyway?
- Is learning teamwork essential? How can we do it?
- I keep hearing about anchored instruction. I still don't understand just what you mean and why it would be good. Please explain!
- Information resources in the world are now at our fingertips, but engineering educators don't use these resources effectively. Should they? If so, how?
- How should engineering design be taught in a new instructional paradigm for engineering education?
- Will things like avatars and knowbots really be used in engineering education?
- Are lectures going to disappear?
- What will faculty do in the future world of engineering education?
- Will textbooks continue to be used? Which alternatives seem most likely.
- Computer technology appears to be taking over engineering and engineering education. Is this bad?
- Can engineering education be taught more quickly and easily than it is now?
- Why is research activity the only way to success in faculty advancement?
- What role will distance education play in the engineering education paradigms a half-century hence?
- How can we know what to do?

Table 1. (continued)

ANSWERS

The following possible answers are offered. We do not see any answer as being "correct," however, most of the answers given below tend to be more liberal and less conservative. That is, we generally prefer dynamic change rather than stagnation.

- **Why change?**

Change is essential to growth. Engineering education has been in a holding pattern for forty years or more. The lecture/laboratory paradigm for pedagogy has worked for us. Student supply, while cyclical, has remained sufficient to fuel the research/teaching machine of our institutions. Will the supply change? Do we need to attract more students or a more diverse student population? If we decide to change, we must understand what the drivers are for change. After all, as pointed out by Mattson, industry has been quite satisfied with our product. If industry, as our customer is satisfied, then should we just keep doing what we are doing now? No! Why do we say this? Dominantly, because a dynamic curriculum needs infusion of new ideas. Students need to learn to function in an increasingly complex world. They need to acquire skill sets from multiple areas and must learn to function in teams. The world of the graduate engineer is no longer most likely to reside in the large industries of America. More of our graduates are likely to be entrepreneurs who have numerous jobs during their lifetimes. Thus, they must learn to twist and turn and adapt. We think that it inadequate to give only an education in the fundamentals. Yes, fundamentals are important. Rigor is important. Yet we don't really know that fundamentals plus rigor equals success, do we? We only assume that it does because this has worked in the past. Yet, it is likely that the shifting nature of the world will surely require a new skill mix that includes fundamentals, dashes of rigor, heavy doses of diverse learning, preparation for lifelong learning and learning to work in teams.

- **Change in what direction?**

Change should fulfill perceived need. A purpose of this study was to generate a look at perceived need. Clear results have emerged as documented in the second section of this book. We should change toward enabling electronically-based education so that institutions can share knowledge and that students anywhere and at anytime can learn. We must begin to understand new learning styles based on anchored instruction and cooperative work groups in order to teach our students how to function in modern society. We must think about enabling tools for education, including electronically-mediated teaching tools, visualization and simulation tools, and access to and seamless natural use of materials located anywhere in the world's information infrastructure.

- **How to change?**

Change may be enabled in many ways. In the best case, change will occur slowly and beneficially as enabled by national federal initiatives. A clear, current example of such change is the current group of engineering colleges, funded by the National Science Foundation, that are leading change in engineering education through the engineering coalitions initiative. In this funding initiative, around fifty colleges of engineering in several coalitions have made significant inroads into understanding how to change. For example, course ware is being created, national networks for sharing curriculum are being constructed, workshops to teach faculty how to build course ware, experiments on effectiveness of team learning, and anchored instruction experiments are being carried out under the aegis of this large federally funded program.

In the worst case, change will occur rapidly due to extreme pressure. At every downturn of numbers of entering engineering first year students, engineering educators imagine the worst - i.e., for example, reductions in staff, lowering of "standards", and introduction of draconian measures to increase funded research. Then, before we know it, there is an upturn in student interest in engineering education and all is forgotten! One might think that this ebb and flow might continue forever. Will it?

To effect change means working hard to shift the paradigm. Even at non-NSF-coalition schools change is possible. Various foundations fund study of innovation in education. Many focus on specific target areas (e.g., the Sloan Foundation funds work on improving distance learning—a core topic in this volume). To enable change, one clear need is for administrations of engineering colleges to note the value of investigating how to educate engineers. We have been locked into the research primacy paradigm for so many years that it has not been possible for faculty to be promoted for scholarly contributions to engineering education. This fact remains true at most institutions in the mid-1990s. Several of the authors in this volume have pointed out the essential need to change the mind set for assessing faculty creativity for promotion and tenure to permit reward for contribution to education and not just for research. There is evidence that in some institutions change in the reward structure has commenced; however, the slight shifts observed are by no means widespread.

- **What is engineering anyway?**

This is not a silly question. We think engineering is "the practice of organizing the design, construction, and operation of any artifice that transforms the physical world around us to meet some recognized need," as stated by Dede in his chapter in this volume. If this is engineering, then engineering education must teach methods for designing, constructing, and using artifacts. We do this, building on an extensive science base. Yet, as pointed out by Dede, we have not traditionally included topics about quality, teamwork, and ways of dealing with incoherent objectives, incomplete data and imperfect understanding. Hence, teaching engineering students how to deal with their future world probably requires reorganizing the above definition. Dede's alternative view of engineering is: "the practice of organizing the design, construction, and operation of any artifice-based systemic innovation that transforms both our physical and psycho-social environment to meet some recognized need." What is the difference in these definitions? The latter definition includes the notion of systemic change, meaning that we should teach engineering students how to deal with large systems. The term psycho-social is added to connote the need for engineers to work in an integrated social environment that doubtless includes teaming, working around disciplinary boundaries in an intensively social work place. It is highly doubtful that engineering will ever return to the "over-the-wall" paradigm of the past. By "over-the-wall," we refer to the disappearing practice of the design team designing an artifact, throwing it "over-the-wall" to the manufacturing team who make the artifact and, in turn, throw it over the wall to the test group. Instead, engineers will work in highly integrated teams that include customers, designers, manufacturers, and test personnel as well as others. Hence, in view of this changing view of engineering, we must change our pedagogical viewpoint and provide educational experiences that will match these future needs.

- **Must engineering education be science based?**

The answer is mostly, yes! Certainly, the changes that moved engineering from a trade-style discipline following World-War II to a science-based curriculum were very successful. It is widely believed that the underpinnings of science are needed to produce successful and knowledgeable engineers. Another question is "how much science and what kind of science is needed?" Or perhaps, can science and engineering learning be integrated into a cohesive whole? As an example, can we link the learning of physics and chemistry directly into our upper-division courses so that students use the scientific knowledge that they learn as first-year students in these courses? Science integration is not done well today in our curriculum since the teaching and learning of science and engineering in an integrated fashion requires breaking down the boundaries between the traditional departments that teach these topics. Some new ground in this area is currently being broken as a result of funding initiatives by the National Science Foundation.

- **Is the curriculum wedded to a four-year tenure?**

  Engineering curriculum plans are officially cast into four-year molds because of the tradition of college requiring four years. However, at many, if not most, engineering education institutions, the typical student requires more than four years to graduate. What does this mean? It simply means that our curriculum are too jam-packed with courses, labs, and other requirements making it difficult for students to make their way through the curriculum in the proscribed four year time allotment. Students do fail courses, they do get out of sequence, they sometimes cannot find the right course at the right time, they change their majors. In short, everything has to be just right from the outset for students to make it through a typical engineering education curriculum in four years.

  What should be done? Should engineering be made a five-year program? Probably not. This would decrease the attractiveness of engineering vis-à-vis other undergraduate programs. Perhaps what should be done is to provide (1) off-line learning[1], for example, using tutorials that are on-line[2], (2) create more integrated learning experiences, (3) reduce duplication, (4) reduce or relegate legacy knowledge to off-line learning , etc.

- **Will students always be predominantly campus-based?**

  "Always" is a long time. For the near-term future (the lifetime of readers of this document), it is likely that campus-based environments for learning will be the norm. One can already see the slow erosion of campus insularity through distance learning efforts such as those described by Mattson and the use of shared materials on the Internet. However, changes would bring about home learning and at work learning coupled with a move away from the residential campus have not yet occurred. It has been predicted for many years that ever higher costs for on-campus education would drive students to other modes of learning and cause traditional residential education paradigms to change. But this has not happened, because there has been no pressure to change. If there are customers (i.e., students) who will pay for traditional education, there is no need to change, even if change would result in a higher quality education. What will be the impetus for off-campus learning? It seems likely that off-campus learning might become widespread as population demographics change, i.e., people routinely work at home to avoid commuting, and as the information infrastructure of the world becomes robust enough to support almost-there interactions. That is, if I could attend a lecture by Dr. Famous at Harvard rather than go to my local community college professor, Mr. Know-it-just-in-time, and the price was about the same to go to Harvard, I might well opt for Harvard. More likely, however, is the possibility that charismatic lectures can be distributed to many sites and local faculty can act as facilitators. Over time, we suspect that excellent lectures, excellent tutorials, excellent research works, will be distributed widely. Easy access via a pay per view mechanism should provide significant improvement in the quality of education.

---

[1] i.e., not in the classroom or laboratory.

[2] i.e., in the computer sense.

- **What about tools for teaching?**

  Tools for teaching will be useful in the upcoming years. In the mid-1990s, various university members of the engineering education coalitions and others are beginning to create tutorial materials for engineering students that can be accessed over the Internet. Multimedia tools, videos, lectures, simulation systems, virtual realities, print materials, intelligent assistants and the like will all have been experimented with by the end of the decade. The aim is to improve comprehension by permitting students to learn at their own pace, to improve experiences by providing expert knowledge that could not otherwise be easily acquired, to reduce human costs (i.e., decrease the use of the biological tape-recorder, the lecturer), and to provide students with learning scenarios based on technology.

- **Is multimedia going to be an effective teaching and learning tool for engineering?**

  Multimedia refers to the mixing of different types of media, predominantly video, graphics, and sound. As indicated in the article by Katz et al., clear strides are being made into the study of the use of multimedia for engineering education. What are the prospective benefits of multimedia in the classroom and/or laboratory? First and foremost, multimedia can improve learning through visualization of complex physical phenomena. Often it is difficult for instructors to link mathematics to the reality of a physical system and how it operates. Very useful visualizations are being created today that provide, for example, slow motion videos of simple mechanical systems that show how physical movements are translated into descriptive waveforms. Video provides clear utility in depicting situations that are best rendered in motion. A testable question is to what degree can information that involves motion can be rendered in animation, by sequentially presenting graphical images? Why would one be interested in this alternative? Animation can potentially provide the same information with a much lower cost than a video presentation. Clearly a problem with video is the high cost of creating videos and the high level of work needed to process video into a teaching and learning system. We suspect that multimedia will develop at a fairly rapid rate and that interesting materials will be developed using multimedia. However, widespread development of materials will be difficult due to the high cost of creating materials.

- **How can we bear the high cost of creating multimedia presentations? How effective will such presentations be?**

  Two likely scenarios for creating multimedia presentations come to mind. First, the government, through the NSF, may well continue funding programs to develop course ware for engineering education. Second, publishers will likely support more and more of the development costs for multimedia instructions. The second possibility will only happen as instructors in colleges of engineering increase the use of multimedia materials in their courses. Changes of this type require a shift in the thinking of faculty about how to teach. As long as the lecture is viewed as more suitable than using technology to support teaching, introduction of multimedia instruction in the classroom is likely to move slowly.

  Can multimedia presentations be effective? In a word, yes! However, it is likely that the most useful multimedia experiences will be interactive and will be created so that students can experience the presentations on their own time, where and when they want to. If this conjecture turns out to be true, it may mean that the traditional lecture should be supplanted with instructor/student interactions and discussions and the rote presentation of materials relegated to out-of-the-classroom multimedia learning experiences.

- **Is simulation a viable paradigm in engineering education?**

  Simulation is the current paradigm for design that is most used by practicing engineers. We, as educators, should widely employ simulation in our teaching. In most engineering colleges, students do learn about using a variety of simulation packages and often use these packages for creating designs in capstone and other more advanced courses. However, simulation has not been explored as a learning tool for augmenting and assisting students in learning basic concepts that are now predominantly discussed in lectures. Most physical phenomena that can be simulated should be, and used as a teaching tool for students to understand how physical artifacts work. Simulation coupled to a visualization modality is a far superior paradigm to presentation multimedia. Here is the contrast. In multimedia presentations, students can see and hear information about a physical system or mathematical construct. Rarely are systems simulated so that students can actually interact with a simulated physical system. An ideal situation would be to create a learning environment in which the real world is almost duplicated. In this type of a "virtual world," learning experience could combine the powerful presentation capabilities of multimedia while providing the student the capability of interacting directly with an almost real environment. We feel there is a significant chance that these types of learning environments would add real value to the engineering education paradigms of the future and that exploration along these lines should commence.

- **What is meant by simulation anyway?**

  In broad brush, simulation is a computer technique that permits observation of the behavior of models of the physical world. Visualization extends simulation by permitting visual rendering of simulations. For example, a mathematical model consisting of a set of differential equations might describe the relationship between the input and output of a physical system. By repetitively "plugging-in" values into the input of the simulation, a computer simulation can predict what the behavior of the real system would be for a given input. Visualization of a simulation can be achieved through computer graphics that permit students to manipulate inputs and observe the output response of a simulated system. Hence, visualization of a simulated model of a physical system can be a very powerful tool for learning, since students can, in effect, manipulate their version of the real world in an artificial environment.

- **Is learning teamwork essential? How can we do it?**

  As discussed by Dally and others in this volume, it should be clear by now that learning in teams is an essential experience for engineering students. Why? Because experiences in teams mimic the experiences that students have when they enter the workplace. A wide variety of experiments are apace in this country on how to best utilize team work in engineering education. Some team work experiences consist of shared design projects, others, shared homework solutions and even shared exams. Some of the most useful learning experiences come from experiences in which students all equally participate in an "awesome" project in which team members work hard and the result is much more than any in the team expected. Unfortunately, such cooperation in student teams is not as universal as might be hoped. Student teams often contain at least one malingerer, if not more. A finding is that team experiences should be highly structured, perhaps even with roles and specific clear duties and milestones for each team member. We (engineering educators) have not traditionally structured team learning experiences in this way; we would probably achieve better results if we did. These assertions are, however, testable questions!

- **I keep hearing about anchored instruction. I still don't understand just what you mean and why it would be good. Please explain!**

  Anchored instruction or situated learning can help students to learn better and more quickly by creating learning environments that mimic the real world. Re-creation of realistic situations by video and computer can be used to cast students into a virtual world in which they must exercise their problem-solving abilities to solve complex problems. One might imagine a virtual world for learning in a chemical engineering laboratory in which computer simulation provides realistic approximations of test tubes, beakers, chemical reactions, and so forth that provide students with a learning experience which closely approximates the learning experience in a physical laboratory. This type of simulated world for anchoring instruction is not currently used extensively in engineering education. Experiments by Mosterman et al. (1994) have shown that introduction of simulated laboratories prior to a physical laboratory saves student and teaching assistant time and improves student comprehension.

- **Information resources in the world are now at our fingertips but engineering educators don't use these resources effectively. Should they? If so, how?**

  Sometimes one feels that the information revolution has touched engineering education perhaps less than other fields, when in fact, engineering education should be in a leadership position to utilize information technology for enhancing learning. Students in pre-college learning experiences are using information resources on a variety of networks. The popular media has continuous discussions of the use of the Internet and the various commercial services that make up the fabric of the "Information Highway." Yet, in colleges of engineering little attention is devoted to using this information infrastructure to augment traditional education. Some motion is noted at numbers of colleges and universities worldwide: clearly, student use of the networks is very high, although not necessarily for useful intellectual activities. If we are to show students what the art of the possible is for learning by means of networks, we, the faculty, must help them to understand. For example, at Vanderbilt, we now introduce the first-year students in engineering to information technology in the first semester. However, this learning is not reinforced in later years since the faculty, by and large, do not use the networks as a ubiquitous learning tool in their classrooms. Anecdotal information from other institutions indicate that in 1995, the situation is similar elsewhere. Certainly, it seems plausible that we should investigate how to make optimal use of this splendid resource to support our teaching and learning activities in colleges of engineering.

- **How should engineering design be taught in a new instructional paradigm for engineering education?**

  This is a difficult question. We know how engineering design has been taught up to now. We understand how teams work in capstone design courses. Is this the best way?

This is unknown at this time. We need to undertake studies to determine if there are other possibilities. We suggest two possibilities next.

Design, anchored experiences. One idea is to build anchored design experiences in which students work as teams to create a complete product, taking the product from concept to market with all the attendant experiences including test, manufacture, marketing, and customer service. Including these latter factors would anchor the design experience in a more "real-life" type of experience that could mimic experiences that would be similar to those that students would experience after they graduate.

Design, distributed experiences. Another idea is the creation of team design experiences that extend across universities and into industry. Given that students in multiple institutions and at multiple industries can easily communicate and share engineering drawings, spreadsheets, and detailed designs, it may well be possible to create learning experiences across multiple institutions that also collaborate with industry, probably all network enabled. Industrial collaboration would significantly improve the design experiences of most engineering students.

- **Will things like avatars and knowbots really be used in engineering education?**

The avatars (simulated humans in computers) mentioned by Dede in this volume, seem to us to be unlikely adjuncts to engineering education in the near-term. While a professorial avatar[3] might have some impact on a student interacting with a computer-based learning system, it appears that the creation of one is difficult and the utility not clear. While personal contact is essential in learning for some people, it is unlikely that faculty-student interaction can be replaced by machine-based creations. It is, furthermore, unclear whether personal avatars[4] will really help students or not.

In contrast to avatars, knowbots will likely become central to engineering education in the future. Already, there is overwhelming information overload for students. What with library resources, journal articles, design documents, experiences of others, there is too much to understand and sift through when trying to learn about a topic. What can help? Your personal knowbot who can find information, package it, and deliver it to you, as needed! We suspect that specialized knowbots of various types will become commonplace as they are invented in the upcoming years. Of course, an avatar knowbot would be interesting, particularly if it were specialized as a professorial figure and linked to a multimedia presentation: "Hi this is Professor Jones, I'm your personal avatar. Can I help you with Chapter 3 homework problems today?"

- **Are lectures going to disappear?**

Lectures are the most efficient way of fulfilling professorial obligations and the perceived student need for securing information. Lectures have been used for presentation of information and for discussion for centuries. Hence, it is unlikely that lectures will disappear any time soon. Nevertheless, let's explore what could be done with faculty members' time if the lecture were not central? Can we imagine an educational world in which faculty act mainly as facilitators, creating an educational

---

[3] i.e., a simulation of a known professor.

[4] Personal avatars are like personal knowbots. In fact, one might think of a personal avatar as an anthropomorphized personal knowbot, tuned for interaction with you.

environment for students that consists of faculty avatars, on-line knowbots, multimedia presentations, interactive simulations, virtual realities for learning, and broad-gauged team-oriented design experiences offered in collaboration with industry and other universities? It would certainly take a paradigm shift to effect such change. Would we want to? Studies need to be done to see if there would be pedagogical gains. Could we afford to? We would probably have to change what we do now if such paradigm shifts were made. We can see a clear role for faculty as learning facilitators, for creation of materials, for assessing materials, for guiding students, etc. The answer to this question is that there will probably be incremental change; for now, however, going to a nonlecture-based educational system for engineering education is unlikely to happen. We must conduct studies to find out what gains are possible through shifting the paradigm!

- **What will the faculty do in the future world of engineering education?**

  As indicated above, we think that it is most likely that faculty will become facilitators and organizers of knowledge. As information grows exponentially, we must try and extract knowledge from the world's knowledge bases and provide the salient features to students. The role of faculty may become much more intellectually based, since there is too much information and so little time to teach students what they need to prepare for the work world. The challenge is to create teaching paradigms that enable faculty to do what they do best—that is, sift and sort information, create new knowledge, organize old knowledge, and impart it to students in a way that the students are able to use.

- **Will textbooks continue to be used? Which alternatives seem most likely?**

  We are just now in 1995 seeing the beginning of changes in the traditional textbook market. Publishers are moving rapidly to approach the multimedia publishing market for education. A significant number of new companies have been created to create and distribute educational software. There is an excellent chance that within five years we will see a shift to computer-based learning modules that are distributed with textbooks (e.g., demonstrations, simulations, tutorials). Then, as the student and faculty populations become more familiar with the medium and as handheld notebook computers become part of the everyday culture of engineering education, the textbook as we know it today may disappear.

- **Computer technology appears to be taking over engineering and engineering education. Is this bad?**

  Yes and no. Yes, because the rank and file faculty member cannot keep up. The typical engineering faculty member eschews using computer technology in the classroom because of the difficulty. Thus, the educational opportunities that exist with computers are not taken advantage of and will not be taken advantage of unless there is real pressure to do so. No, because there is the opportunity for very significant change. We have the opportunity for very significant change that was never possible before, due to the immaturity of technology. We must take advantage of the possibilities and move forward.

- **Can engineering education be taught more quickly and easily than it is now?**

  The answer from the composite answers to the questions given above is obviously, yes. However, it should also be clear that we don't yet know how to do it. "Quickly" may be possible through the use of multimedia and simulation tools. We have been locked into the lecture/laboratory paradigm for so long that we cannot imagine what gains might be possible if learning could take place anywhere and anytime rather than in the constrained lecture/laboratory sessions that we now employ (it is indeed arguable that learning occurs there). "Easily" is highly doubtful. Lectures are easy and getting teaching assistants to organize laboratories is easy to facilitate. Changing to a paradigm that is highly technology oriented will make teaching more difficult, yet, we think, exciting!

- **Why is research activity the only way to success in faculty advancement?**

  Research is a sticking point. Currently, collecting research grants and writing scholarly articles is the only way for assistant professors to get promoted in the university system of America. Untenured faculty cannot afford to participate in educational change. Hence, the source of some of our keenest energy and best ideas goes untapped because of a promotion and tenure paradigm that has outlived its usefulness. For administrators reading this, we suggest you consider shifting the paradigm for promotion and tenure to include new research on educational methods and publication.

- **What role will distance education play in the engineering education paradigms a half-century hence?**

  A half-century ago, engineering education wasn't very different from today. At that time, change was just moving engineering education from a trade orientation to a science orientation. Lectures and laboratories were given. The professoriate was much less highly skilled in research and facilities were primitive by today's standards. If similar change occurs in the upcoming half century, we may expect vast improvements in learning styles and methods discovered that will support learning about engineering. Perhaps most significant will be the capability for having machine-based learning and support assistants. While no one can accurately guess the future, this prediction has some face validity. First, information is increasing at an exponential rate. We will be unable to cope with this information within a few short years. Thus, one can predict that knowledge-assistants will have to be invented to deal with the information explosion. Second, our curricula are bursting at the seams. We cannot continue to learn more and more literal knowledge, that is, just knowledge about what things are. Learning must shift to how to learn about things—how to make inferences, how to apply knowledge, and how to evaluate situations. Even within the span of fifty years such changes may come slowly.

- **How can we know what to do?**

  You can't be sure! However, there are certainly indications about the directions in which to go. We hope that the contents of this volume have evoked ideas about how to change what you and your colleagues are doing. There are certainly many things that would be classified as part of the "grand challenges" for engineering education, including ways of using technology to facilitate education; ways of creating more viable team learning experiences; techniques for enhancing learning with others at a distance; methods for exchanging, accessing, and using information across networks; and ways to improve pedagogy through multimedia and simulation, among others discussed. If change is to come about, we must all work on threads of inquiry until we are able to see the paradigm begin to shift. Early on, we probably won't even notice a shift since change will be incremental. We suggest that you think about what you have read here, choose a direction for making a contribution, and go for it!

REFERENCES

1. **Mosterman, Pieter J., Marcel A.M. Dorlandt, J. Olin Campbell, Craig Burow, René Bouw, Arthur J. Brodersen and John R. Bourne**, Virtual Engineering Laboratories: Design and Experiments, *ASEE Journal*, July, 1994.

Part II

THE INFLUENCE OF TECHNOLOGY ON ENGINEERING EDUCATION: ANALYSIS AND RECOMMENDATIONS

The materials presented in the second part of the book are summaries of surveys and results from the focus group meeting held in the summer of 1993.

Part II:

TABLE OF CONTENTS

FIGURES

TABLES

I. INTRODUCTION

A. THE PROBLEM

Engineering education is changing. The way engineering educators teach and the way students learn is being altered by the availability of new and inexpensive computer and communications technology. Engineering educators can now provide students with experiences which could not be even considered in the past. For example, access to on-line libraries, robust calculation capabilities on laptop computers, visualization tools, technology-facilitated communications among peers (e.g., electronic mail) and information resources are among the technology-based learning facilitators that were not in existence only a few short years ago. Due to the existence of new communications and computer technology, it is clear that the paradigms for teaching engineering education must and will change. How and in what way will the standard lecture/laboratory paradigm shift? Figure 1 captures the idea of transforming engineering education from where it is now to a future state. Advances in technology will enable curriculum transformations that will change both how education is delivered as well as what is taught. This study has the modest ambition of assessing how computer and communications technology might change engineering education and to help provide guidance to engineering educators, administrators and governmental agencies about possible future directions. These two parts attempt to examine in detail what changes are possible in a variety of different areas. The first part in this two part set is a compendium of review papers which analyze major areas within engineering education This second part contains data collected from focus group meetings, surveys and provides an analysis of this information.

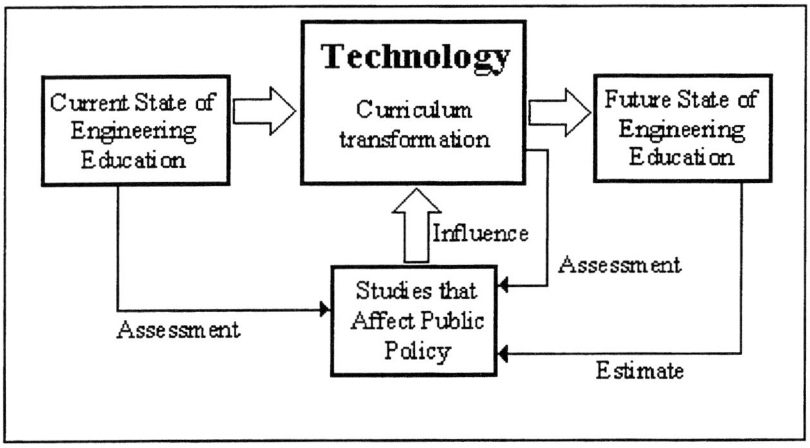

Figure 1. Transforming Engineering Education.

The audience for this two part set is (1) engineering education administrators, (2) engineering educators, and (3) governmental policy and decision makers Some of the material created may be of use to the science education community, as well.

B. THE OPPORTUNITIES AND PRESSURES

At the outset of this study, it was recognized that there was a very significant opportunity to improve engineering education using computer and communications technology. Work reported until 1992, only sporadically addressed the overall picture of what changes might come about due to the availability of inexpensive technology. We viewed this omission as an opportunity to attempt to discern how engineering education could be improved by making computer and communications technology an integral part of the engineering education infrastructure. Likewise, pressures exist. For example, our culture has become more technology oriented. Students arriving in colleges of engineering today virtually expect to be entertained - a product of their video nurtured generation! These cultural changes practically mandate that our teaching and learning paradigms in engineering education shift as well.

C. METHODS AND METRICS

This study employed a systematic process to go about uncovering information, collecting opinions, organizing thoughts and making recommendations. This second of Part II describes in detail the process by which the study was run.

1. The Process

Figure 2 shows the process adopted for the study. As shown, the study was divided into three phases. Phase I was devoted to planning and organization and lasted for about three months. Phase II was the core of the study devoted to making assessments and meeting with the consultant group. This phase lasted approximately a year. The final phase (Phase III) was dedicated to organizing the material generated from the previous phases and creating these two parts of material. Phase III had an approximate duration of nine months.

The study was conducted from July, 1992 and was largely complete by September of 1994.

2. Surveys

Three surveys were conducted during the course of the Study to assess the state-of-mind of engineering educators and other interested parties. Surveys were designed to assess the use of technology in the classroom and laboratory, electronic media for teaching, the engineering curriculum, the influence of internal and external factors on engineering education, and participant information. Three survey's were administered - (1) to the attendees of the Innovations in Engineering Education Conference held in Santa Barbara, CA during the summer of 1992, (2) to the attendees of the Frontiers in Education Conference held in Nashville, TN in November of 1992 and (3) to engineering educators worldwide using Internet during 1993. Over 100 replies to the survey were received.

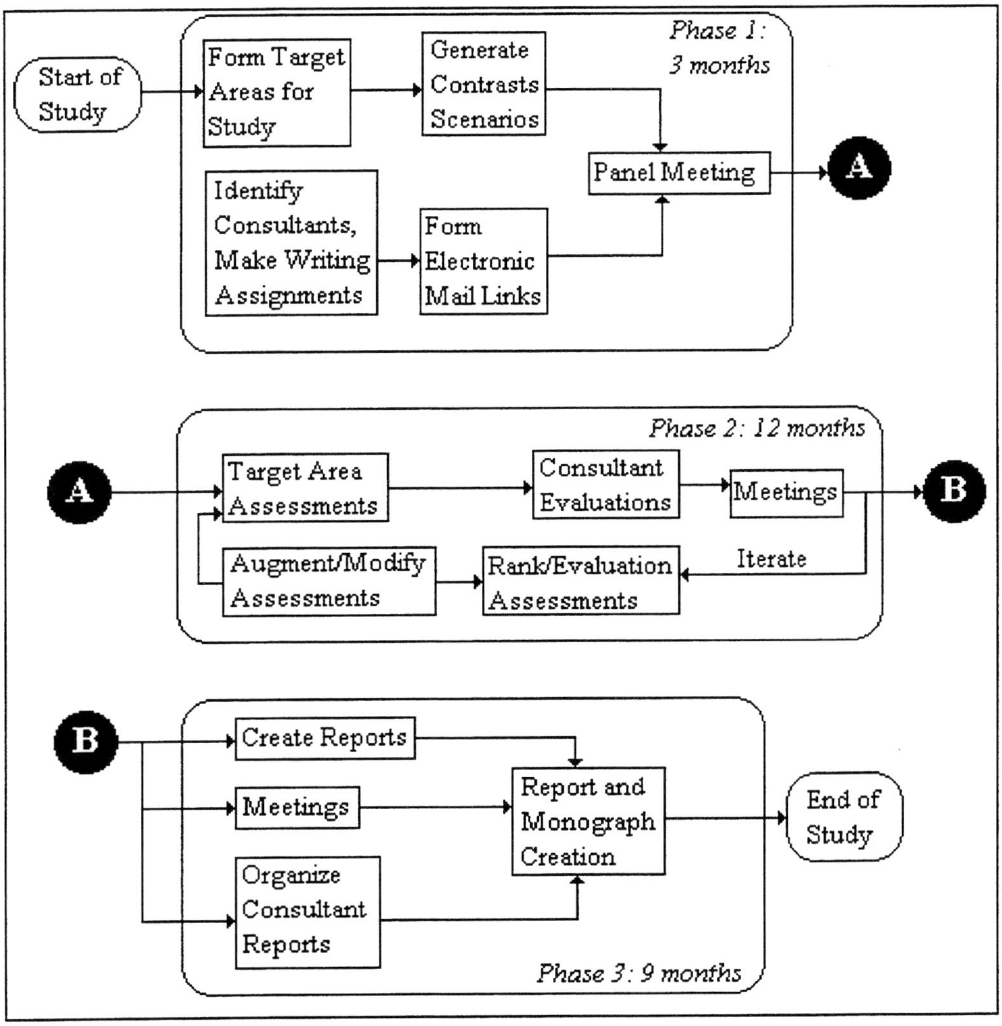

Figure 2. The Study Process.

3. Study Group Formation and Meetings

The study group was selected based on a number of criteria; (1) different facets of engineering education should be represented, including those with interests in lecture and laboratory instruction, (2) both graduate and undergraduate concerns should be included, (3) both practical thinkers and visionaries should be part of the study team, (4) a mix of institutions and members of the engineering education coalitions should be included, (5) the participants should not be made up entirely of engineers and should include educators, economists, and futurists as well as engineering educators. The group, as finally constituted, did conform to all the above criteria. For the purpose of focusing, this group was divided into three subgroups to deal with the study of (1) distributed learning, (2) curriculum paradigms and (3) delivery methods.

A first meeting of the study group was held in November 1992 in which the methods of procedure were reviewed and the computer tools to be used were reviewed. In particular, the first meeting reviewed the use of Kepner-Tregoe[1] analysis techniques in which various alternatives are selected by rating the effect of various criteria on the alternatives. Study participants were asked to complete a set of Kepner-Tregoe tradeoff analyses about curriculum concerns, tools, delivery methods, classroom use, laboratories, learning outside the classroom and multimedia during the winter and spring of 1993.. A second three day meeting was held during the following summer of 1993 to conduct detailed analyses. The methodology described below was used to organize this meeting. A final discussion meeting was held at the November, 1994 at the Frontiers in Education Conference held in Washington, DC.

4. Study Group Methodology

In a three day meeting, the three study groups mentioned above worked separately and together as a larger team using the process described below:

a. A list of pressures and opportunities for each designated target area (i.e., (1) distributed learning, (2) curriculum paradigms and (3) delivery mechanisms) were listed and rank ordered by each group. Examination of pressures and opportunities was selected as a first step to give study participants something to relate to when studying the selection of alternatives for each area later in the meeting. Forming and prioritizing a list of pressures and opportunities helped organize the thinking of a group about the topics under consideration.

b. A Kepner-Tregoe tradeoff analysis sheet was constructed by each group and individually evaluated by group members, a vision statement was written by each group and a set of recommendations for future emphasis areas in engineering education created. After each of the items above was created by each small focus group, the tradeoffs were discussed by the larger group. Each group debated and collected a select of alternatives for their specific issue. Next, a set of criteria for differentiating between the alternatives was collected and refined. Qualitative scales were defined and used to make decisions about how each criterion affected the alternatives. Scales such as "agree!, agree, neutral, disagree, disagree!" or "excellent, very good, good, fair, poor" were used. The computer tool used to aggregate scoring converted qualitative scales to quantitative scales on a linear basis for scoring purposes. Once the alternative/criteria/scale selection was complete, individual ballots were printed and participants scored each alternative/criteria item according to the scale. Finally, these individual scores were amalgamated, printed and discussed with the total study group. An example will help to clarify the Kepner-Tregoe tradeoff methodology. Consider:

Suppose you are attempting to differentiate between three alternatives called, for example, {A, B, and C} and that you have several criteria $\{C_1, C_2$ and $C_3\}$ that can be used to differentiate between the alternatives. Further, suppose that each of the criteria can be weighted by weights W_i and that you have a qualitative scale that extends from "excellent" to "poor" as indicated above. One can form these items into a matrix and places a rating at each element of the matrix using the rating scale in order to evaluate the criteria for each

[1]See Section III of this volume on trade-off analyses for a complete description of Kepner-Tregoe trade-off analysis methods. A brief introduction to the method is given in this section.

alternative. While there are various ways that one could organize the scoring of a matrix of this sort, for descriptive purposes imagine that excellent is assigned a numerical score of 5,

Criteria	Weights	Alternatives A	B	C
C_1	W_1	excellent	good	poor
C_2	W_2	fair	good	poor
C_3	W_3	very good	good	good

very good = 4, good = 3, fair = 2 and poor = 1. We can also assign weights which rate the relative importance of the criteria when compared against each other. In our example, if we assign $W_1 = 3$, $W_2 = 4$ and $W_3 = 5$ then our matrix would look like:

Criteria	Weights	Alternatives A	B	C
C_1	3	5	3	2
C_2	4	2	3	2
C_3	5	4	3	3

The weights, as shown in the example, indicate that we felt that W_3 was more important than W_2 and that W_2 was more important than W_1. Now if we multiple each weight times the numerical representation of the rating and sum each alternative column we get:

		43	37	29

Using this ranking scheme, alternative A is best, followed by B and C. The concept of organizing alternative ranking in this way is not difficult to understand, nor is it very precise. However, it does have the virtue of helping organize information and providing a way for groups to understand how they are thinking about alternatives and criteria. A commercial computer tool (Apian, 1992) was used to facilitate the organization of the data.

c. Each focus group created a vision statement in its area of interest.

d. Each focus group created a set of recommendations which would assist in achieving the vision described in the vision statement.

e. Finally, the study group created an outline for the two parts that would become the major output of this study. Two monograph outlines were created: (1) a set of papers on selected topics prepared by the study group participants and (2) a compendium of data representing the information collected during the study.

D. A GUIDELINE TO THIS DOCUMENT

This part first reviews the surveys and the results of the three surveys conducted. Second, the detailed Kepner-Tregoe tradeoff analyses conducted at the outset of the Study are explained and reviewed. Third, the results from the focus groups are presented and analyzed. Fourth, recommendations secured from the focus groups are presented and discussed. Finally, a road map for the future is given.

The information in this document is probably best read after reading the material given in the papers in Part I. However, the information presented in this part does stand alone and can be digested without reading the critical review articles. It should be noted that the recommendations made in the last part of this part represent the thinking of the participants in the Study groups and, in no way, should be interpreted as policy of the National Science Foundation.

For the casual reader of this document, a quick scan of the survey results in Section II will confirm almost predictable responses. A review of how the Kepner-Tregoe tradeoffs work in Section III will help readers who are unfamiliar with the methodology to understand the subsequent sections.

We have constructed this Part almost as a chronicle of our experiences in this Study. The materials presented move according to a process that was determined at the outset and rigorously followed during the course of the Study. We made mistakes from time to time in how we created survey and tradeoff information. However, even the results of how these mistakes influenced our thinking turned up interesting results and ultimately resulted in the refinement of our analytical instruments. Thus, read this part almost as a chronicle — an evolution of our thinking about the future of engineering education. It was a learning experience!

REFERENCE

Apian Software, Decision Pad., San Mateo, CA, 1990-1994

II. THE SURVEY

At the outset of the study, an extensive survey was undertaken to assess the current state of engineering education and measure the extent of the use of computer and communication technology in the delivery of engineering education. This section presents the survey methods and the results which were obtained.

A. SURVEY METHOD

A survey instrument was designed to explore the teaching methods and various perceptions of engineering educators with a particular emphasis on the influence of computer and communication technology on the current practices. A commercial software tool (Survey Pro by Apian Software) was used to create the survey. The survey was divided into the following topical areas:

- *The Classroom:* The first section consisted of 5 questions and explored teaching practices in the classroom.
- *The Laboratory:* Seven questions explored the use of the laboratories and their relationship the classroom in engineering education.
- *Computers and Other Electronic Media in the Classroom:* This section surveyed the current uses of computers and communication technology and was divided into two sections. (1) Current Use of Computers in the Classroom and (2) Electronic Media.
- *The Engineering Curriculum:* This section surveyed the current state of engineering education and attempted to explore the pressures and opportunities for change in the curriculum. Thirteen questions were posed in this section.
- *Teaching and Research:* Three questions were ask to explore the respondent's perceptions about the relative importance of teaching and research at their institutions.
- *Demographic Information:* Two questions were included to determine the teaching experience and position of the respondents.

B. THE SURVEY AND RESULTS

This section presents the survey and the results. The survey included fifty questions. The survey was distributed through various methods. The attendees at the Innovations in Engineering Conference, Santa Barbara, CA, August 1992 were the first group to complete the survey. This small conference was attended by engineering educators who are at the forefront of the changes in engineering education. The survey was also distributed to the attendees of the Frontiers in Engineering Education, Nashville, TN, November 1992. This conference was attended by over 450 people and included a cross section of engineering educators from engineering and technology programs. The survey was also distributed through the Internet. A notice was posted in various pertinent news groups and list servers inviting participation in the survey. The survey was distributed and returned using electronic mail.

The survey is shown in Table 1. On the left side are the questions and the possible responses. On the right side the responses are shown both in tabular and graphical form. The responses for The Classroom indicate that lectures are the predominate delivery mechanism in engineering education today with the overhead projector and transparencies being the most popular teaching aid. Less than 10% of those surveyed use computer with presentation graphics and LCD panel in the classroom.

Engineering laboratories play the traditional role of linking theory and practice in the curriculum. The surprising result of this section is the perception that the equipment in the laboratories is not dangerously obsolescent. Over 60% of the respondents either *disagreed* or *disagreed strongly* with the statement "Our laboratory equipment is old and outdated." In a somewhat related statement, "Laboratories should always have the latest state-of-the-art equipment," only 60% either *agreed* or *strongly agreed.* One conclusion that could be drawn is that the most modern equipment in the laboratories is not necessarily a prerequisite for good pedagogy. The respondents generally agreed that computer based training can play an important role in laboratory instruction.

In the section on current use of computers in the classroom, the results indicate that computers are being accepted as a useful adjunct in the classroom. Approximately three-quarters of the respondents have used computers in the classroom and find them useful. IBM PC compatible computers are used for undergraduate instruction in 75% of the institutions surveyed. The use of computer packages rather than programming languages is perceived to be more important for the student to learn. Productivity packages such as word processing, math packages, and spreadsheets are the most important packages.

In contrast, video in the classroom is not as well accepted. Nearly half of the respondents have not used video in the classroom. Most faculty have not taped their lectures or engaged in distance learning through a video network. Electronic mail has become commonplace among engineering faculty and is used to some extent to communicate with undergraduate students.

The respondents to this survey felt strongly about the need for change in the engineering curriculum. Of course, that is not surprising since the sample for this survey was drawn from those faculty who are typically the agents for change at their respective institutions. There was less unanimity for the specific changes which are desirable. The top factors here are those that have traditionally driven engineering education such as problem solving and design. There is a consensus that the pipeline problem needs to be addressed and that communication technology is a possible way to reach into the K-12 stream.

The faculty that responded found teaching to be an important priority in their universities. Only half agreed with the statement that "Research is the only way to get promoted at my institution." However, 40% of the respondents were more than twenty years from their highest degree and have passed the stage in their careers where tenure and promotion are their primary concern.

This survey revealed no dramatic results from what we perceive to be the current state in engineering education. It does show that computers, in particular, are playing an important role in the education of our undergraduate students. Less clear is the influence of communication technology. Distance education has found an important role in graduate education, but currently appears to have little impact on the undergraduate.

REFERENCE

Survey Pro, Apian Software, San Mateo, California, 1993.

Survey Results

The Classroom

1. Which of the following do you agree with?

1 My classes are strictly lecture	6.1 %
2 My classes are mostly lecture and some discussion	67.8
3 My classes are about half lecture and half discussion	23.5
4 Other	6.3
5 No answer	0.9

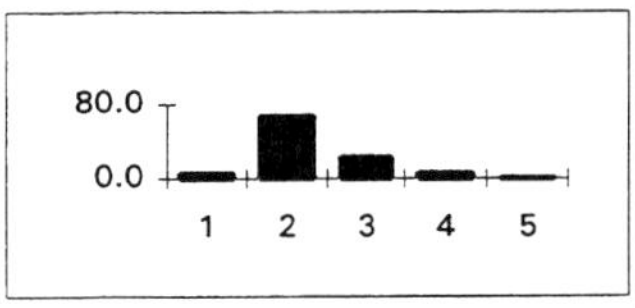

2. In the courses in your department, what are the different types of instruction that are used?

1 Traditional lecture	95.7 %
2 Traditional laboratory	88.8
3 Teaming experiences	60.0
4 Industrial linkages/experiences	32.4
5 Cross-disciplinary experiences	32.3
6 Just-in-time engineering education	8.8
7 Distance education	18.4
8 Other (Capstone Design, etc.)	41.0
9 No answer	0.9

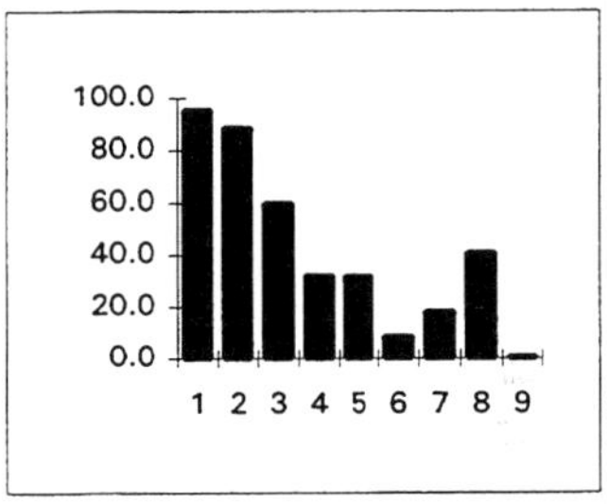

3. I typically use the following classroom presentation aids

1 Chalkboard	89.6 %
2 Transparencies with overhead projector	85.2
3 Video projector/VCR player	23.6
4 Computer and LCD Panel (black & white)	19.1
5 Computer and LCD Panel (color)	16.6
6 Computer presentation at each student's desk	9.7
7 Other	7.9
8 No answer	0.0

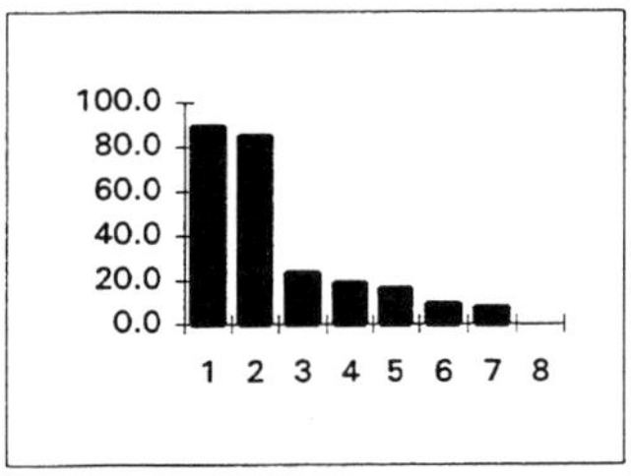

4. What do most of your colleagues at your institution use for presentation aids?

1 Chalkboard	89.5 %
2 Transparencies with overhead projector	81.5
3 Video projector/VCR player	21.5
4 Computer and LCD Panel (black & white)	26.5
5 Computer and LCD Panel (color)	19.6
6 Computer presentation at each student's desk	7.8
7 Other	6.1
8 No answer	2.6

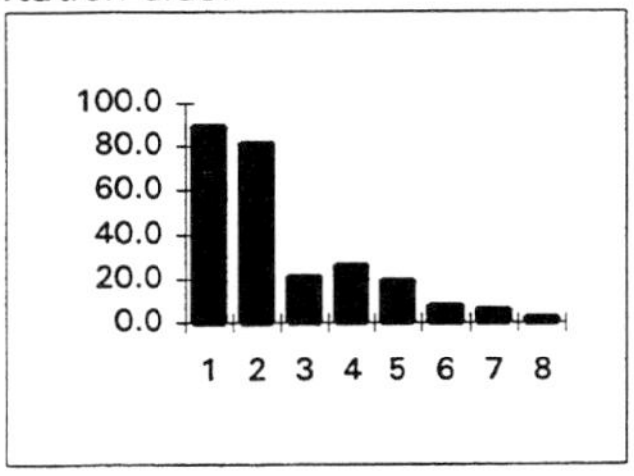

5. Which of the following materials are used in your classes?

1 Textbook	83.9 %
2 Course notes, handed out	75.0
3 Periodic tests given on several pages	70.5
4 Homework sheets	68.9
5 Copies of lecture transparencies or slides	54.8
6 Remedial exercices	21.7
7 Library materials	40.8
8 Other	10.5
9 No answer	0.0

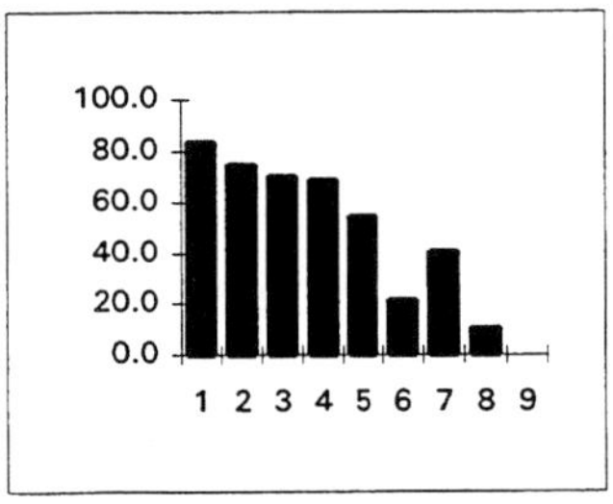

Table 1. Survey Results.

The Laboratory

6. Are laboratories in your curriculum separated or directly linked to the lecture?

1 Both	43.5 %
2 Tied to the lecture	34.2
3 Separate from the lecture	20.8
4 No answer	1.7

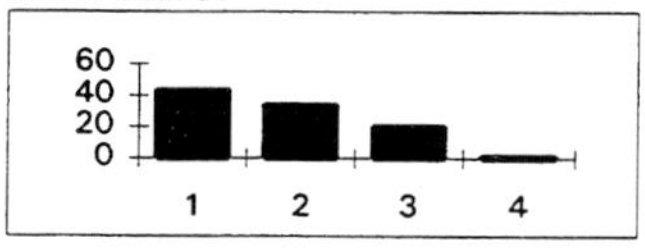

7. Students clearly see the relevancy of the laboratories

1 Strongly agree	25.3 %
2 Agree	60.8
3 Disagree	13.1
4 Strongly disagree	0
5 No answer	0.85

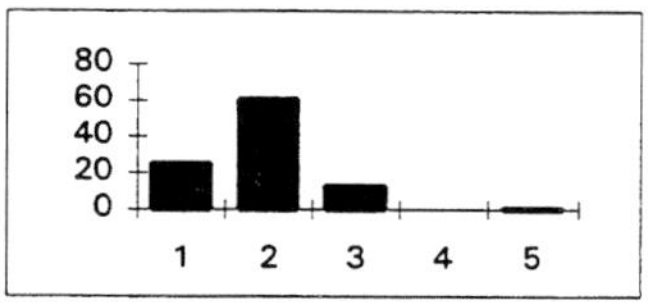

8. Laboratories should, in some way, be linked to professional practice

1 Strongly agree	27.3 %
2 Agree	60.7
3 Disagree	11.3
4 Strongly disagree	0.85
5 No answer	0

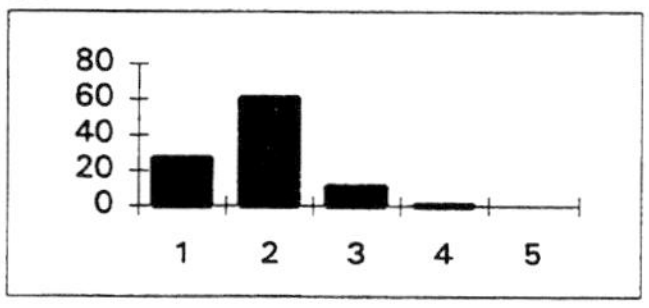

9. Our laboratory equipment is old and outdated

1 Strongly agree	12 %
2 Agree	33.8
3 Disagree	40.2
4 Strongly disagree	13.2
5 No answer	0.85

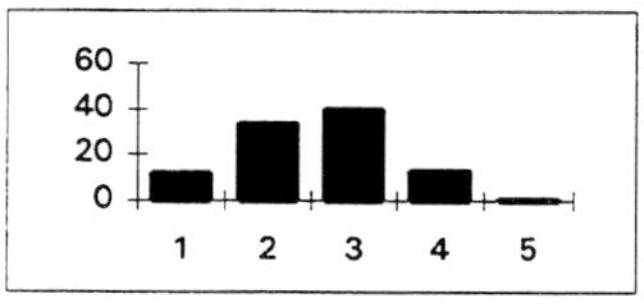

10. Computer-Based Training (CBT) can be an effective means for instruction in the laboratory

1 Strongly agree	23.4 %
2 Agree	55.7
3 Disagree	20
4 Strongly disagree	0
5 No answer	0.9

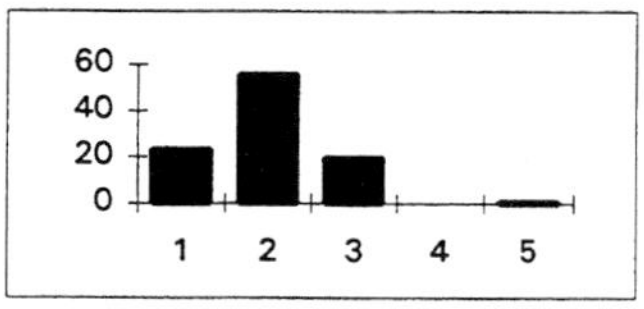

11. Laboratories should always have the latest state-of-the-art equipment

1 Strongly agree	8.7 %
2 Agree	48.9
3 Disagree	39.8
4 Strongly disagree	2.6
5 No answer	0

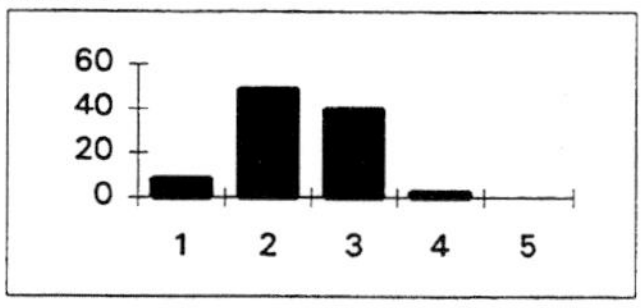

12. It is practical to always have state-of-the-art equipment

1 Strongly agree	1.7 %
2 Agree	13.2
3 Disagree	66.2
4 Strongly disagree	19
5 No answer	0

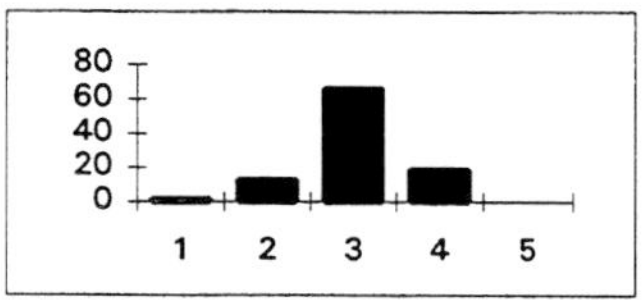

Table 1. Survey Results. (continued)

Computers and Other Electronic Media in Instruction

Current Use of Computers in the Classroom:

13. Have you ever used a computer in the classroom?

	%
1 Yes	75
2 No	24.2
3 No answer	0.85

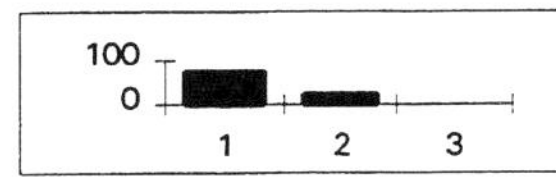

14. Do you think computers are useful in the classroom?

	%
1 Yes	77.5
2 No	2.7
3 Uncertain	19
4 No answer	0.85

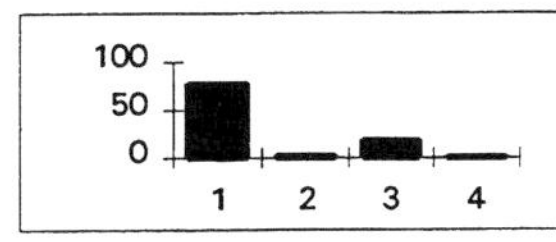

15. For undergraduates, do you predominantly use:

	%
1 PCs	75.6
2 Macs	22.9
3 Unix workstations	17.6
4 Others (Mainframe computers, etc.)	10.4
5 No answer	4.3

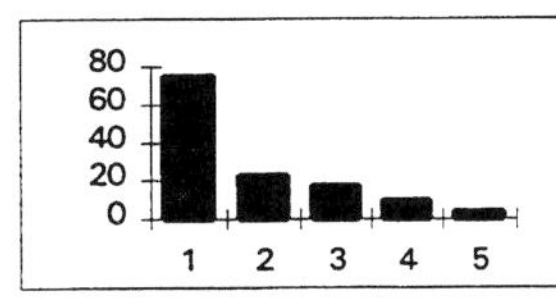

16. What approximate percentage of your students have their own computers?

	%
1 0%	0
2 20%	23.4
3 40%	28
4 60%	25
5 80%	8.7
6 100%	2.65
7 Other	7.95
8 No answer	4.35

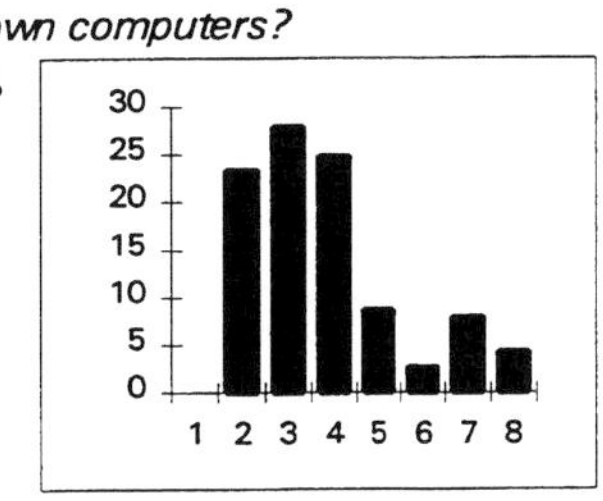

17. Engineering students should learn a wide variety of programming languages.

	%
1 Strongly agree	4.4
2 Agree	30.2
3 Disagree	50.5
4 Strongly disagree	7
5 No answer	3.45

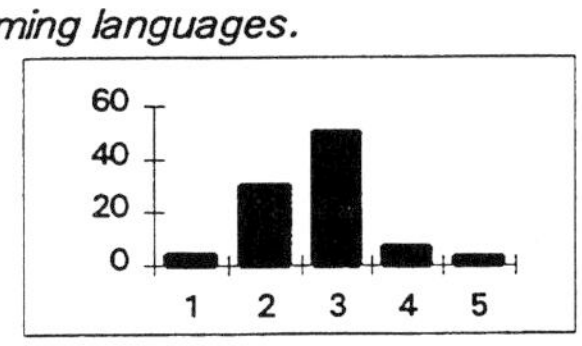

Table 1. Survey Results (continued).

18. Check which languages you feel are important:

1	Fortran	59.9 %
2	C	68.5
3	C++	45.6
4	Pascal	25.3
5	Ada	11.4
6	Basic	17.3
7	Prolog	6.25
8	Smalltalk	5.25
9	Visual Basic	7.8
10	Cobol	0.9
11	Other	16.6
12	No answer	4.4

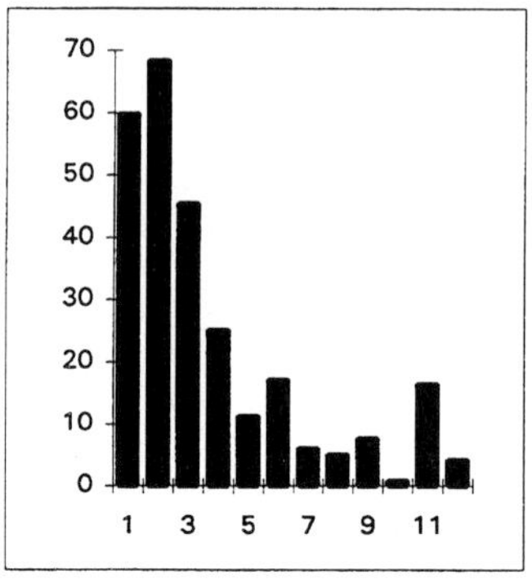

19. Is it more important for engineering students to learn programming languages or to understand how to use computer packages?

1	Computer packages	65.3 %
2	Programming languages	47.7
3	No answer	2.6

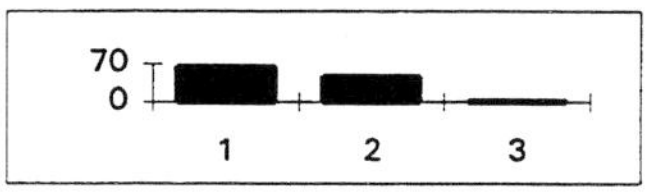

20. Check which packages you feel engineering students should learn:

1	WordPerfect/Word/other word processors	91.2 %
2	Math packages: MATlab/Mathematica/etc.	78.4
3	Excel/Quattro/other spreadsheets	75.7
4	Simulation packages such as SPICE or ASPEN	34.7
5	DBase/other database packages	28.8
6	Other (Presentation graphics)	35.4
7	No answer	3.45

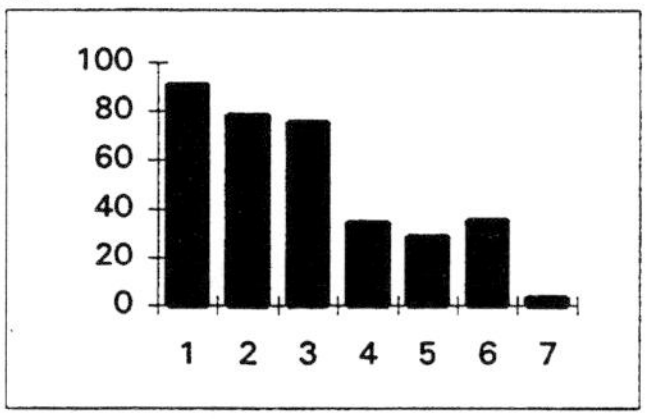

Electronic Media

21. Do you use video presentation in the classroom?

1	Never	46.1 %
2	Sometimes	48
3	Frequently	5.15
4	Always	0
5	No answer	0.9

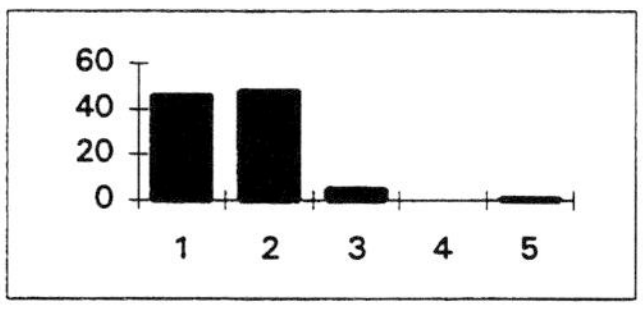

22. If yes, where do these videos come from?

1	Commercial distributors	28.6 %
2	Other universities	12.3
3	Internal or state system video	6.95
4	NTU	7.05
5	Other	22.4
6	No answer	49.7

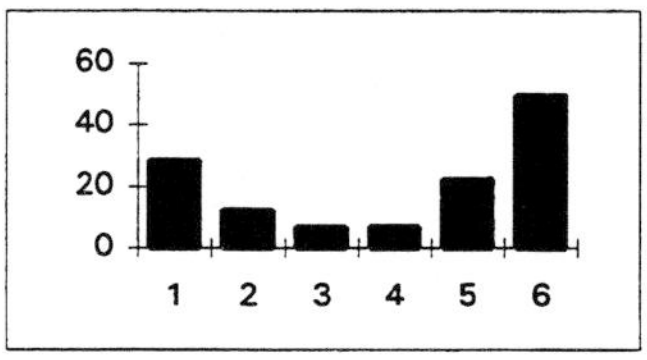

Table 1. Survey Results (continued).

23. Do you video tape your own lectures?

1 Never	68.7 %
2 Sometimes	28.8
3 Frequently	1.75
4 Always	0.85
5 No answer	0

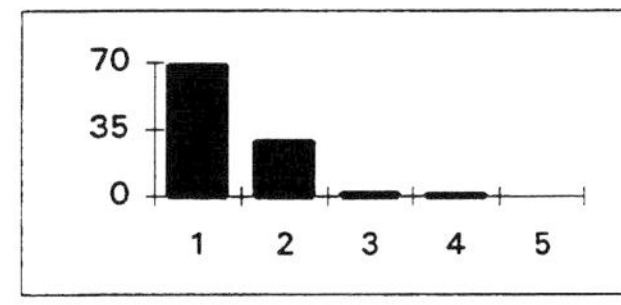

24. If you record your lectures, can students view the lectures at another time?

1 Yes	24.7 %
2 No	18.2
3 No answer	57.2

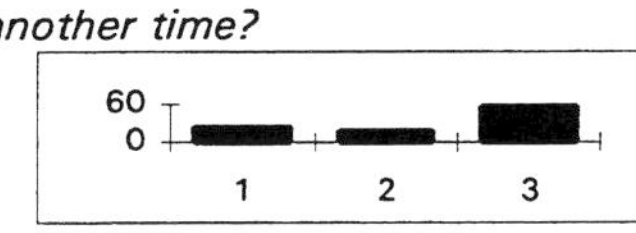

25. Do you broadcast your lectures?

1 Yes	8.8 %
2 No	74.8
3 No answer	16.4

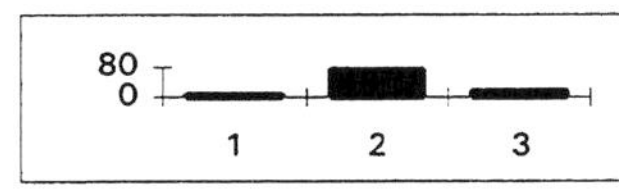

26. If you or your colleagues broadcast your lectures, where are they broadcast?

1 Within the state system	21.8 %
2 Within my institution	8.8
3 National Technological University (NTU)	15
4 Global system analysis and simulation network	0.85
5 Other	6.95
6 No answer	62.5

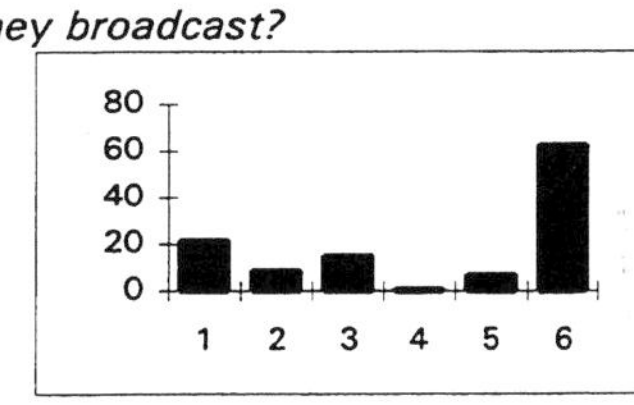

27. Do you use electronic mail in communicating with your colleagues?

1 Yes	84.7 %
2 No	15.3
3 No answer	0

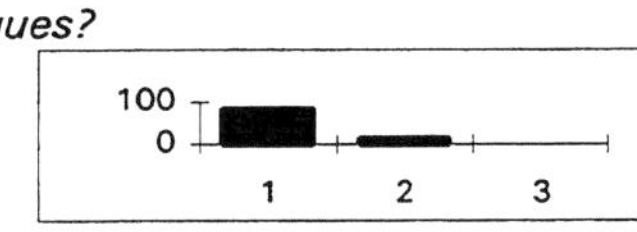

28. Do you communicate with students in your undergraduate classes?

1 Yes	54.2 %
2 No	39
3 No answer	6.8

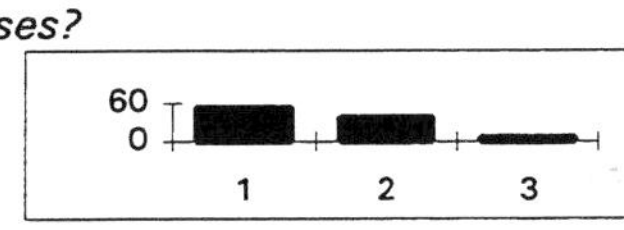

29. Would computer-assisted workgroups be an assistance in the management of group projects?

1 Never	3.4 %
2 Sometimes	45.8
3 Frequently	22
3 Always	6.8
4 No answer	22

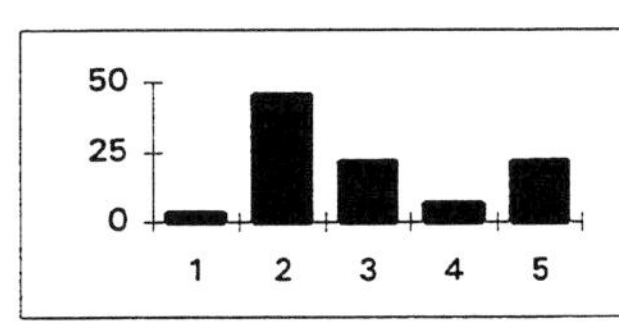

Table 1. Survey Results (continued).

The Engineering Curriculum

30. Do you think that there is any need for change in engineering education?

1	Yes	85.4 %
2	Uncertain	11.1
3	No	2.7
4	No answer	0.9

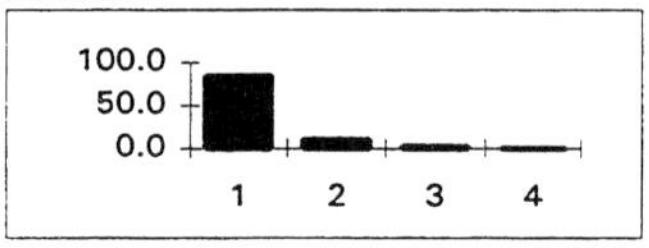

31. ABET is a barrier to innovation in the engineering curriculum.

1	Strongly agree	18.5 %
2	Agree	33.9
3	Disagree	35.6
4	Strongly disagree	13.6
5	No answer	5.1

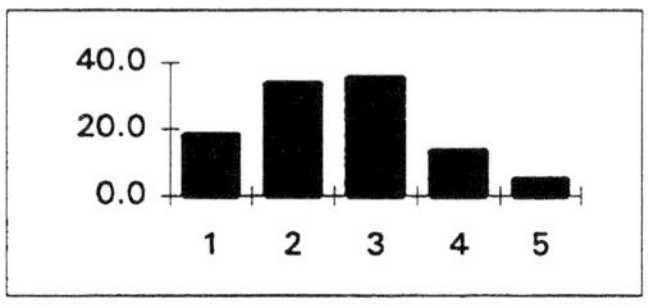

32. In your opinion, what are the most important curriculum-related issues facing engineering education?

1	Making the curriculum relevant to current engineering practice.	69.0 %
2	Creating innovative learning methods.	48.8
3	Keeping students interested in courses.	40.8
4	Ensuring that students "know the basics."	52.0
5	Producing students that are widely accepted in industry.	41.7
6	Maintaining a flow of students into engineering schools.	13.8
7	Providing the basis for lifelong learning.	86.4
8	Presenting the history of engineering to students.	3.5
9	Giving students a broad background in solving problems.	72.9
10	Other.	15.1
11	No answer.	0.0

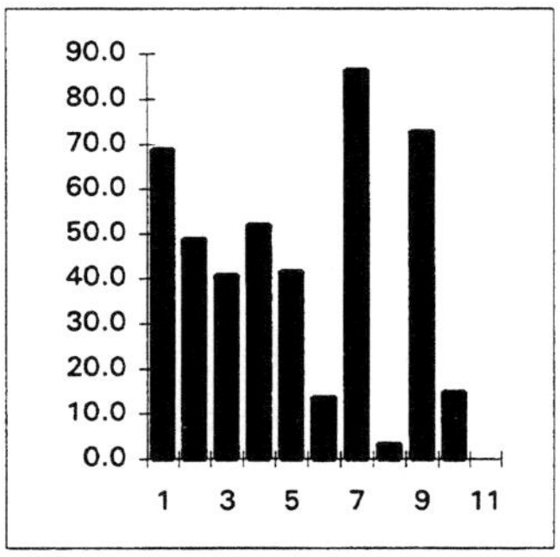

33. Which of the following facets of engineering education are strongly influenced by the introduction of technology into the engineering curriculum?

1	Mathematics	39.4 %
2	Problem solving	75.2
3	Analytical techniques	41.1
4	Learning basic engineering science	36.7
5	Computer methods and programming	54.2
6	Learning design methodologies	61.1
7	Capability for lifelong learning	35.6
8	Other	12.4
9	No answer	4.3

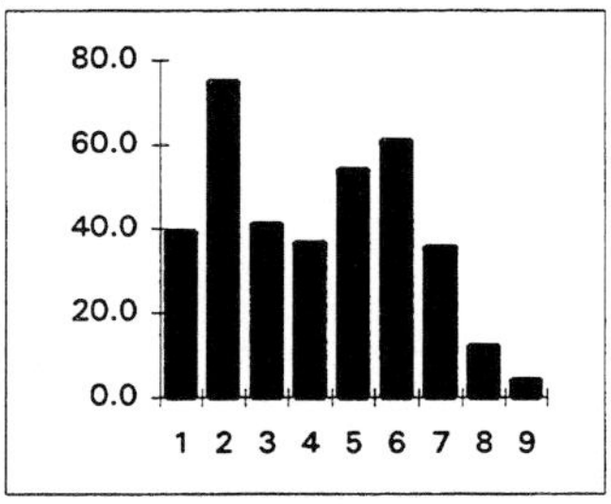

34. The average engineering educator believes that computer and telecommunications technology will change the way we teach.

1	Strongly agree	13.1 %
2	Agree	53.9
3	Disagree	30.5
4	Strongly disagree	1.8
5	No answer	0.9

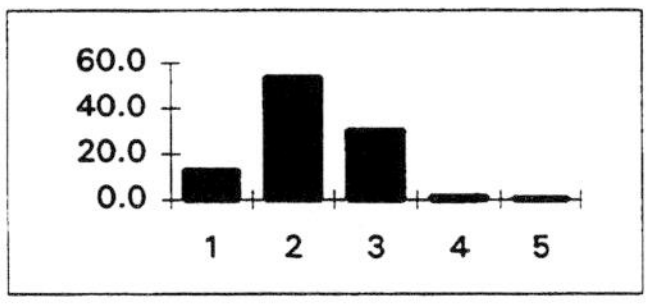

Table 1. Survey Results (continued).

35. *What do you perceive as the drivers for change?*

		%
1	Decreasing cost	36.0 %
2	Student demand	36.5
3	Improved learning environments	52.2
4	Improved information transfer rate	41.8
5	Improved lecture efficiency	27.9
6	Industrial demand	33.6
7	Administrative demand	12.0
8	Home/Dormitory learning	22.5
9	The economic cost of higher education	28.8
10	ABET	15.9
11	Other	16.6
12	No answer	4.3

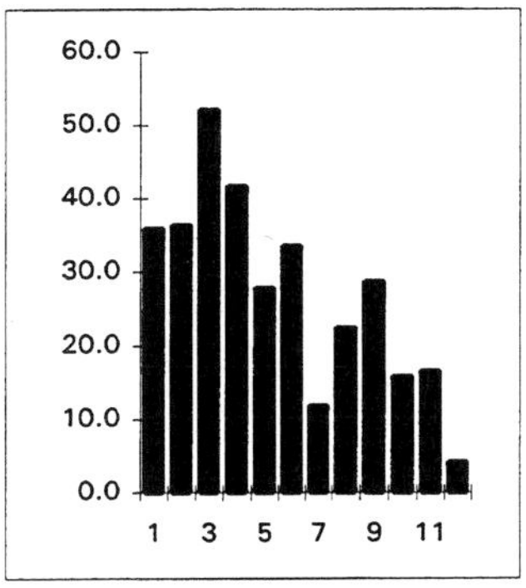

36. *Engineering societies play an important role in education of the students through student branches of professional societies.*

		%
1	Strongly agree	5.3 %
2	Agree	40.7
3	Disagree	42.8
4	Strongly disagree	8.7
5	No answer	2.6

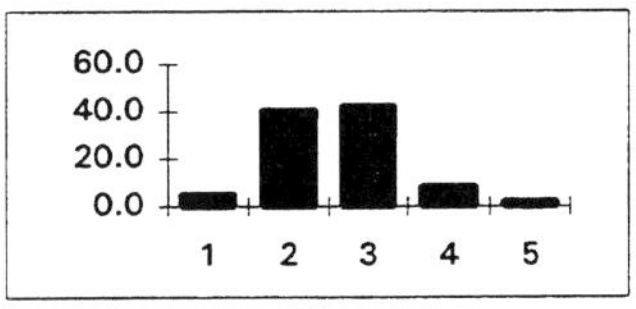

37. *Is it important for students to spend part of their undergraduate years in an industrial experience?*

		%
1	Yes	57.6 %
2	Uncertain	29.4
3	No	13.1
4	No answer	0.0

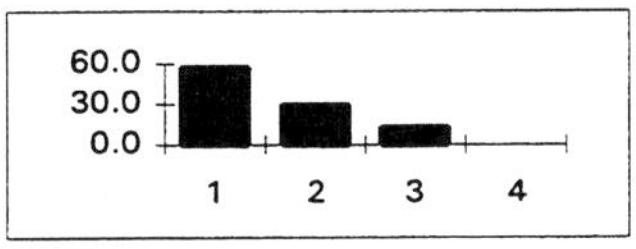

38. *Can an industrial experience be provided without sending students to actually work in industry?*

		%
1	Never	13.1 %
2	Sometimes	75.6
3	Frequently	8.8
4	Always	0.9
5	No answer	0.9

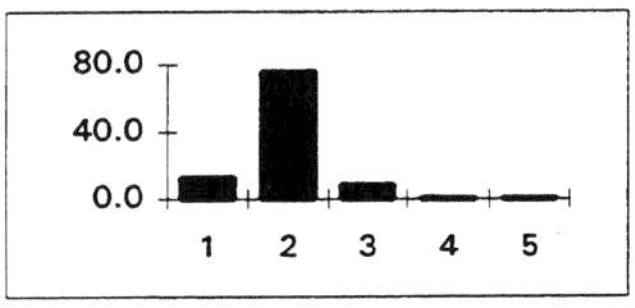

40. *Does your institution have any formal education component that gives students an industrial experience?*

		%
1	Yes	52.4 %
2	No	41.5
3	No answer	6.2

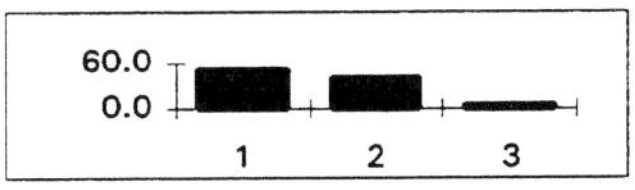

42. *How important is it for engineering colleges to reach into K-12?*

		%
1	Very important	56.6 %
2	Somewhat important	30.5
3	Slightly important	16.9
4	Not needed	0.0
5	No answer	0.0

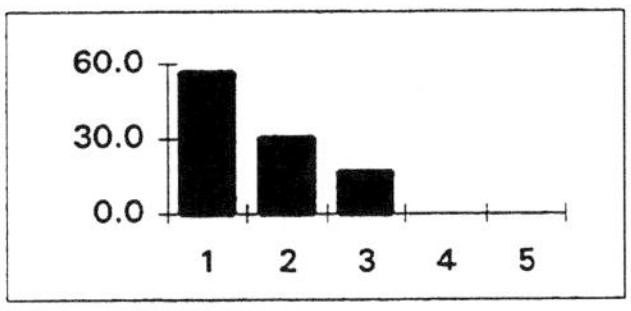

Table 1. Survey Results (continued).

43. If important, what types of computer and communications technology might help?

1	Giving old computers to high schools	25.1 %
2	Providing computer services to high schools	25.3
3	Engineering college lectures transmitted to high schools	12.4
4	Giving high schools video tapes of faculty lectures	13.2
5	Providing ways for univ. students to interact with K-12 students.	71.2
6	Faculty visits to give lectures and/or demonstrations	67.8
7	Providing instructional materials to K-12 teachers	54.2
8	Other	20.2
9	No answer	13.1

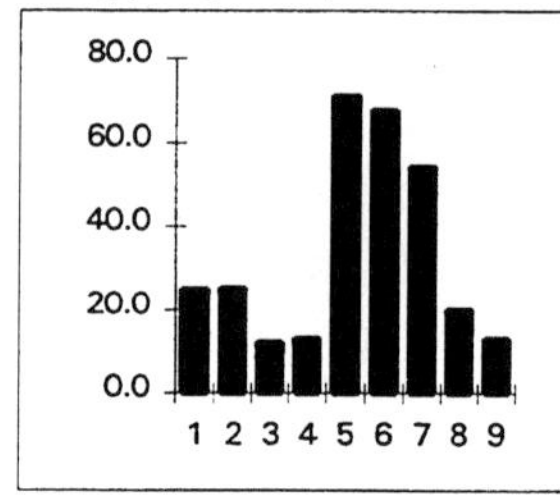

44. Technology can help disadvantaged students.

1	Strongly agree	35.0 %
2	Agree	48.6
3	Disagree	6.0
4	Strongly disagree	0.9
5	No answer	3.4

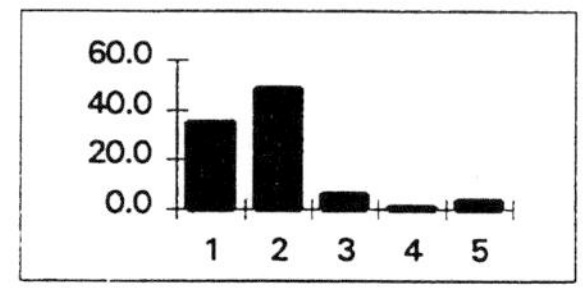

Table 1. Survey Results (continued).

Teaching and Research

46. I like teaching.

		%
1	Strongly agree	62.5 %
2	Agree	34.9
3	Disagree	0.9
4	Strongly disagree	0.0
5	No answer	1.7

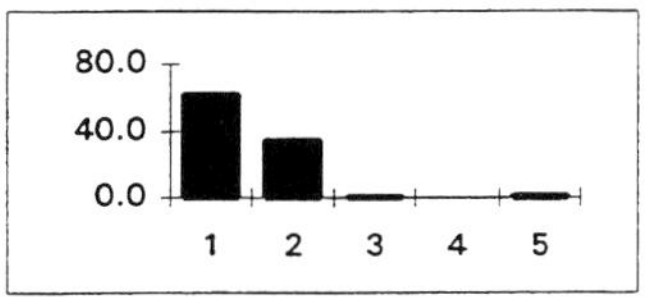

47. Teaching is not very important at my institution.

		%
1	Strongly agree	8.0 %
2	Agree	20.1
3	Disagree	50.3
4	Strongly disagree	20.8
5	No answer	0.9

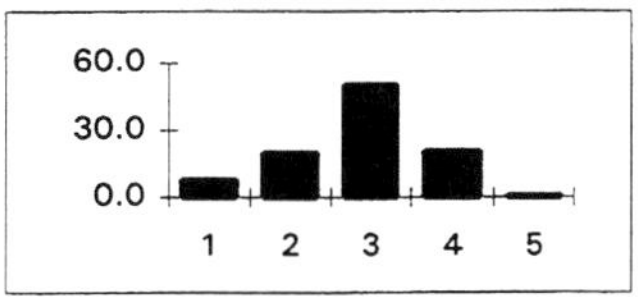

48. Research is the only way to get promoted at my institution.

		%
1	Strongly agree	24.4 %
2	Agree	32.2
3	Disagree	30.5
4	Strongly disagree	7.9
5	No answer	3.4

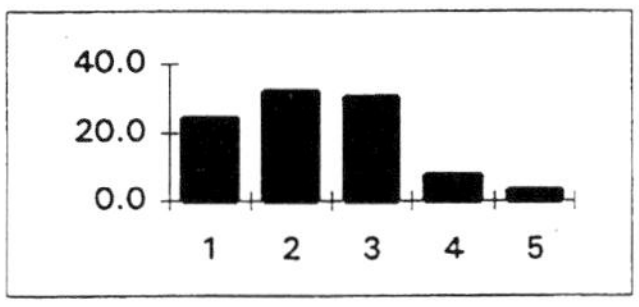

Information about You and Comments:

49. How many years since you received your highest degree?

		%
1	Less than 5 years	19.4 %
2	5 to 10 years	16.3
3	10 to 20 years	22.7
4	20 and more	40.9
5	No answer	1.7

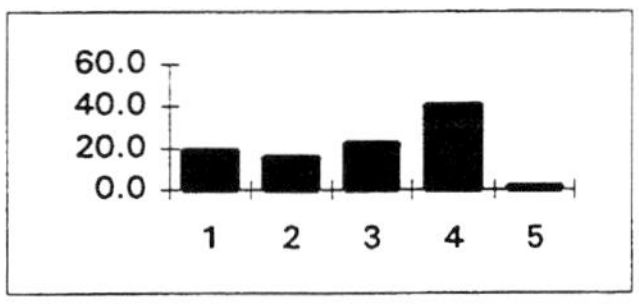

50. What is your position?

		%
1	Instructor	2.6 %
2	Assistant Professor	21.8
3	Associate Professor	26.1
4	Professor	40.0
5	Dean, Associate Dean or Assistant Dean	4.3
6	Other	14.9
7	No answer	0.9

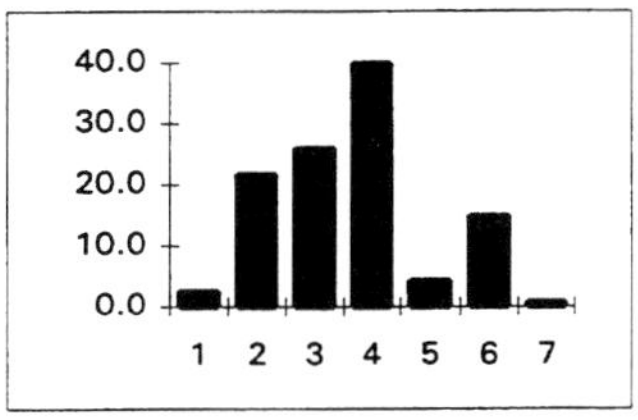

Table 1. Survey Results (continued).

III. THE TRADEOFF PARADIGMS

A. INTRODUCTION TO THE TRADEOFF PARADIGM

During the first six months of the study, a series of Kepner-Tregoe (KT) tradeoff analysis skeleton forms were created. The rationale for using the KT methodology was that it was that (1) we needed a quantitative way to deal with understanding the variety of alternatives that were possible in the different areas being studied, (2) we needed a tool that was relatively easy to understand, and (3) we needed a tool for which computer tools already existed. The KT tradeoff method fit all these criteria; a computer implementation of the analysis created by Apian Software was selected (Decision Pad) and purchased. We felt that the software provided an excellent way of keeping track of the decisions that we studied. Of particular value was the capability of projecting the computer screen to a group analyzing a problem. We used this capability during our focus group meetings.

To begin to acquaint our consultant group with KT concepts, we built a group of 7 decisions for the group to examine. We created, with the data entry software provided by Apian, a set of 7 tradeoffs ballots that we mailed to our consultants for review. These ballots were also available on computer disks which were sent to the consultant group. An example of the type of paper ballot used is shown in Table 2. The consultant group completed these ballots and returned them to us for tabulation. The results of this tabulation are shown later in this section. In our view, a primary function of having the consultant group fill out these KT ballots was to acquaint the group with the use of this tool. Our intent from the outset was to use the KT analysis method for shaping the thinking of the focus group at later meetings. Hence, this initial exercise helped make the use of the tools familiar to the group. In addition, the analysis of the resulting ballots provided excellent thinking and background information in the 7 target areas selected. Table 3 shows the areas of the tradeoffs scenarios:

CURRICULUM PARADIGMS
COMPUTER-FACILITATED EDUCATIONAL EXPERIENCES
DELIVERY OF EDUCATIONAL MATERIALS
THE CLASSROOM OF TOMORROW
THE ENGINEERING EDUCATION LABORATORY OF TOMORROW
LEARNING OUTSIDE THE CLASSROOM AND LABORATORY
THE ROLE OF MULTIMEDIA

Table 2. The Seven Tradeoff Scenarios Ballot

Ballot for Curriculum Delivery

Input from ________________________________ Date __________

Circle entry for Student Acceptance:	
Lectures	V-High, High, Medium, Low, V-Low
Lecture + Technology	V-High, High, Medium, Low, V-Low
Lectures + Rec/Lab	V-High, High, Medium, Low, V-Low
Lec+Rec/Lab +Technology	V-High, High, Medium, Low, V-Low
Cooperative Workgroups	V-High, High, Medium, Low, V-Low
Anchored Instruction	V-High, High, Medium, Low, V-Low
Case Studies	V-High, High, Medium, Low, V-Low
Integrated Curriculum	V-High, High, Medium, Low, V-Low
Tutorials	V-High, High, Medium, Low, V-Low
Internships	V-High, High, Medium, Low, V-Low
Circle entry for Faculty Acceptance:	
Lectures	V-High, High, Medium, Low, V-Low
Lecture + Technology	V-High, High, Medium, Low, V-Low
Lectures + Rec/Lab	V-High, High, Medium, Low, V-Low
Lec+Rec/Lab +Technology	V-High, High, Medium, Low, V-Low
Cooperative Workgroups	V-High, High, Medium, Low, V-Low
Anchored Instruction	V-High, High, Medium, Low, V-Low
Case Studies	V-High, High, Medium, Low, V-Low
Integrated Curriculum	V-High, High, Medium, Low, V-Low
Tutorials	V-High, High, Medium, Low, V-Low
Internships	V-High, High, Medium, Low, V-Low
Circle entry for Industry Acceptance:	
Lectures	V-High, High, Medium, Low, V-Low
Lecture + Technology	V-High, High, Medium, Low, V-Low
Lectures + Rec/Lab	V-High, High, Medium, Low, V-Low
Lec+Rec/Lab +Technology	V-High, High, Medium, Low, V-Low
Cooperative Workgroups	V-High, High, Medium, Low, V-Low
Anchored Instruction	V-High, High, Medium, Low, V-Low
Case Studies	V-High, High, Medium, Low, V-Low
Integrated Curriculum	V-High, High, Medium, Low, V-Low
Tutorials	V-High, High, Medium, Low, V-Low
Internships	V-High, High, Medium, Low, V-Low

Table 3. A Portion of an Example Ballot

We specifically wanted to cover the seven areas listed in Table 3. First, we wondered whether the curriculum as presented in traditional classroom and laboratory experiences found in current Colleges of Engineering would be replaced by some type of new teaching experience? From the array of possible alternative ways of delivering the curriculum, which might be likely candidates for more intensive study? Second, since this study was concerned with the impact of computers and communications technology on engineering education, we attempted to discern which educational experiences could be facilitated by computer were most important. Third, since computers and networks will permit delivery of education materials, we wondered which methods might prove most useful? Fourth, we were interested in what the classroom of tomorrow might contain. Likewise, fifth, we wished to explored the impact of technology on the laboratory of tomorrow. Learning outside the classroom seemed to be an important dimension to learning that technology should facilitate. Finally, we explicitly included a tradeoff about multimedia since it has the potential to influence most of the technology facilitated education approaches studied.

B. KEPNER-TREGOE TRADEOFF ANALYSES

The next paragraphs describe the use of the computer tool Decision Pad created by Apian Software. To aid in the explanation, we will use a familiar example — that of trying to select which automobile to purchase. This example is the only one given that doesn't directly relate to the materials studied — it is presented simply because the material is easily understood. The intent is to have the reader completely understand the tradeoff methodology prior to attempting to interpret the results of the study that utilize the tradeoff methodology. Table 1 displays a sample Kepner-Tregoe tradeoff analysis for a three alternative problem. Imagine that you are using the tool to decide whether to purchase "Car 1", "Car 2" or "Car 3" - in this case three different automobiles. Our first step was to select the alternatives among which we want to choose. Next, we chose a set of criteria which will help us distinguish between the alternatives. For example, some clear criteria might be: cost, fuel efficiency, luxury, sporty, longevity, repair costs, trade-in value, etc. As you can clearly see some of these criteria are important to one person but not to another. The tradeoff tool permits adaptation to each person's preferences through a weighting of each criteria. The alternatives, criteria and weights are formed into a matrix as follows:

		Alternatives		
Criteria	Weights	Car 1	Car 2	Car 3
Cost	W_1	?	?	?
Fuel Efficiency	W_2	?	?	?
Luxury	W_3	?	?	?
Sporty	W_4	?	?	?
Repair Costs	W_5	?	?	?

Table 4. Example Kepner-Tregoe Tradeoff Matrix

Suppose we insert weights to emphasize criteria that are important to us and create scales for rating each alternative. For example, we might fill out the matrix as shown below if we are interested in a luxurious, sporty car. Then, we would fill in each of the cells for each car according to the scale that is provided for each criteria. These criteria scales could be, for example, from 'high' to 'low' for Cost, with 'low' being best, or from 'low' to 'high' for luxury and sporty with 'high' being best. These types of quantitative scales can be created with any desired, yet meaningful, names to rate the criteria for each alternatives. A feature of the program is the conversion of the qualitative scale into a quantitative value according to the definitions given by the user.

		Alternatives		
Criteria	Weights	Car 1	Car 2	Car 3
Cost	5	low	medium	high
Fuel Efficiency	2	high	medium	low
Luxury	8	high	medium	low
Sporty	10	high	low	high
Repair Costs	5	medium	medium	medium

In the above example, the ratings inserted in each cell could be secured from a car rating journal and from the prospective owner's own ideas (e.g., is it sporty?). To illustrate the methodology, for the scale shown, we could assign a numerical score, for example: very high = 4, high = 3, medium = 2, low = 1 for the Fuel Efficiency, Luxury, and Sporty categories and the inverse for the Cost and Repair Costs categories (i.e., low = 4. . . very high = 1). Then, if we multiply the weights for each criteria times the numerical scales we would get:

		Alternatives		
Criteria	Weights	Car 1	Car 2	Car 3
Cost	5	20+	10	15+
Fuel Efficiency	2	6	6	2
Luxury	8	24++	16--	16+
Sporty	10	30++	10-	30++
Repair Costs	5	15	15	15

If we now sum the columns we get:

		95	57	78

As shown, for our choice of weights, Car 1 is the best. It is the most sporty and luxurious of the group. The +'s and -'s show where there was the most impact in the decision, both positively and negatively. A different set of weights would have yielded a completely different answer to our tradeoff example.

Decision Pad, the software tool that was used in the study, does the basic type of calculations shown above but is considerably more sophisticated. Many types of scales are provided, including the ability for the user to create their own scales. The scoring is automatic and graphical setting of weights is provided. In addition, various analytical tools such as sensitivity analysis is included. For the study, perhaps the most important feature that Decision Pad added was the capability of collecting ballots from multiple evaluators who could complete tradeoffs of their own including their individually set weights. The program then amalgamates multiple ballots and provides a summary. The summaries from multiple ballots are presented in this section. In order to make the results somewhat clearer, we took the results from Decision Pad and entered the results into Excel matrices and create bar graphs from these matrices.

C. THE SEVEN TRADEOFF SCENARIOS

The instructions shown in italics below were given to individuals that participated in the initial tradeoff analyses. These individuals were members of the study group and others who acquired our tradeoff from our Internet FTP[2] computer directory. It should be noted at the outset that these tradeoffs were created as tests and as prototypes to help the study team understand how to utilize tradeoff tools. They are not perfect contrasts and have redundancies in them. The reader perhaps will be able to select more appropriate and better contrasts after reading this material. However, the study team did not refine these tradeoffs, instead choosing to move to focus groups with specific tradeoffs for facilitating the understanding of the particular issues in the focus groups. Even with imperfections, however,

[2]file transfer protocol

these seven tradeoffs contain quite interesting results. The imperfections tend to wash out as the ratings are amalgamated.

> *The purpose of the following tradeoff analyses is to provide a mechanism to evaluate the importance of various criteria that distinguish between alternatives posed in each tradeoff. Alternatives in each tradeoff are to be contrasted using the criteria given, rated using the scales provided. The outcome of this effort will be a ranking of the importance of each alternatives based on the criteria selections made. These same tradeoffs will be given to each person who participates and a composite team score will be the final result of this part of the study.*

Tables 5 through 11 display the seven tradeoff matrices supplied to each participant in the study. An explanation of each of the tradeoffs is given next. The general information given to each person who completed the tradeoff is given first in italics, followed by an explanation of each term used.

1. Curriculum Paradigms

> ***The purpose of this tradeoff analysis is to secure a collective viewpoint of which curriculum enhancements might be plausible and which factors influence the decisions to favor particular alternatives.***

This tradeoff analysis scenario contrasts:

Traditional teaching and learning paradigms currently employed in colleges of engineering. This is the current situation as it exists in the early to mid-1990s; that is, the use of lecture and laboratories. Assume typical lecture sizes of 30 to 40 students and laboratories of students that cluster around instruments.

Traditional teaching and learning paradigms enhanced by technology. This alternative is the current alternative enhanced by technology such as the use of computer projectors, etc. For example, use of email for enhancing student-to-student and student-to-faculty communications, using computer tutors or classroom projectors would be examples of simple enhancements to the way we do things now.

Reorganizing the curriculum in a just-in-time manner. This alternative refers to an experimental curriculum delivery paradigm designed to introduce new information only when it is needed, not just-in-case it is needed.

Experiential learning. Experiential learning refers to making student centered learning involve acquiring experiences. In engineering education, this might be extensive design experiences. At some institutions, experiential learning is manifested in dissection experiences. At others, analysis and problem solving and design. The intent is to give students more control of the learning experience by providing longer experiences which require synthesis of knowledge rather than rote problem solving.

Distance education. The final alternative is distance education; i.e., delivery of lectures over satellite or via some other medium (video tape, electronic mail, etc.).

Distance education exists at many institutions. Would it improve our undergraduate curriculum to have this type of delivery mechanisms used more broadly?

Tradeoff Matrix for Curriculum Paradigms

	WEIGHT	Traditional	Traditional+	Experiential Learning	Just-in-Time
Student Acceptance	1	?	?	?	?
Faculty Acceptance	1	?	?	?	?
Industrial Accept.	1	?	?	?	?
Prep. for Profession	1	?	?	?	?
Cost of Implementing	1	?	?	?	?
Cost of Running	1	?	?	?	?
Process Knowledge	1	?	?	?	?
Problem-Solving Abil.	1	?	?	?	?
Broader Knowledge	1	?	?	?	?
Deeper Knowledge	1	?	?	?	?
Improve Design Abil.	1	?	?	?	?
Learning Team Work	1	?	?	?	?
Learn Fundamentals	1	?	?	?	?
SCORE	13	?	?	?	?
RANK		?	?	?	?

	WEIGHT	Distance Education
Student Acceptance	1	?
Faculty Acceptance	1	?
Industrial Accept	1	?
Prep. for Profession	1	?
Cost of Implementing	1	?
Cost of Running	1	?
Process Knowledge	1	?
Problem-Solving Abil.	1	?
Broader Knowledge	1	?
Deeper Knowledge	1	?
Improve Design Abil.	1	?
Learning Team Work	1	?
Learn Fundamentals	1	?
SCORE	13	?
RANK		?

Table 5.

Tradeoff Matrix for Computer-Facilitated Education

	WEIGHT	Symbolic Math Packages	Computa-tional Math Packages	CAD Tools	Conventional CBT
Student Acceptance	1	?	?	?	?
Faculty Acceptance	1	?	?	?	?
Industry Acceptance	1	?	?	?	?
Prep. for Profession	1	?	?	?	?
Cost of Implementing	1	?	?	?	?
Cost of Running	1	?	?	?	?
Process Knowledge	1	?	?	?	?
Problem-Solving Abil.	1	?	?	?	?
Broader Knowledge	1	?	?	?	?
Deeper Knowledge	1	?	?	?	?
Improve Design Abil	1	?	?	?	?
Learning Team Work	1	?	?	?	?
Learn Fundamentals	1	?	?	?	?
SCORE	13	?	?	?	?
RANK		?	?	?	?

	WEIGHT	Intelligent Tutor System	Computer Simulation	Anchored Instruction	Virtual Realities
Student Acceptance	1	?	?	?	?
Faculty Acceptance	1	?	?	?	?
Industry Acceptance	1	?	?	?	?
Prep. for Profession	1	?	?	?	?
Cost of Implementing	1	?	?	?	?
Cost of Running	1	?	?	?	?
Process Knowledge	1	?	?	?	?
Problem-Solving Abil.	1	?	?	?	?
Broader Knowledge	1	?	?	?	?
Deeper Knowledge	1	?	?	?	?
Improve Design Abil.	1	?	?	?	?
Learning Team Work	1	?	?	?	?
Learn Fundamentals	1	?	?	?	?
SCORE	13	?	?	?	?
RANK		?	?	?	?

Table 6.

Tradeoff Matrix for Computer-Facilitated Education

	WEIGHT	Laboratory Simulation	Mathematics Visualization
Student Acceptance	1	?	?
Faculty Acceptance	1	?	?
Industry Acceptance	1	?	?
Prep. for Profession	1	?	?
Cost of Implementing	1	?	?
Cost of Running	1	?	?
Process Knowledge	1	?	?
Problem-Solving Abil.	1	?	?
Broader Knowledge	1	?	?
Deeper Knowledge	1	?	?
Improve Design Abil.	1	?	?
Learning Team Work	1	?	?
Learn Fundamentals	1	?	?
	===		
SCORE	13	?	?
RANK		?	?

Table 6 (continued).

Tradeoff Matrix for the Delivery of Education Materials

	WEIGHT	Current Method of Delivery	Video Use	Knowledge Pipes	Moderated Bul. Boards
Student Acceptance	1	?	?	?	?
Faculty Acceptance	1	?	?	?	?
Industry Acceptance	1	?	?	?	?
Cost of Implementing	1	?	?	?	?
Cost of Use	1	?	?	?	?
Knowledge Access	1	?	?	?	?
Effectiveness	1	?	?	?	?
Feasibility	1	?	?	?	?
Effect on Teaming	1	?	?	?	?
SCORE	9	?	?	?	?
RANK		?	?	?	?

	WEIGHT	Computer books
Student Acceptance	1	?
Faculty Acceptance	1	?
Industry Acceptance	1	?
Cost of Implementing	1	?
Cost of Use	1	?
Knowledge Access	1	?
Effectiveness	1	?
Feasibility	1	?
Effect on Teaming	1	?
SCORE	9	?
RANK		?

Table 7.

Tradeoff Matrix for the Classroom of Tomorrow.

	WEIGHT	Current classroom model	Computer Projector	Interactive Classrooms	Multimedia Classroom
Student Acceptance	1	?	?	?	?
Faculty Acceptance	1	?	?	?	?
Industrial Accept.	1	?	?	?	?
Prep. for Profession	1	?	?	?	?
Cost of Implementing	1	?	?	?	?
Cost of Running	1	?	?	?	?
Process Knowledge	1	?	?	?	?
Improve Prob. Solv.	1	?	?	?	?
Broader Knowledge	1	?	?	?	?
Deeper Knowledge	1	?	?	?	?
Improve Design Abil.	1	?	?	?	?
Learning Team Work	1	?	?	?	?
Learn Fundamentals	1	?	?	?	?
SCORE	13	?	?	?	?
RANK		?	?	?	?

	WEIGHT	Hyperteaching/ Teaming	Interactive TeleEducation
Student Acceptance	1	?	?
Faculty Acceptance	1	?	?
Industrial Accept.	1	?	?
Prep. for Profession	1	?	?
Cost of Implementing	1	?	?
Cost of Running	1	?	?
Process Knowledge	1	?	?
Improve Prob. Solv.	1	?	?
Broader Knowledge	1	?	?
Deeper Knowledge	1	?	?
Improve Design Abil.	1	?	?
Learning Team Work	1	?	?
Learn Fundamentals	1	?	?
SCORE	13	?	?
RANK		?	?

Table 8.

Tradeoff Matrix for the Engineering Education Laboratory

	WEIGHT	Current Model	Problem Solving	Virtual Reality Laboratory	Computer-Simulated Laboratory
Student Acceptance	1	?	?	?	?
Faculty Acceptance	1	?	?	?	?
Industry Acceptance	1	?	?	?	?
Prep. for Profession	1	?	?	?	?
Cost of Implementing	1	?	?	?	?
Cost of Running	1	?	?	?	?
Process Knowledge	1	?	?	?	?
Problem-Solving	1	?	?	?	?
Broader Knowledge	1	?	?	?	?
Deeper Knowledge	1	?	?	?	?
Improve Design Abil.	1	?	?	?	?
Learning Team Work	1	?	?	?	?
Learn Fundamentals	1	?	?	?	?
SCORE	13	?	?	?	?
RANK		?	?	?	?

Table 9.

Tradeoff Matrix about Learning Outside the Classroom

	WEIGHT	Current Model	Electronic Lecture Access	Structured Team Learning	CSCW
Student Acceptance	1	?	?	?	?
Faculty Acceptance	1	?	?	?	?
Industry Acceptance	1	?	?	?	?
Prep. for Profession	1	?	?	?	?
Cost of Implementing	1	?	?	?	?
Cost of Running	1	?	?	?	?
Process Knowledge	1	?	?	?	?
Problem Solving	1	?	?	?	?
Broader Knowledge	1	?	?	?	?
Deeper Knowledge	1	?	?	?	?
Improve Design Abil.	1	?	?	?	?
Learning Team Work	1	?	?	?	?
Learn Fundamentals	1	?	?	?	?
SCORE	13	?	?	?	?
RANK		?	?	?	?

	WEIGHT	Simulated Laboratories	Computer-based Tutorials
Student Acceptance	1	?	?
Faculty Acceptance	1	?	?
Industry Acceptance	1	?	?
Prep. for Profession	1	?	?
Cost of Implementing	1	?	?
Cost of Running	1	?	?
Process Knowledge	1	?	?
Problem Solving	1	?	?
Broader Knowledge	1	?	?
Deeper Knowledge	1	?	?
Improve Design Abil.	1	?	?
Learning Team Work	1	?	?
Learn Fundamentals	1	?	?
SCORE	13	?	?
RANK		?	?

Table 10.

Tradeoff Matrix about Multimedia

	WEIGHT	No Computer Facilitation	Routine Video Use	Sound	Animation
Student Acceptance	1	?	?	?	?
Faculty Acceptance	1	?	?	?	?
Industry Acceptance	1	?	?	?	?
Prep. for Profession	1	?	?	?	?
Cost of Implementing	1	?	?	?	?
Cost of Running	1	?	?	?	?
Process Knowledge	1	?	?	?	?
Problem Solving	1	?	?	?	?
Broader Knowledge	1	?	?	?	?
Deeper Knowledge	1	?	?	?	?
Improve Design Abil.	1	?	?	?	?
Learning Team Work	1	?	?	?	?
Learn Fundamentals	1	?	?	?	?
SCORE	13	?	?	?	?
RANK		?	?	?	?

	WEIGHT	Interactive Simulation	Multi-modalities
Student Acceptance	1	?	?
Faculty Acceptance	1	?	?
Industry Acceptance	1	?	?
Prep. for Profession	1	?	?
Cost of Implementing	1	?	?
Cost of Running	1	?	?
Process Knowledge	1	?	?
Problem Solving	1	?	?
Broader Knowledge	1	?	?
Deeper Knowledge	1	?	?
Improve Design Abil.	1	?	?
Learning Team Work	1	?	?
Learn Fundamentals	1	?	?
SCORE	13	?	?
RANK		?	?

Table 11.

2. Computer-Facilitated Educational Experiences

The purpose of this scenario is to evaluate the importance of several possible technology-based methods for improving engineering education. The outcome will be a ranking.

Alternatives:

Mathematics teaching packages (symbolic). This alternative includes tools such as Mathematica and Maple.

Mathematics teaching packages (computational). This alternative refers to packages such as MatLab and MatCad. The alternative is included to permit the comparison of symbolic and computational packages.

CAD tools. Computer-aided design tools. CAD tools are already used in many courses in colleges of engineering.

Conventional CBT. Conventional computer-based training tools which march students lock-step through computer-presented lessons.

Intelligent tutoring systems. Intelligent tutoring systems are an alternative that adds some "intelligence" to conventional CBT; i.e., an ITS would alter the course of instruction based on the "perception" of the computer delivering the lessons.

Simulation exercises./computer simulation tools for simulating engineering problems in the real world. These simulations could also be of physical phenomena such as particle flow or electron flow.

Anchored instruction. Instructional experiences centered around the solving of a problem - i.e. anchoring the instructional experience in a real, easily understandable surroundings. Could be a design experience.

Virtual realities. A make-believe environment that looks and feels like the real world, but isn't. In instruction, this could be an "engineering game" in which virtual companies vie in a make-believe world to create competing products, for example.

Laboratory simulation. This alternative refers to replacing the physical laboratory that is traditional in engineering education with a simulation of the physical laboratory.

Mathematics visualization. The final alternative was concerned with tools for visualizing mathematics. This type of tool might be a Mathematica or a MatCad but would have the explicit capability of helping the student visualize the meaning of mathematical representations.

3. Delivery of Educational Materials

The purpose of this tradeoff analysis is to attempt to determine which of various methods for delivery of educational materials are useful.

Alternatives:

Current methods of delivery. This alternative is the way that we deliver engineering education currently. Assume a lecture section with about 30 students. The only aids available are the blackboard (and chalk) and an overhead projector.

Video use. Video use would be an increased reliance on the use of video materials; for example libraries of video lectures that are shared.

Knowledge pipes. This alternative refers to the concept of providing educational materials over satellite and/or fiber optic delivery infrastructures. We are thinking of much more than the standard distance education paradigm in which information is viewed at specific times. In this alternative, a high bandwidth "knowledge pipe" should be easily accessible and browsable as needed; anyplace, anywhere, anytime.

Moderated bulletin boards. This alternative is the "poor man's" knowledge infrastructure. It uses email bulletin boards that can be moderated. For student and faculty use, this alternative is usable today using Internet facilities.

Computer books. Computer books are books with computers built-in that have the capability of providing simulations and hypertext. Any computer book would have a large and rich knowledge infrastructure.

4. The Classroom of Tomorrow

The purpose of this analysis is to determine what are the most desirable attributes of the classroom of tomorrow.

Current model (chalk, lecture, overheads). This is way we teach in the classroom now.

Computer projector. This simple alternative simply adds a computer projector as a substitute for the overhead projector.

Interactive classrooms. Interactive classroom have computer capabilities, including the ability for students to interact with the computer and for instructors to view and utilize computer-acquired information about how the student use their own computers during a classroom session. Students may ask questions and receive answers via computer.

Multimedia classrooms. Multimedia classrooms employ video, audio, animation, and sound tightly linked with computers to create a multi-sensory learning environment.

Hyperteaching/teaming. Hyperteaching/teaming refers to hypermedia facilitated classrooms combined with team workgroups of students engaged in problem solving.

Interactive TeleEducation. This alternative refers to a classroom environment composed of students freely mixed that are both collocated and remotely located in a seamless electronic computer-based electronic classroom multimedia environment.

5. The Engineering Education Laboratory of Tomorrow

The purpose of this analysis is to determine the features of tomorrow's laboratories for engineering students.

Current model. The current model is the engineering education laboratory. Students are normally formed into small groups that conduct experiments using equipment. There is virtually no exposure to multimedia, organized teaming efforts or to large-scale problem posing and solving.

Problem solving model. This alternative for laboratory instruction is to give students a large problem to solve, complete with detailed information. Their task is to create plans for solution and experimentally attempt to solve the posed problem. The concept is to give students a realistic problem which will require use of concepts learned in class and the complete use of the laboratory facilities to solve the problem.

Virtual reality laboratory. A virtual reality laboratory is an experience which appears real to students. This alternative is the same as the problem solving alternative, except the problem solving space is cast into an almost real (yet virtual) reality that helps students imagine that they are really doing the exercise. VRLs might be large design projects.

Simulated laboratory. In a computer simulated laboratory, physical instrumentation disappears or is minimized. Instead, laboratory materials and instruments are simulated on the computer.

6. Learning Outside the Classroom and Laboratory

The purpose of this analysis is to understand whether it is plausible to suggest that more learning can take place outside the traditional classroom and laboratory.

Current model. The current model is that students learn outside the classroom through a variety of means: e.g., through reading, searching libraries, using electronic mail, etc.

Electronic lecture access. Electronic lecture access would permit on-demand access of lectures from dormitories, home and elsewhere. More learning may occur away from the lecture hall in the future. Is it important to permit the shifting of when and where learning occurs?

Structured team learning. Student learn from each other. Should this type of learning outside the classroom be structured in some way? For example, formalizing study groups would be one way.

Computer-supported cooperative work. CSCW can serve as an infrastructure for supporting communities of learners who are electronically exchanging ideas, information, problem discussions, etc. This alternative should include both collocated individuals and persons located at a distance that collaborate in this medium.

Simulation laboratories. Should laboratory experiences be created that can be used by students at home or in their dormitory rooms?

Computer-based tutorials. CBT materials can be used outside the classroom for students engaging in self-paced learning. Significant amounts of learning that is now accomplished in classrooms and laboratories might be replaced through clever CBT systems.

7. The Role of Multimedia

Multimedia potentially has a role in facilitating engineering education; the purpose of this tradeoff is to analyze alternative multimedia futures.

No multimedia. This is the situation in most classrooms today.

Video. What value does video add to multimedia presentations?

Sound. Is sound an essential ingredient?

Animation. Is animation useful? Will animation substitute for video?

Interactive simulations. Do interactive simulations help learning?

Combined sound, animation, video, simulation. Is the combination of these elements more useful than any alone?

C. THE CRITERIA

A set of criteria were selected that were thought to help distinguish among the various alternatives. A common set of criteria were selected to use across the seven tradeoffs examined. The choice of criteria was based on a 'quality' theme; that is, criteria should affect time, cost, customer satisfaction and learning. The four main metrics that are measured in attempting to improve quality are time, cost, customer satisfaction and errors made. We modified these metrics for educational objectives to be time, cost, customer satisfaction (both student and industry) and learning gains achieved.

The definitions for the criteria given to the team evaluating the tradeoff alternatives are shown below:

> *Student acceptance.* This criteria refers to how well students are likely to accept the alternative. This criteria is particularly concerned with whether more students are likely to enter engineering colleges and ultimately the engineering profession because of adopting the alternative under consideration.
>
> *Faculty acceptance.* This criteria is concerned with whether the faculty will accept the alternative.
>
> *Industrial acceptance.* Will industry accept the alternative?
>
> *Knowledge retention.* Is the alternative going to assist in knowledge retention?
>
> *Engineering process knowledge.* Does the alternative assist in understanding the engineering process?
>
> *Improved problem solving capabilities.* Can students solve problems better using the selected alternative?
>
> *Broader knowledge.* Will students secure a broader base of knowledge?
>
> *Deeper knowledge.* ...or deeper knowledge?
>
> *Improved design ability.* Will student's design ability be improved?
>
> *Learning team work.* Does the alternative favor team work?
>
> *Learning fundamental/basic science and engineering.* Does the alternative assist in learning fundamentals?
>
> *Preparation for the profession.* Does the alternative assist in preparing for the profession of engineering?
>
> *Speed of learning.* How about speed of learning?
>
> *Cost of implementing*_ What will it cost to implement the chosen alternative?
>
> *Cost of running,*_ What are the continuing costs of running the alternative chosen?

As shown in the tables showing the skeleton forms for each of the tradeoffs, there is some variability in the criteria for the different tradeoffs. For example, the criteria were modified for delivery of engineering education somewhat as shown.

The next section analyzes the results obtained from the study group.

D. ANALYSIS OF THE TRADEOFFS

Figure 3 displays the results for the tradeoff conducted for the curriculum paradigms. As shown, experiential learning rated first among the five alternatives, followed by improving the traditional modes of instruction. Distance education finished last in this evaluation. Examination of Table 12 shows why these ratings occurred as they did. In this table, "++" indicates a high positive influence, a "+" a positive influence, a "-" is a negative influence and a"--" is a very negative influence. The values shown were computed for this and subsequent tables for the composite of all the evaluators of each tradeoff. As shown, experiential learning scored high in a number of categories according to the evaluators. Only the "cost of running" introduced a negative influence. At the alternative end of the spectrum, distance education rating suffered most in two categories, "student acceptance" and "broader knowledge". These ratings were likely secured because students are not captivated by a "talking head", but in lectures appear to resonate better to live lecturers. However, a caveat must be introduced here - if one considers mature industrial learners seeking a graduate degree, the rating will be much higher than for undergraduate engineering students. Why? Because mature learners appreciate the targeted specific information that can be obtained by watching an expert lecturer on broadcast TV. Note also, the positive rating secured by traditional education in terms of cost of running and implementing.

Figure 4 displays the results of the tradeoff for computer-facilitated education. The results were close for the top alternatives. The top ranked alternative was the use of CAD tools, followed by computational math packages and a tie for third place between computer simulation and anchored instruction. CAD tools and computation math packages and symbolic math packages score well on the implementation metric since there are many relatively inexpensive tools in the marketplace which can assist student learning in these areas. Very importantly, there would be very high faculty acceptance of the CAD tools, math packages and the math visualization alternatives. Also, the packages reinforce and assist in teaching of design, one of the major features of an engineering education. Interestingly, virtual realities lose in this ranking due to lack of faculty acceptance and high cost.

Figure 5 shows the results of the tradeoff among different delivery methods for educational materials. The group felt that moderated bulletin boards would be the most useful mechanism and knowledge pipes the least. Moderated bulletin boards are certainly the simplest mechanism to implement today for exchanging information world-wide. Students and faculty alike do not currently make sufficient use of this resource that is available today. The definition of "knowledge pipes" is simply a much advanced version of the bulletin board concept, yet the tradeoff rated it last. This ranking came about due to the effect of cost on the rankings. Bulletin boards can be implemented, viewed and used for a very low cost with a high payoff (yet to be proven) while knowledge pipes will require much higher investments. Computer books and our current methods of delivery tied in this tradeoff, but interesting enough, for different reasons as shown in Table 14. Current delivery methods secure a tremendous boost because of the extremely low cost of implementing the traditional lecture. Also, faculty accept the traditional delivery approach. These findings are not much of a surprise. It is much more work to deliver education in new modalities. New modalities of instruction require much more preparation time than the standard lecture. However, computer books rank tied for first in this analysis for a different reason. The panel felt that computer books would be effective! Why effective? Probably because the adding of in-book simulations and portrayals of mathematical representations of physical phenomena was perceived to be an important learning tool.

In the tradeoff concerned with the classroom of the future, one alternative stands out, as shown in Figure 6. This alternative is the hyperteaching/team alternative. Recalling the definition given the panel of consultants - hyperteaching/teaming referred to hypermedia facilitated classrooms combined with team workgroups of students engaged in problem solving. No wonder this alternative ranked high - it combines the best features of the other alternatives! Interactive TeleEducation ranked low once more, confirming the perceptions of inadequacy currently surrounding distance education. As shown in Table 15, the high ranking for the hyperteaching/teaming alternative came even though costs were perceived to be negative or neutral. The very high scores secured on improving learning and acceptance made this alternative very appealing even though the cost would be moderate to high.

Figure 7 displays the results for the tradeoffs concerned with the engineering laboratory. "Problem solving" ranked first because of acceptance and knowledge improvement. Cost of running laboratories was acknowledged as a negative factor. In second place, computer simulation laboratories were found, only slightly higher ranked than virtual realities. Note in Table 16, that the latter alternative secured more positive ratings than the former. However, the ranking was reversed due to the weighting of the cost factor. The virtual reality score for cost, however, seems unrealistic since virtual realities will have high costs for implementation (see Dede, Bourne and Brodersen in Part I for a discussion of Virtual realities).

Figure 8 presents the data for the tradeoff concerned with learning outside the classroom. The results are striking! Structured team learning is in first place by a considerable margin earning that rating (see Table 17) by having high scores in teaming, preparation for the profession, industrial acceptance and improving design knowledge and deeper knowledge. Interesting, no negative factors were noted for this alternative. The current model was negatively rated, except for cost. The remaining alternatives were not strongly rated, except for electronic access to lectures which ran dead last in these ratings.

Finally, Figure 9 and Table 18 display the results for the tradeoff about multimedia. Here the clear winner is interactive simulation. The ranking is driven by high scores throughout except for cost. The fact that this alternative came in first even with negative valuations in both cost categories makes this alternative worth considering strongly. The categories animation and multiple modalities actually facilitate the interactive simulation category since interactive simulation would likely employ both animation and multiple types of media.

Figure 10 summarizes the seven tradeoffs by reproducing the bar chart ratings in one figure. Table 19 lists the seven top alternatives listed in the left column. In the right column is an interpretation of the category in which these top-ranking alternatives fall.

As shown, there are really only three areas of major concern pin-pointed by these analyses: problem solving, teaming, and learning facilitation by use of tools.

	A	B	C	D	E
Names	Traditional	Traditional +	Experiential learning	Just-In-Time	Distance Education
1	5.0	6.6	7.9	5.1	4.8
2	5.9	5.9	6.9	6.4	1.9
3	5.7	5.5	8.5	4.7	3.3
4	4.9	7.1	7.2	5.3	3.8
5	5.3	6.0	8.2	6.7	2.9
6	6.3	7.4	7.9	7.8	4.1
7	6.6	5.9	7.0	4.7	4.3
8	4.6	6.1	5.9	5.2	3.8
Summary	*5.4*	*6.2*	*7.3*	*5.9*	*3.7*
Rank	*4*	*2*	*1*	*3*	*5*

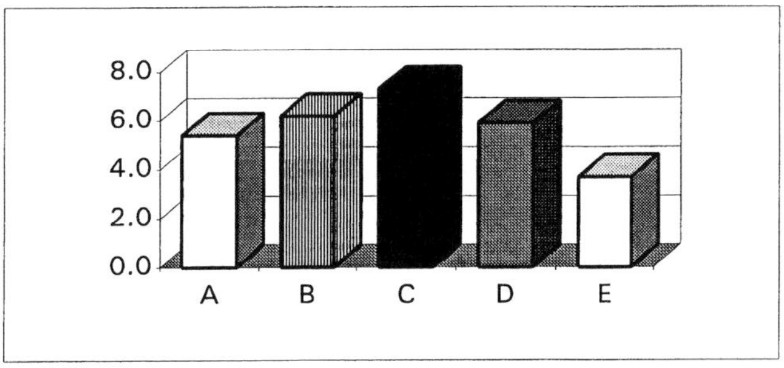

Figure 3. Analysis of Results for Tradeoff Among Curriculum Paradigms

	Traditional	Traditional +	Experiential Learning	Just-In-Time	Distance Education
Student Acceptance			+		- -
Faculty Acceptance	+	+		-	-
Industry Acceptance					
Prep. for Profession			+ +		-
Cost of Implementing	+ +				
Cost of Running	+		-		
Process Knowledge			+		
Problem-Solving Abil.			+		
Broader Knowledge					- -
Deeper Knowledge			+		-
Improve Design Abil.	-		+ +		-
Learning Team Work			+ +		-
Learn Fundamentals					

Table 12. Factors Influencing Result for Curriculum Paradigm Tradeoff

	A	B	C	D	E	F	G	H	I	J
Names	Symbolic Math Package	Comput. Math Package	CAD Tool	Conven-tional CBT	Intelligent Tutor System	Computer Simulation	Anchored Instruction	Virtual Realities	Lab. Simul.	Math Visual-ization
1	7.2	7.2	8.8	5.5	5.4	7.5	8.0	7.1	6.5	7.1
2	5.0	7.1	6.6	5.5	4.8	4.6	6.7	6.2	5.1	5.9
3	7.5	7.5	7.1	4.7	4.7	6.7	4.7	4.3	7.1	6.7
4	6.9	8.0	5.5	6.4	6.7	7.6	5.2	5.4	7.3	6.4
5	3.9	7.0	8.0	6.5	7.9	8.1	7.8	7.0	6.0	4.5
6	5.9	5.2	6.4	6.3	7.3	8.1	8.2	8.2	8.4	6.9
7	5.6	7.3	7.3	4.3	5.3	7.2	7.4	4.7	5.7	6.7
8	7.6	6.4	5.1	3.8	5.7	6.1	6.3	3.9	7.2	?
Summary	*5.9*	*6.8*	*6.9*	*5.4*	*5.7*	*6.7*	*6.7*	*5.3*	*6.2*	*6.3*
Rank	*8*	*2*	*1*	*9*	*8*	*3*	*3*	*10*	*6*	*5*

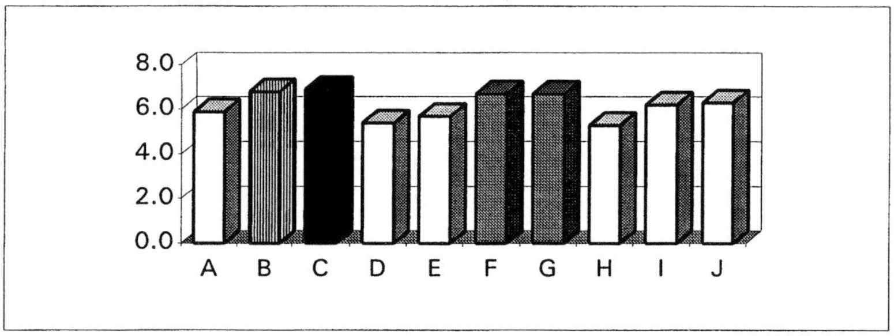

Figure 4. Analysis Results for Computer-Facilitated Education Experiences

	Symbolic Math Packages	Computational Math Packages	Experiential CAD Tools	Conventional CBT	Intelligent Tutor System
Student Acceptance				- -	- -
Faculty Acceptance	-	+ +	+ +	-	-
Industry Acceptance	-		+ +	-	-
Prep. for Profession		+	+	- -	
Cost of Implementing	+ +	+ +	+ +	+ +	-
Cost of Running				+ +	
Process Knowledge	-	-		-	
Problem-Solving Abil.					
Broader Knowledge			-	-	
Deeper Knowledge			-	-	
Improve Design Abil.	- -	+	+ +	- -	- -
Learning Team Work	-			-	
Learn Fundamentals			-		

	Computer Simulation	Anchored Instruction	Virtual Realities	Laboratory Simulation	Mathematics Visualization
Student Acceptance					
Faculty Acceptance	+ +	-	- -	-	+ +
Industry Acceptance			-		-
Prep. for Profession	+	+			
Cost of Implementing	-	-	- -	-	-
Cost of Running			- -		
Process Knowledge					
Problem-Solving Abil.					
Broader Knowledge					
Deeper Knowledge					
Improve Design Abil.	+	+	+	+	- -
Learning Team Work		+ +			
Learn Fundamentals				-	+ +

Table 13. Factors Influencing Result for Computer Facilitated Education Experiences Tradeoff

	A	B	C	D	E
Names	Current Method of Delivery	Video Use	Knowledge Pipes	Moderated Bul. Boards	Computer Books
1	4.2	5.3	6.2	5.9	5.8
2	6.4	4.6	3.3	3.9	5.6
3	4.9	5.7	3.9	8.3	5.9
4	5.2	4.8	5.7	6.6	6.2
5	4.9	5.3	4.4	5.6	7.6
6	5.4	5.2	4.7	4.5	5.4
7	6.4	3.8	4.7	5.4	5.0
8	4.9	5.0	6.6	6.4	4.8
Summary	*5.5*	*4.9*	*4.5*	*5.7*	*5.5*
Rank	*2*	*4*	*5*	*1*	*2*

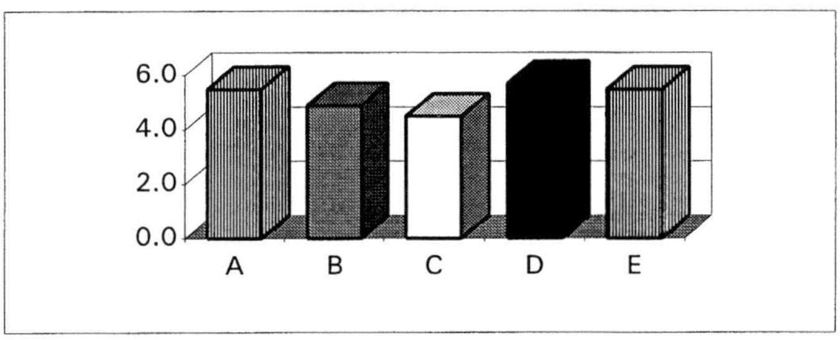

Figure 5. Results for the Delivery of Educational Materials

	Current Methods of Delivery	Video Use	Knowledge Pipes	Moderated Bulletin Boards	Computer Books
Student Acceptance					+
Faculty Acceptance	+				
Industry Acceptance					
Cost of Implementing	+ +		-		-
Cost of Use	+		- -	+	
Knowledge Access!	- -	-	+		+
Effectiveness					+ +
Feasibility			-		-
Effect on Teaming				+	

Table 14. Factors Influencing Result for Delivery of Educational Materials Tradeoff

	A	B	C	D	E	F
Names	Current Classroom Model	Computer Projector	Interactive Classrooms	Multimedia Classroom	Hyper-teaching /Teaming	Interactive Tele Education
1	5.0	5.5	6.4	6.1	7.3	6.1
2	5.8	7.0	6.9	6.6	7.1	6.6
3	4.4	6.9	6.7	7.3	8.5	6.1
4	5.4	6.6	6.3	7.2	6.4	6.4
5	5.6	6.2	6.7	7.0	8.4	4.4
6	6.8	6.7	8.4	8.5	8.5	6.5
7	6.5	6.4	5.5	5.0	5.2	4.6
8	5.3	6.1	5.3	5.7	6.2	3.5
Summary	*5.4*	*6.2*	*6.4*	*6.2*	*7.4*	*5.6*
Rank	*6*	*3*	*2*	*3*	*1*	*5*

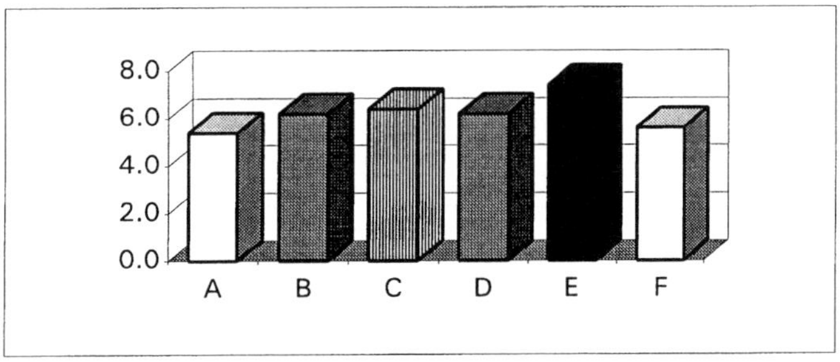

Figure 6. Results for the Classroom of Tomorrow

	Current Classroom	Computer Projector	Interactive Classroom	Multimedia Classroom
Student Acceptance	-			+
Faculty Acceptance	+	+		
Industry Acceptance	-			
Prep. for Profession	-			
Cost of Implementing	+ +	+		- -
Cost of Running	+ +	+ +		-
Process Knowledge	-	-		
Problem-Solving Abil.	-	-		
Broader Knowledge	-			+
Deeper Knowledge	-			
Improve Design Abil.	-	-		
Learning Team Work				
Learn Fundamentals				

	Hyperteaching /Teaching	Interactive TeleEducation
Student Acceptance	+	
Faculty Acceptance		-
Industrial Accept.		
Prep. for Profession	+	
Cost of Implementing		
Cost of Running	-	-
Process Knowledge	+	
Problem-Solving Abil.	+ +	
Broader Knowledge		
Deeper Knowledge		
Improve Design Abil.	+ +	
Learning Team Work	+ +	
Learn Fundamentals		

Table 15. Factors Influencing Result for Classroom of Tomorrow Tradeoff

	A	B	C	D
Names	Current Model	Problem Solving	Virtual Reality Laboratory	Comp.-Simulated Laboratory
1	4.9	8.1	7.6	8.0
2	4.3	7.9	7.2	3.3
3	5.5	7.7	6.2	7.5
4	5.3	6.0	5.2	5.8
5	4.6	5.9	7.4	6.1
6	5.9	8.0	7.5	8.8
7	5.7	8.4	5.0	5.4
8	4.9	5.9	4.7	7.2
Summary	*4.9*	*7.2*	*6.2*	*6.5*
Rank	*4*	*1*	*3*	*2*

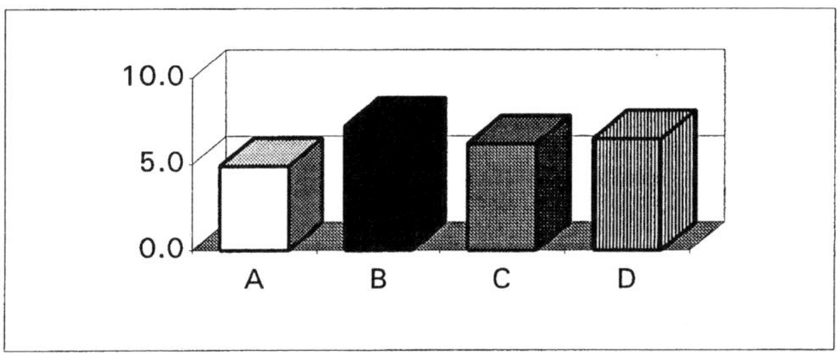

Figure 7. Results for the Engineering Education Laboratory

	Current Model	Problem Solving	Virtual Reality Laboratory	Comp.-Simul. Laboratory
Student Acceptance		+	+	
Faculty Acceptance	+	+ +		- -
Industry Acceptance			+	- -
Prep. for Profession	-	+ +	+ +	
Cost of Implementing	- -			- -
Cost of Running	- -	- -	-	
Process Knowledge	- -		+	+ +
Problem-Solving Abil.				-
Broader Knowledge	+ +	+ +	+ +	+ +
Deeper Knowledge		-	- -	
Improve Design Abil.		+	-	
Learning Team Work		+		
Learn Fundamentals				

Table 16. Factors Influencing Result for Engineering Education Laboratory Tradeoff

	A	B	C	D	E	F
Names	Current Model	Electronic Lecture Access	Structured Team Learning	CSCW	Simulated Laboratories	Computer-based Tutorials
1	4.9	5.1	5.9	6.5	5.3	5.5
2	4.9	5.7	6.6	4.1	5.2	5.7
3	5.5	6.6	8.5	6.4	7.5	5.7
4	5.2	5.6	6.1	6.3	5.8	6.8
5	6.1	4.6	8.6	7.5	7.5	7.7
6	6.2	6.5	8.1	7.8	8.3	8.9
7	5.8	5.7	7.3	6.3	5.8	5.6
8	5.1	6.1	5.5	5.1	6.8	4.9
Summary	*5.4*	*5.1*	*7.1*	*6.0*	*6.3*	*6.5*
Rank	*5*	*6*	*1*	*4*	*3*	*2*

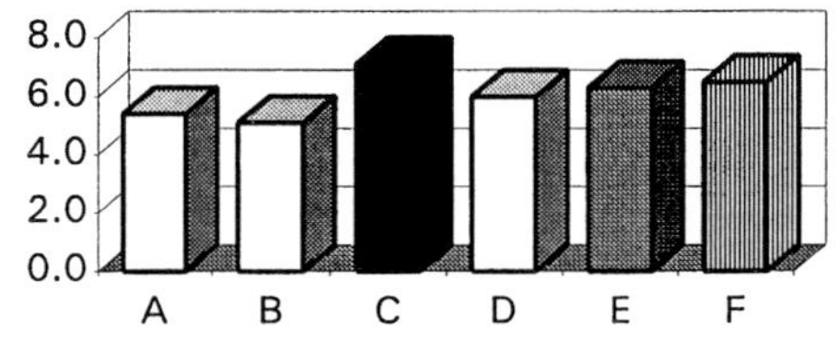

Figure 8. Results for Learning Outside the Classroom

	Current Model	Electronic Lecture Access	Structured Team Learning	CSW
Student Acceptance	-	-		
Faculty Acceptance				
Industry Acceptance			+	
Prep. for Profession	-	-	+ +	
Cost of Implementing	+ +			-
Cost of Running	+ +			
Process Knowledge	-	-		
Problem-Solving Abil.	-	-		
Broader Knowledge	-	-		
Deeper Knowledge	-	-	+	-
Improve Design Abil.	- -	- -	+ +	
Learning Team Work	-	-	+ +	+
Learn Fundamentals				

	Simulated Laboratories	Computer-Based Tutorials
Student Acceptance		
Faculty Acceptance	+	+
Industry Acceptance		+
Prep. for Profession		
Cost of Implementing	-	-
Cost of Running		
Process Knowledge		
Problem-Solving Abil.		
Broader Knowledge		
Deeper Knowledge	+	+
Improve Design Abil.		
Learning Team Work	-	-
Learn Fundamentals		

Table 17. Factors Influencing Result for Learning Outside the Classroom Tradeoff

	A	B	C	D	E	F
Names	No Computer Facilitation	Routine Video Use	Sound	Animation	Interactive Simulation	Multi-Modalities
1	4.9	5.5	5.3	6.1	6.4	6.3
2	5.4	5.0	5.2	6.8	7.6	?
3	3.6	6.2	5.7	7.2	8.5	5.7
4	5.2	6.4	6.2	7.2	7.7	7.6
5	5.8	5.7	4.9	5.9	6.5	6.5
6	4.3	6.2	6.3	8.6	9.1	9.2
7	5.7	5.2	5.2	5.2	6.1	5.4
8	4.9	4.9	5.3	5.6	6.7	6.1
Summary	*5.0*	*5.5*	*5.1*	*6.4*	*7.3*	*6.5*
Rank	*6*	*4*	*5*	*3*	*1*	*2*

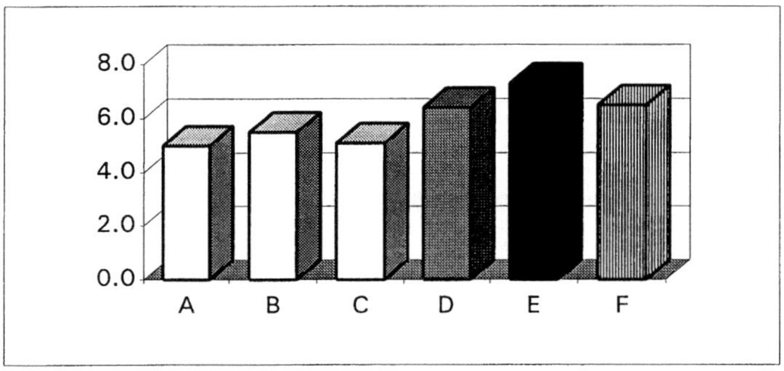

Figure 9. Multimedia Results

	No computer Facilitation	Routine Video Use	Sound	Animation
Student Acceptance	-	-	-	
Faculty Acceptance				+
Industry Acceptance	-			
Prep. for Profession	- -			
Cost of Implementing	+ +			-
Cost of Running	+	+	+	
Process Knowledge				
Problem Solving Abil.	-	-	-	
Broader Knowledge	-		-	
Deeper Knowledge	-		-	
Improve Design Abil.				
Learning Team Work				
Learn Fundamentals	-	-	-	

	Interactive Simulation	Multi-Modalities
Student Acceptance	+ +	
Faculty Acceptance	+	
Industry Acceptance		
Prep. for Profession	+ +	+ +
Cost of Implementing	-	-
Cost of Running	- -	- -
Process Knowledge		
Problem Solving Abil.	+ +	
Broader Knowledge		
Deeper Knowledge	+	
Improve Design Abil.	+	+
Learning Team Work		
Learn Fundamentals	+	+

Table 18. Factors Influencing Result for Multimedia Tradeoff

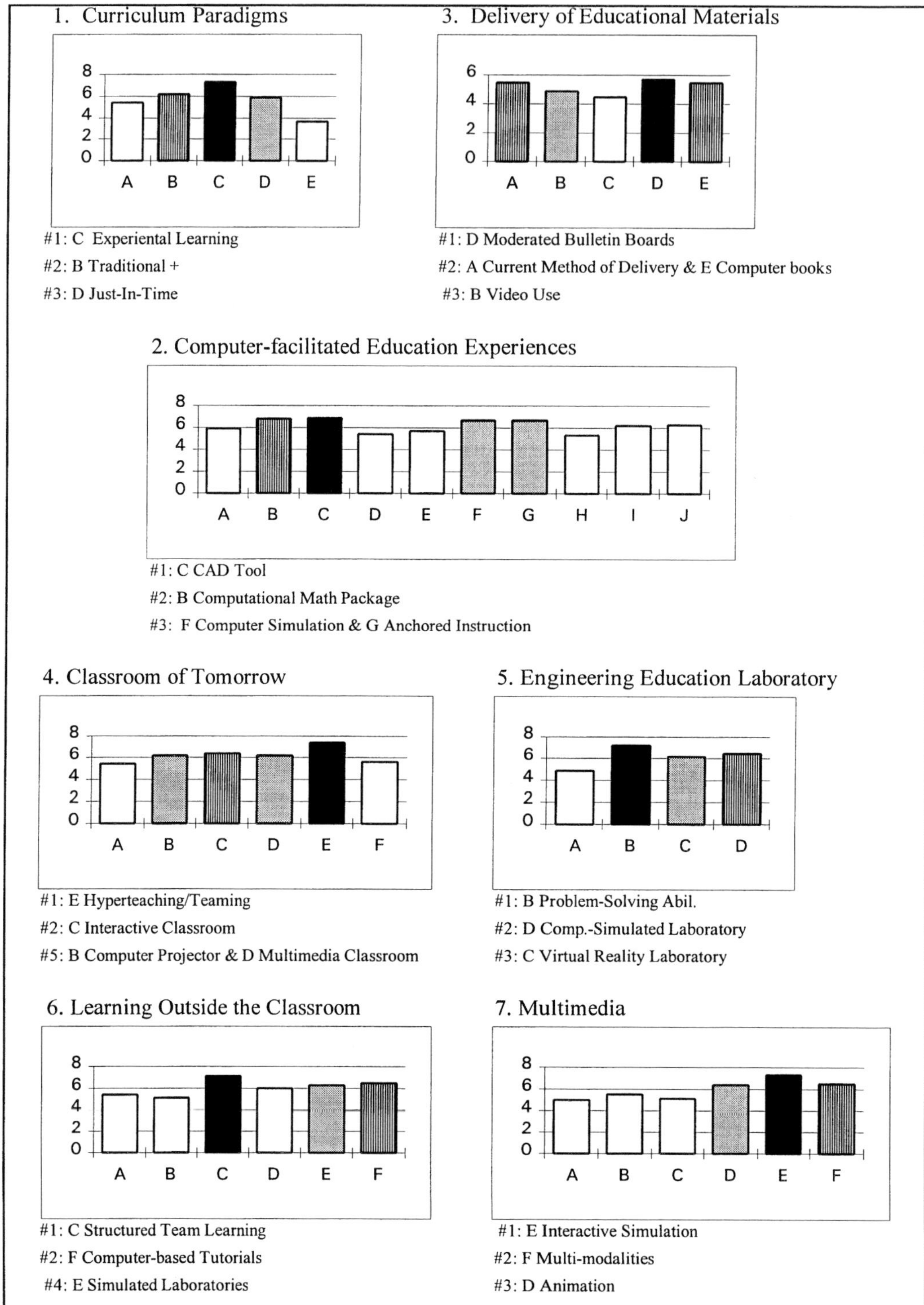

Figure 10. Composite of Analysis of Tradeoffs.

Alternative	Alternative Classification
Experiential Learning	Problem solving, teaming
CAD Tools	Learning facilitation via tools
Moderated Bulletin Boards	Learning facilitation via tools, teaming
Hyperteaching/ Teaming	Teaming, problem solving
Problem Solving	Problem Solving
Structured Team Learning	Teaming
Interactive Simulation	Learning facilitation via tools

Table 19. The Top Seven Alternatives and Their Classifications

IV. FOCUS GROUP RESULTS

A. INTRODUCTION

The individuals shown in Table 3 met in Steamboat Springs, Colorado from July 16, 1993 until July 21, 1993 to discuss and debate the impact of advances in computer and communications technology on the future of engineering education. The group shown in this table was selected to represent various perspectives on education, in general, and engineering education, in particular.

Name of Study Participant	**Area of Expertise**
John Bourne, Vanderbilt University, Principal Investigator of Study	Intelligent Systems, Computers, Engr. Education, Tutoring Systems
Arthur Brodersen, Vanderbilt University, Co-Principal Investigator of Study	Intelligent Systems, Electronics, Engr. Education , Tutoring Systems
Malcolm Getz, Vanderbilt University, Investigator	Economics, Libraries, Teaching with Technology
Charles Kinzer, Vanderbilt University, Investigator	Education, Reading, Learning Technology
Chris Dede, George Mason University	Future of Education
Bruce Black, Rose-Hulman	Mathematics, Computers and Engineering Education
Robert Thomas, Cornell University	Multimedia, Infrastructures
Randy Katz, Univ.of Calif., Berkeley	Information Infrastructures, Electronics, Electronic Classrooms
Roy Mattson, National Technological University	Distance Education
Francis Kulacki, University of Minnesota	Economics and Demographics of Engineering Education
James Avery, University of Colorado	Laboratory Education
James Dally, University of Maryland	Anchored Instruction, Design

Table 20. Study Participants

B. THE FOCUS AREAS

Three focus areas were chosen for examination by the study participants. These focus areas were:

1. Distributed Learning
2. Curriculum Paradigms
3. Delivery Mechanisms

Distributed Learning was initially defined as learning that takes place without regard to physical location. The topic includes distance education including delivery of education over high bandwidth communications media, group learning on networks and individual learning[3]. *Curriculum Paradigms* were defined as different ways of teaching and learning.

Delivery Mechanisms were defined as the way that we deliver engineering education to the students. Ambiguity among these three focus areas was purposely built in. There is considerable overlap between the areas - differences are mainly those of perspective and definition.

Tables 21, 22 and 23 show sample pressure and opportunity lists that were supplied to the participants to seed their discussions about pressure and opportunities at the meeting.

Pressures
Education at Home Education in the Work Place - continuing life-long learning Collaborative Distributed Education - among students; among students and faculty Creation and Use of Distributed Resources Commercial Sourcing of Learning Materials Curriculum Reform Curriculum Delivery Methods Need to Reduce Cost

Opportunities
NREN Capabilities Vast Market for Life-Long Learning Collaborative Engineering Education as a Paradigm Electronic Resources: Tutoring Systems, Library Materials Knowledge Sharing Across Institutions - lectures - materials: electronic and otherwise - faculty; students

Table 21. Prototype Pressure and Opportunity List for Distributed Learning

[3]The careful reader may wonder why we chose distributed learning as a category for inquiry since distance education almost always came in last in the tradeoff that were done by the consultant group at the outset of the study. The reason is that we chose to define distributed learning in a much broader sense than the distance learning of today. That is, we attempted to look forward a decade and predict that computer and communications technology would make it possible for us to do things that cannot be done today. Hence, we selected this area that could (and probably will) have dramatic change and may be an area into which national inquiries should focus.

Pressures
To incorporate computer techniques into the engineering curriculum
To educate the student for the needs of industry
To attract and retain students from a diverse and non-traditional pool (minority and women)
To make engineering education relevant to the societal problems in health-care, environment and urban problem areas

Opportunities
To shift learning to student-centered techniques such as active learning
To use emerging software tools such as Mathematica or Mathcad to relieve the tedium of education and enrich problem solving
To take advantage of visualization tools and simulation tools to leverage education opportunities
To use intelligent tutoring systems to level the playing field for non-traditional student populations

Table 22. Prototype Pressure and Opportunity Lists for Curriculum Paradigms

Pressures
Rising student-faculty ratios.
Higher education is pressed to contain costs.
Because skills are valued in the market, the quality of education must meet a market test.
Students read less and see more videos.

Opportunities
Improve the quality of individual delivery mechanisms.
Change the mix of delivery used in the four year curriculum.
Make computers and networks ubiquitous.

Table 23. Prototype Pressure and Opportunity List for Delivery Mechanisms

Most of the initial pressures for distributed learning centered about the perceived need to shift the time and place at which learning occurs - i.e., to enable students to learn at home or in the workplace. There was perceived to be a pressure from industry for students to learn to work together at a distance, for example in engineering design teams that are distributed globally. Various initial opportunities were perceived such as the prospective availability of the National Research and Education Network (NREN) which would enable connectivity to learners everywhere. Knowledge sharing among institutions was perceived to be an opportunity linked to the availability of NREN.

Pressure on curriculum paradigms were initially seen as pressures requiring a shift from the current lecture/laboratory paradigm of engineering education. The specific proposed external pressures were to incorporate computers into the curriculum, to educate students for the needs of industry, attract and retain students and to make engineering education relevant to social needs. Likewise, the prototypical opportunities given to the team revolved around the prospect of creating future curriculum paradigms that are student-centered, rather than faculty-centered.

Finally, the pressures on delivery mechanisms were seeded to be increases in student-faculty ratios, cost containment and market-driven tests. The opportunities were seen as improving quality of instruction, the way we deliver instruction and the opportunity to make computers and communications media ubiquitous.

Tables 24, 25 and 26 show the three prototype Kepner-Tregoe tradeoffs that were supplied to the participants as a place to commence their discussions. These tradeoffs were formulated as simplified versions of the tradeoffs that the participants in the workshop had already completed that are described in Section III, above. Since the detailed tradeoffs described above required considerable time to complete, it was felt that seeding the focus group discussions with simple skeletal outlines that could be modified was reasonable to initiate discussion. As we shall see, all the focus groups significantly changed their criteria and alternatives.

The prototype tradeoff matrix for distributed learning had the following alternatives: distributed team learning, shared learning modules, networked individual learning, lectures by experts via video, interactive distance education modules, delivery of education by commercial entities to the home, traditional distance education and a combination of the feasible alternatives. Distributed team learning referred to teams of non-collocated students working on problem solving. Networked individual learning referred to enhancing the capability of providing learning materials to individuals over high bandwidth networks.

The prototype curriculum paradigm tradeoff matrix contained the alternatives: Cooperative workgroups, anchored instruction , just-in-time curriculum delivery, case studies, traditional lectures and enhancement of traditional lectures with technology. Cooperative workgroups were team learning environments. Anchored instruction dealt with learning by dealing with real issues. Just-in-time curriculum delivery was defined as providing student knowledge when they need it, rather than providing knowledge just-in-case they need it.

The prototype delivery mechanisms alternatives were more extensive, consisting of: team projects, seminars, tutorials, electronic classrooms, business school style lectures, classroom instruction (e.g., sections of 30 students), self-paced instruction, traditional laboratory instruction and tradition lecture hall instruction. Business school style classes differ from traditional engineering education lectures in the intense degree in which students are involved in interacting with the lecturer.

	WEIGHT	Combination of Feasible Alternatives	Distrib. Team Learning	Shared Learning Modules	Networked Individual Learning
Faculty Acceptance	5	V-Good	V-Good	V-Good	Fair
Student Acceptance	18	Excell++	V-Good	V-Good	Good
Industry Acceptance	2	Excell	V-Good	V-Good	V-Good
Cost of Implementing	8	Medium	Low+	Medium	Low+
Cost of Running	15	Medium	Low+	Low+	Low+
Improve Education	2	AGREE!++	Agree	Agree	Agree
Fac./Stu. Contact Impr.	13	Neutral+	Neutral+	DISAGREE!-	DISAGREE!-
Feasibility	17	V-Good	V-Good	V-Good	V-Good
SCORE	103	7.8	7.4	6.6	6.2
RANK		1	2	3	4
	WEIGHT	Video Expert	Interactive Distance Ed. Lectures	Commercial Ed. Home Delivery	Traditional Distance Education
Faculty Acceptance	5	Fair	Fair	Poor	Fair
Student Acceptance	18	V-Good	V-Good	Good	Fair--
Industry Acceptance	2	V-Good	Excell	V-Good	Good
Cost of Implementing	8	Medium	V-High-	Medium	V-High-
Cost of Running	15	Medium	High-	Medium	High-
Improve Education	25	Agree	Agree	Neutral--	Neutral--
Fac./Stu. Contact Impr.	13	DISAGREE!-	Neutral+	DISAGREE!-	DISAGREE!-
Feasibility	17	Good	Fair--	Fair--	V-Good
SCORE	103	5.7	5.3	4.3	3.9
RANK		5	6	7	8

Table 24. Kepner-Tregoe Tradeoff Prototype for Distributed Learning

	WEIGHT	Traditional Lectures	Traditional Lecture +	Cooperative Workgroups	Anchored Instruction
Student Acceptance	10	Medium	Medium	High+	High+
Faculty Acceptance	10	High	High	Medium	High
Industry Acceptance	5	Medium	Medium	High	Medium
Improve Problem-Solving Skills	9	Disagree--	Disagree--	Agree+	Agree+
Improve Design Skills	8	Neutral	Neutral	Agree	Agree
Improve Communicat. Skills	7	Disagree	Disagree-	Agree+	Neutral
Improve Process Knowledge	6	Disagree-	Disagree-	Agree+	Agree+
Enhance Team-Building Skills	8	DISAGREE!--	DISAGREE!--	AGREE!++	Neutral
Master Basic Math & Science	10	Neutral	Neutral	Neutral	Agree
Cost of Implementing	9	V-Low++	Medium	Medium	Medium
Cost of Running	9	V-Low++	Medium	Medium	Medium
Preparation for Profession	9	Medium-	Medium-	High	High
SCORE	100	5.2	4.3	6.8	6.6
RANK		5	6	1	2

	WEIGHT	Case Studies Curriculum	Just-in-Time
Student Acceptance	10	Medium	Medium
Faculty Acceptance	10	Medium	Medium
Industry Acceptance	5	Medium	Medium
Improve Problem-Solving Skills	9	Agree+	Agree+
Improve Design Skills	8	Neutral	Neutral
Improve Communicat. Skills	7	Neutral	Neutral
Improve Process Knowledge	6	Agree+	Disagree-
Enhance Team-Building Skills	8	Disagree	Disagree
Master Basic Math & Science	10	Neutral	AGREE!++
Cost of Implementing	9	Medium	High-
Cost of Running	9	Medium	Medium
Preparation for Profession	9	High	High
SCORE	100	5.4	5.4
RANK		3	3

Table 25. Kepner-Tregoe Tradeoff Prototype for Curriculum Paradigms

	WEIGHT	Team Project	Seminar	Tutorial	E. classroom
Faculty Acceptance	6	Medium	High+	Medium	Medium
Std. Accepts 1st year	10	High	V-High+	High	High
Std. Accepts 2-3 year	5	High	High	High	High
Std. Accepts 4th year	10	V-High+	V-High+	V-High+	High
Prep for Career	7	High	High	V-High+	High
Cost of Implementing	10	Medium	High	High	High
Cost of Running	16	Medium	High--	High--	High--
Learn Fundamentals	12	Good	V-Good+	Good	V-Good+
Problem-Solving Abil.	10	Excell++	Good	V-Good+	V-Good+
Team Work	8	Excell++	Good	Fair	Fair
Improve Design Abil.	6	V-High+	Medium	High	Medium
==					
SCORE	100	7.4	6.3	6.1	5.7
RANK		1	2	3	4

	WEIGHT	Law/B Style	Classroom	Self-paced	Lab
Faculty Acceptance	6	Low	Medium	Low	Medium
Std. Accepts 1st year	10	Medium	High	Medium	Medium
Std. Accepts 2-3 year	5	Medium	Medium	Low	Medium
Std. Accepts 4th year	10	Medium	Medium	Low-	Medium
Prep for Career	7	Medium	Medium	Low-	Medium
Cost of Implementing	10	High	Medium	Medium	High
Cost of Running	16	Low++	Medium	Low++	High--
Learn Fundamentals	12	Good	Good	V-Good+	Fair-
Problem-Solving Abil.	10	Fair-	Fair-	Good	Fair-
Team Work	8	Good	Poor	Unacc-	Fair
Improve Design Abil.	6	Low	Medium	Low	Medium
==					
SCORE	100	5.0	5.0	4.6	4.1
RANK		5	5	7	8

Table 26. Kepner-Tregoe Tradeoff Prototype for Delivery Mechanisms

C. PROBLEM SOLVING METHODS

The process flow for the study group was to first list pressures, then opportunities, conduct the Kepner-Tregoe tradeoffs, analyze and rank the alternatives, create vision statements and finally produce and prioritize recommendations for which alternatives are most important to carry out.

D. PRESSURES AND OPPORTUNITIES FOR EACH FOCUS AREA

Tables 27, 28 and 29 display the pressures and opportunities lists that each focus group completed during the meeting discussions.

Pressures
External Pressures:
Cultural Change
Financial Pressure
Changing Demographics/Societal Morphology
Competition: Economic, Intellectual
Internationalism
Technological Change
Policies and Intellectual Rights
External:
Faculty Effectiveness/Productivity
Value Added by Distributed Learning
Financial Pressure of Accessing Knowledge on Network

Opportunities
Faculty Cultural Change
Quality Improvement
Outreach: Community Colleges, Extension Programs, K-12
Curriculum support
Access to Learning Materials
Dynamic Staffing

Table 27. Completed Pressure and Opportunity List for Distributed Learning

Pressures
Develop the student as a creative, lifelong learner.
Be responsive to the needs, goals and learning styles of students to address issues of matriculation, retention, and diversity.
Incorporate computer techniques into the curriculum as a learning and engineering tool.
Make engineering education relevant to the changing needs of industry and society.
Involve all faculty in the continuous improvement of the educational enterprise

Opportunities
Improve the ability to teach design and to consider larger, more realistic engineering problems.
Build intuition and judgment in the student through the use of visualization and simulation tools.
Enrich problem-solving skills.
Shift to student-centered models of instructions
Allow the exploration of knowledge within and across domains by each student.

Table 28. Complete Pressure and Opportunity List for Curriculum Paradigms

The delivery mechanism group started with nine different alternatives and ended with seventeen. The initial group includes alternate ways of organizing classroom instruction from seminars to lectures with associated variation in student-faculty ratios and costs. Among these were large lectures, self-paced learning, seminars, tutorials, the more intense, large classes typical of law and business schools, a class making extensive use of assignments of projects to be completed by teams of students, an electronic classroom where each student has a computer at each class meeting, and a standard classroom with about thirty-five students.

Pressures
The workplace is changing. Individual engineers will collaborate more by network. Engineers will work in teams, allowing integration of design with manufacturing. Engineers are concerned with the success of the organization rather than simply focusing on the design of artifacts. With markets for more good becoming global, designs must meet the needs of many cultures and be increasingly cost conscious. More people will use computers more of the time for more purposes.
Some fundamentals of the workplace endure. The workplace will continue to value a balance of mathematics and scientific analysis with hands-on design skills. Engineers must be sensitive to the organizational context in which they work, including the economic and political forces that shape the organization. An engineer's success depends on oral and written communication.
Higher Education is costly Higher education overall is cost constrained with rising student-faculty ratios, limitations on the budgets of state governments, limits on increases in tuition, increased interest in accountability, and a very competitive marketplace. Engineering education costs more than many other disciplines because of higher faculty salaries, lower student-faculty ratios, and the importance of laboratories and equipment.
Students are changing. Students will come from increasingly diverse backgrounds. Students will continue to enjoy design experiences Students will be concerned about how individuals are evaluated when they work in teams. Most students will own computers that are portable, connect to networks, employ more easily mastered software, and contain larger storage devices. Students will be interested in life-long learning.

Opportunities
Engineering schools have opportunities to design new courses and curricula to take advantage of computers to take advantage of networks to promote teamwork to provide more design experiences to encourage more creative, self-directed activity to use more self-paced tutorials, peer-to-peer interactions, and electronically mediated learning regimes.

Table 29. Completed Pressure and Opportunity List for Delivery Mechanisms

The original list of course options included a laboratory, but the discussion group thought it important to distinguish several kinds of labs. The conventional lab allows students to follow a prescribed regime to demonstrate known results. In contrast, a design lab allows students to work as a team in meeting some design objective. In the design lab, students will set their own agendas, learn what they need to know in order to achieve a particular design goal, and work with others in a team to achieve a collaborative result. A third laboratory regime offers laboratories as modules of varying length that can be more easily interspersed with other courses.

The group added a lecture/recitation/lab where the course combines a large lecture with small recitation sections and a laboratory.

The final tradeoffs added an electronic infrastructure as an investment option for enhancing the delivery of educational services. The infrastructure category reflected the role that on-line resources, deliverable by network to classrooms, faculty, and particularly students may play in a variety of teaching strategies. The digital library of materials, including multi-media documents, data sets from laboratories, and discussion groups may in fact be decentralized around the network but be conveniently accessible from everywhere on the network.

Another added category was a course that integrates several pedagogical strategies and uses the multi-media infrastructure. "Peer-to-peer virtual learning" wherein individuals may use the network services interactively, includes consulting the digital library and participating in bulletin boards and electronic discussion groups.

The group added a faculty training category, recognizing that faculty will need time and support to learn to use new tools and to become proficient with new teaching strategies.

E. ALTERNATIVES AND CRITERIA IN EACH FOCUS AREA

Each focus group significantly changed the alternatives and criteria for their areas as shown in Tables 30, 31, and 32.

After discussion, the focus group on distributed learning changed the alternatives to those shown in Table 30. A significant addition was the alternative of the electronic library (a.k.a. - the digital library). Distributed team learning was retained as an alternative. Structured delivery of materials to individuals was a change in name from shared learning modules. The key in this change was the concept of providing curriculum structure to materials that are available to distant learners. The idea was that students learn better when given a structured environment. The interactive distance education category was retained.

The distributed learning focus group conducted two ratings: one considering graduate education and the other undergraduate education. This dichotomization is probably important from the perspective that distance learning now serves many graduate learners who are located in industry or in government.

The curriculum paradigm group also modified the alternatives which were used in the preliminary tradeoff analysis. The category of Lecture was expanded to include the alternative of Lecture plus recitation/laboratory. This distinction is necessary since these two modalities exist in the current engineering curriculum. Two additional alternatives were created by adding the use of technology to the traditional presentations and laboratories. The alternative of Just-in-Time Curriculum was changed to Integrated Curriculum as being more representative of models of instruction being pursued by institutions such as Rose-Hulman University. The alternatives of tutorial, the British model; and internships, the cooperative education model were added for completeness. Additional criteria were added by this group. Mastery of math and science and improving communication skills were added as measures of curriculum effectiveness. These criteria can be found by examining Table 31.

The delivery mechanism group started with nine different alternatives and ended with seventeen. The modifications in the final tradeoff was to change the alternatives and criteria to match the pressures and opportunities formulated by the group. A variety of interesting ideas were added including delivery infrastructure building, peer-to-peer virtual learning, and others as shown in Table 32.

Undergraduate Ratings

Names	A Electronic Library/Ref. Individual	B Video Expert Lectures	C Traditional Distance Education	D Distributed Team Learning	G Structured Individual Delivery	H Interactive Distance Education
1	6.6	4.1	5.7	7.3	4.3	7.0
2	7.3	6.3	4.2	6.6	4.2	3.6
3	9.6	7.9	6.4	8.6	3.0	4.7
4	8.2	6.0	1.9	1.8	2.4	1.1
Summary	*7.5*	*5.8*	*4.3*	*5.7*	*3.7*	*4.1*
Rank	*1*	*2*	*4*	*3*	*6*	*5*

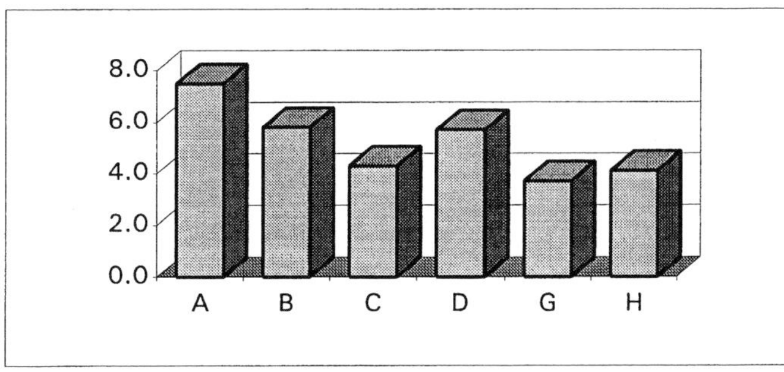

Graduate Ratings

Names	A Electronic Library/Ref. Individual	B Video Expert Lectures	C Traditional Distance Education	D Distributed Team Learning	G Structured Individual Delivery	H Interactive Distance Education
1	6.6	5.6	5.9	7.4	5.9	6.8
2	7.2	6.2	5.3	5.2	5.7	5.3
3	9.6	8.2	7.3	7.2	5.1	8.0
4	8.2	7.0	7.6	3.0	5.7	3.9
Summary	*7.5*	*6.2*	*6.0*	*5.8*	*5.7*	*5.6*
Rank	*1*	*2*	*3*	*4*	*5*	*6*

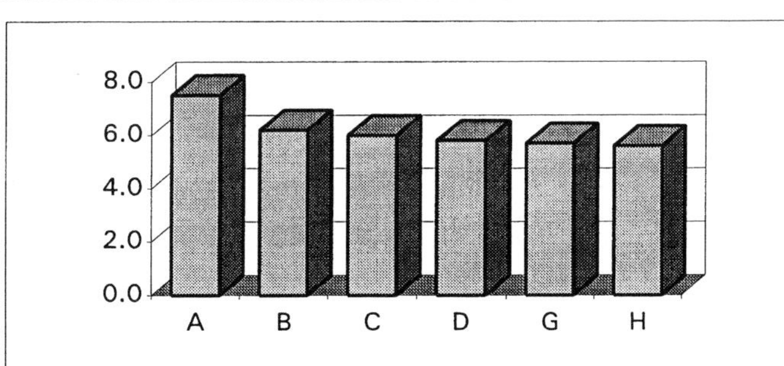

Table 30. Completed Tradeoffs for Distributed Education.

	A	B	C	D	E	F	G	H	I	J
Names	Lectures	Lectures + Technol.	lectures + Rec./Lab.	Lec. + Rec/Lab + Technol.	Coop. Work--groups	Anchored Instruction	case Studies	Integrated Curriculum	Tutorials	Intern-ships
1	3.9	4.0	5.7	6.0	6.3	7.4	7.1	6.1	6.7	7.0
2	6.6	6.2	7.2	7.4	5.0	6.4	4.8	5.0	5.6	4.4
3	4.4	3.5	6.2	5.8	8.2	7.2	6.4	5.1	4.6	5.8
4	4.9	5.1	6.3	6.7	7.4	7.4	6.3	6.2	6.3	7.3
5	5.4	5.4	5.3	5.0	6.9	6.8	6.1	4.0	4.9	6.2
Summary	*5.0*	*4.9*	*6.3*	*6.1*	*6.4*	*7.1*	*6.0*	*5.0*	*6.0*	*6.6*
Rank	*8*	*10*	*4*	*5*	*3*	*1*	*6*	*8*	*6*	*2*

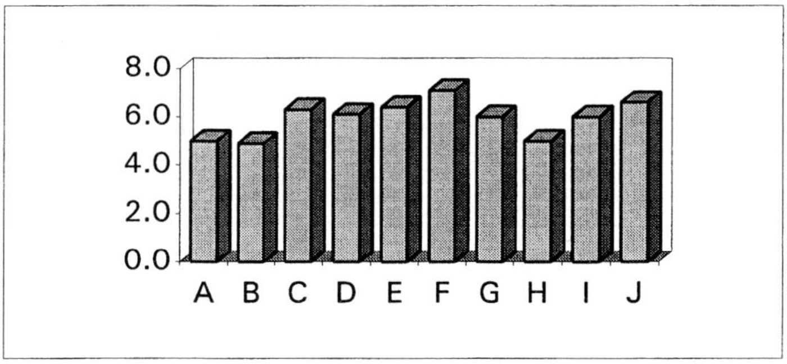

Table 31. Completed Tradeoffs for Curriculum Paradigms.

	A	B	C	D	E	F	G	H
	Infrastruct.	Intg./ Mediat.	Team Project	Peer Virtual	Lab./ Design	Cooperat. Learning	Self Paced	Modules/ Lab.
Summary	7.0	6.6	6.5	5.4	5.3	5.1	5.1	4.7
Rank	1	2	3	4	5	6	6	8

I	J	K	L	M	N	O	P	Q
Faculty training	Elec. Classroom	Lec./rec./ lab.	Seminar	Tutorial	Lab./ Demo	Law/ Busin.	Class- room	Lecture
4.7	4.4	4.1	3.9	2.9	2.8	1.9	1.8	1.4
8	10	11	12	13	14	15	16	17

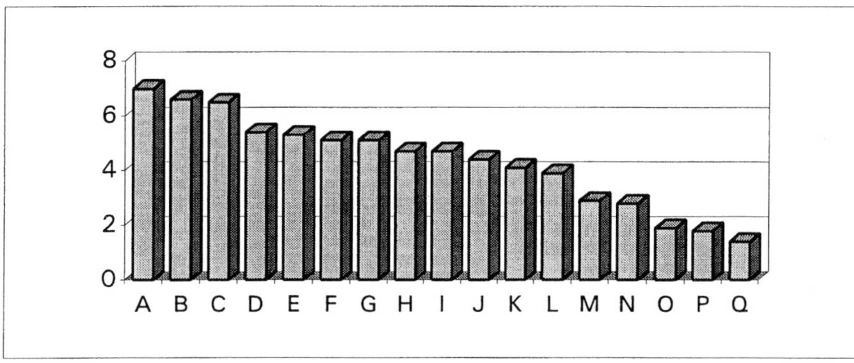

Table 32. Completed Tradeoffs for Delivery Mechanisms .

F. RESULTS OF FOCUS GROUP RATINGS

1. Distributed Learning.

Table 33 shows the KT matrix completed with the addition of sensitivity markers that show which factors were significant in the rating of the alternatives in the distributed learning focus group. A "++" indicates that the criteria was highly significant, a "+" less significant. "-" and "--" were negative factors in deciding the ranking of the alternatives. By simply scanning the matrix, one can pick out the positive and negative factors that impacted the ranking of the alternatives. In this printout, the alternatives are ranked from best to worst.

Analyzing the KT matrix for distributed learning, the top ranked alternative is the electronic library. This alternative was thought to potentially have very high faculty acceptance and to be quite feasible. An electronic library would have learning materials of all types, including interactive multimedia and simulations of all types. Lectures by experts over video ranked a distant second primarily because of excellent feasibility (i.e., this is done now). The rationale for this alternative is that excellent lectures do have a way of engaging students - some lecturers should be captured on video and widely distributed.

2. Curriculum Paradigms

The KT matrix for curriculum paradigms is shown in Table 34. The top ranked alternative is anchored instruction. This alternative was judged by the study group to have high student acceptance, to enhance design skills and to build team skills. The only negative was the implementation cost. Internships or cooperative education was second. The cost of running such programs was judged to be among the highest of the alternatives; however, cooperative education was judged to be the best for preparation for the engineering profession. The lowest ranked alternative was the traditional lecture enhanced with technology. In contrast with traditional lectures the group felt that faculty were slow to adapt to the technology and the costs of the technology were naturally more that the blackboard and chalk of the traditional lecture.

3. Delivery Mechanisms

The Delivery Mechanisms group revised the criteria for evaluating the alternatives to prominence to attracting and retaining the best students, providing experience at hands-on design, increasing faculty effectiveness, and to enhanced professional relevance.

With the weighting of each criteria, the infrastructure investment category received the highest priority. It scored very high on increasing faculty effectiveness, usefulness in life-long learning, providing ready access to information resources, using modern technology, and emphasizing fundamentals. The infrastructure idea carried the day in part because it seems useful with so many different classroom strategies and kinds of students.

Second came the integrated classroom with multi-media, a pedagogical approach that itself makes extensive use of the infrastructure. This classroom approach score very high at professional relevance, on imparting fundamentals, and at promoting new models of teaching.

Third came team projects with very high scores for promoting team work and communication skills as well as promoting hands on design experiences, attracting the best students, and new models of teaching.

The "Peer-to-peer" virtual learning reached four priority, with its dependence on modern technology, emphasis on communication skills, and low cost. Similarly, self-paced computer-based learning scored in the top half because it may be low cost yet use high technology.

Design labs scored well because of their professional relevance.

A group of options occupied the middle of the set, including modular labs, faculty training, and the electronic classroom.

The lowest priority fell to rather conventional delivery mechanisms, including the lecture/recitation/lab, seminars and tutorials, demonstration labs, and the lecture-classrooms. The latter group scored low in professional relevance and in promoting communication skills, very low in using modern technology, and very low in creating a student centered education.

Undergraduate Ratings

	Electronic Library/Ref. Individual	Video Expert Lectures	Traditional Distance Education	Distributed Team Learning	Interactive Distance Education	Structured Individual Delivery
Faculty Acceptance	++	-	-		-	-
Student Acceptance	+		- -	+	-	-
Industry Acceptance						
Cost of Implementing						
Cost of Running	+					
Improve Education	+	+		+		- -
Fac./Stu. Contact Improv.						
Feasibility	++	++	-		-	-

Graduate Ratings

	Electronic Library/Ref. Individual	Video Expert Lectures	Traditional Distance Education	Distributed Team Learning	Structured Individual Delivery	Interactive Distance Education
Faculty Acceptance	++			-		
Student Acceptance	+	+	-			
Industry Acceptance						
Cost of Implementing				-		-
Cost of Running	++	-		-		-
Improve Education				++		
Fac./Stu. Contact Improv.						++
Feasibility		+	+	-	-	-

Table 33. Factors Contributing to Alternative Rankings in Distributed Learning Tradeoffs

	Anchored Instruction	Intern-ships	Cooperat. Workgroups	Lectures + Rec./Lab.	Lec.+Rec./Lab. + Technology
Student Acceptance	++				
Faculty Acceptance	+			+	
Industry Acceptance		+			
Improve Probl.-Solving Skills					
Improve Design Skills	++			-	
Improve Communicat. Skills	+	+	+		
Improve Process Knowledge					
Enhance Team Building Skills	++	++	++		
Master Basic Math & Science					
Cost of Implementing	-	-			
Cost of Running		- -		+	
Preparation for Profession		++			

	Case Studies	Tutorials	Integrated Curriculum	Lectures	Lectures + Technology
Student Acceptance	-		-	-	-
Faculty Acceptance				+	
Industry Acceptance					
Improve Probl.-Solving Skills			-	-	-
Improve Design Skills			-	- -	-
Improve communicat. Skills		+		- -	- -
Improve Process Knowledge			-	-	-
Enhance Team Building Skills			-	- -	- -
Master Basic Math & Science					
Cost of Implementing		-	-	++	
Cost of Running		- -		++	+
Preparation for Profession				- -	

Table 34. Factors Contributing to Alternative Rankings in The Curriculum Paradigm Tradeoffs

	Infratruct.	Intg./ Mediat.	Team Project	Peer Virtual	Lab./ Design	Coop. Learning
Prof Relevance		+			+	+
Teamwork			+		+	++
Comm. Skills			++	+		
Fundamentals	+	+		-		
Hands-On Design			++		++	
Organiz. Context			+			
Compete Glbl. Mkt.						
Analysis						
Use Modern Technlgy	++	+	-	+	-	
Access To Info. RSC	++			+		
Student-Centered Ed.		+	+			
Attrt./Rtain. Best	+	+	++		+	
New Models Teaching	+	++	++	+	+	+
Lifelong Learning	++	+		+		
Meet Resrc. Pressure	+					
Cost of Running						
Faculty Effectivens	++		-		- -	
Faculty Recruiting		+				
Fac Prof Developmnt						

	Self Paced	Modules /Lab.	Fac Training	Elec. Classroo	Lec./Re /Lab.	Seminar
Prof Relevance	-					
Teamwork	- -					
Comm. Skills						+
Fundamentals			- -			
Hands-On Design						
Organiz. Context						
Compete Glbl. Mkt.						
Analysis			-			
Use Modern Technlgy	++				-	-
Access To Info. RSC	+	- -	++	+		
Student-Centered Ed.	+	+				
Attrt./Rtain. Best	+	+				-
New Models Teaching				+	-	
Lifelong Learning	+			-	-	-
Meet Resrc. Pressure						
Cost of Running						
Faculty Effectivens	++					-
Faculty Recruiting						
Fac Prof Developmnt						

Table 35. Factors Contributing to Alternative Rankings in The Delivery Mechanisms Tradeoffs

	Tutorial	Lab./Demo	Law/Busin.	Classroo	Lecture
Prof Relevance	-		- -	-	- -
Teamwork					-
Comm. Skills	+			- -	- -
Fundamentals	-	+			-
Hands-On Design			-		-
Organiz. Context					
Compete Glbl. Mkt.					
Analysis					-
Use Modern Technlgy	-	-	-	-	-
Access To Info. RSC	- -	- -	- -	- -	- -
Student-Centered Ed.		-	-	-	-
Attrt./Rtain. Best	-	-	- -	- -	- -
New Models Teaching	-	-	- -	- -	- -
Lifelong Learning	-	-	-	-	-
Meet Resrc. Pressure					
Cost of Running					-
Faculty Effectivens	-	-			++
Faculty Recruiting					
Fac Prof Developmnt					

Table 35. (continued) Factors Contributing to Alternative Rankings in The Delivery Mechanisms Tradeoffs

Build electronic libraries that are accessible over high speed networks
Create a national infrastructure for delivery of learning materials
Anchored instruction
Cooperative group learning
Lectures plus recitation/laboratories: the traditional model

Table 36. Summary of Top Ranked Alternatives from Focus Group Meetings.

G. VISION STATEMENTS

Tables 37, 38, 39 and 40 display the vision statements of each focus group that were created and the overall vision statement.

Vision Statement for Distributed Learning. The vision for distributed learning is to provide engineering education to anyone, anywhere, and at anytime. A major capability offered via highly interconnected knowledge highways will be easily accessible knowledge, learning modules and structured educational materials that will permit students, faculty and practitioners to find information and knowledge they need. This knowledge will be accessible via taxonomies that permit common vocabularies to be used. Global networked knowledge sources that include reference materials, hypertext-style linked materials, browsable expert lectures, simulations, and linkages to multimedia-facilitated tutorials will dramatically enhance the richness of information available to students and will leverage faculty in their changing role as facilitator and knowledge arbitrator. Distributed learning will include the capability of students and faculty interacting in teams working together to create designs and solve problems. Knowledge highways will provide the capability for students to (1) learn on their own, (2) work together and (3) engage in structured learning experiences. Shifting of both time and place when knowledge is acquired and learning occurs will be a significant outcome of the implementation of distributed learning technology. Interactive capabilities to permit seamless communications among learning groups will be a feature of distributed learning. Implementation is expected to be via advanced telecommunications networks including Internet, NREN and their future derivatives. Bandwidth sufficient to permit building of high throughput knowledge pipes will be essential to the wide scale success of distributed learning.

Table 37.

Vision Statement for Curriculum Paradigms. Computer and communications technology will facilitate the use of new instructional models that have been recognized by cognitive psychologists and learning theorists as leading to higher order learning. These models include constructivist ("learn by doing") and anchored ("learn through connection to real issues") instructional models. Technology will influence the engineering curriculum by improving the teaching of design and problem solving. It will enable the student to consider larger, more realistic engineering problems and explore design alternatives. Visualization and simulation tools will allow the student to move beyond a focus on calculation methods to develop deeper intuitions and judgments. Technology makes possible the shift from a curriculum that is teacher-centered to one that is centered on the student's needs, goals and learning style. Digital libraries, multimedia, and hypertext will allow students and teachers to explore and link knowledge within and across discipline domains. Motto: "Better engineers through better technology"

Table 38.

Vision Statement for Delivery Mechanisms.

Delivery mechanisms for engineering education should promote student centered education by (1) serving more diverse students, attracting and retaining the best students, building on new ideas about teaching and learning and by providing mechanisms for life-long learning. Delivery costs should be reduced and faculty effectiveness increased. Professional relevance of engineering education should be ensured by using modern tools, employing teamwork, hands-on design experiences, improving communications skills and providing breadth in the educational experience.

To enhance the educational delivery mechanisms of the engineering education community, the delivery infrastructure should be improved. Topics should be investigated such as: electronically mediated education, team projects, virtual peer learning, design laboratories, cooperative learning environments, self-paced instruction and modular laboratories.

The faculty should be trained so that new developments in electronically mediated instruction can be taken advantage of.

Table 39.

The Overall Vision Statement.

Computer and communications technology will enable engineering education to adapt to new and varied curricular structures and learning models that will be far more student-centered than present teaching and learning paradigms.

For either evolutionary or revolutionary change in the engineering curriculum, computer and communications technology offers the potential to enrich the learning experience for students via constructivist and anchored instructional modalities. Faculty effectiveness and the faculty's ability to deal with complex, trans-disciplinary problems will be also enhanced through the use of these technologies.

Computer and communications technology will especially influence the engineering curriculum by facilitating the teaching of engineering design. Visualization and simulation tools will allow students to move beyond a focus on calculation methods toward the development of engineering judgment through the rapid explanation of alternative solutions and problem formulation.

Computer and engineering technology will make possible the use of digital libraries, multimedia, hypertext, and distance learning modalities that will shift both the place and time when learning takes place. Flexible, highly responsive, and multidimensional learning environments will result from the application of such technologies for both institution-centered and dispersed groups of students.

Computer and communications technology will ultimately enable engineering schools to provide education to everyone, anywhere and at anytime. The development of highly connected electronic "knowledge highways" will offer students, faculty, and practitioners the capability to freely access knowledge they need. This knowledge will be available via taxonomies that permit the use of common materials, browsable expert lectures, and simulations. Linkages to multimedia-facilitate tutorials will dramatically enhance the breadth and richness of information available to students and faculty. The faculty will evolve new roles as learning facilitators and knowledge arbitrators within the context provided by these technology and techniques derived from them.

Table 40.

H. THE RECOMMENDATIONS

The recommendations secured from each focus group are shown in Tables 41, 42 and 43.

• Build electronic libraries that can be accessed by remote learners. • Study the need and projected use for distributed learning. • Create engineering knowledge taxonomies to provide coherent access to distributed learning materials. • Develop software tools for a supporting distributed learning infrastructure. • Create a process for developing expert lectures that can be delivered on a distributed learning educational network. • Conduct a study leading to standards for use in distributed learning. • Conduct research on computer supported cooperative work in a distributed learning environment.

Table 41. Recommendations for Distributed Learning

Seven main recommendations, as shown in Table 41, were produced by the distributed learning focus group. First, the need for digital electronic libraries was apparent in the discussion. Digital libraries could provide information to anyone, anywhere, and at anything once high speed networks are in place worldwide. While this is true to some degree in the mid-1990s, more work is needed to further develop the capabilities of a global knowledge network and the attendant digital libraries that can be access via the network. Second, it was felt that evaluation of the projected utilization of distributed learning systems in this country in the upcoming years should be undertaken. While it appears clear that electronically distributable materials will become the norm in the future, it is not so clear just how people and institutions will use these materials. To make digital libraries useful, there needs to be a focus on taxonomy of engineering knowledge that will help organize knowledge that exists on the networked digital libraries of tomorrow. The consultant group suggested that engineering might follow the lead of the field of medicine in attempting to define common vocabularies that can be used to access information in distributed knowledge nets. In addition to knowledge structures, software tools that will permit people to find what they need rapidly will need to be developed. Such tools should include personal knowbots that adapt to the user and assist in sorting through masses of globally-available data and information rapidly. Expert lecturers will play a role in engineering education. It was felt that there should be a systematic way of locating this type of expertise and capturing presentations that can be viewed by anyone with access to the educational infrastructure of the future. Standards should be developed. Finally, research on computer-supported cooperative work should be supported, since it appears that future networks will facilitate the working of individuals in teams in the future. There is little knowledge currently available to suggest how the state of knowledge in America can be more rapidly advanced by provide supporting group infrastructures to facilitate learning and knowledge and information interchange in a more coherent and cohesive way.

Five recommendations emerged from the group on curriculum paradigms as shown rank-ordered in Table 42. The first recommendation acknowledges the need for additional understanding of the learning styles in engineering education. The effective use of student-centered paradigms such as anchored instruction and cooperative work groups needs further elucidation. Many interested projects are currently being explored (See the paper in Part I by Dally and Kinzer) with encouraging results. How such paradigms can be incorporated into

- Research on learning styles in engineering education including: anchored instruction and cooperative workgroups.
- Research, measurement and evaluation of visualization/simulation techniques.
- Support for dissemination of successful methods and training for faculty.
- Infrastructure support: Research on knowledge storage and access as part of digital library initiatives.
- Research on facilitating acceptance of technology-induced change by faculty and students.

Table 42. Recommendations for Curriculum Paradigms

extended more broadly into upper division engineering course needs additional study. The third recommendation stresses the need for metrics to evaluate the success or failure of computer-based visualization and simulation in education. The computer has allowed the extension of the design space that students can explore so that much more realistic problems can be addressed. As yet, there is no quantifiable way that these pedagogical techniques can be evaluated. A clear road map is needed to evaluate the results of the research which would result from first recommendation.

The next two recommendations address the need to disseminate the success in engineering and incorporated these techniques into engineering education. Technology changes so rapidly that many faculty have neither the time nor the resources to adapt their current teaching styles to the advances that will be rapidly emerging by the year 2000. The recommendation on infrastructure support addresses this same concern. Interestingly, this recommendation also emerged as the first recommendation from the group on distance education. Clearly the INTERNET provides the basis to disseminate instructional material to a wide audience. The very nature of the INTERNET makes this information, even when made available, is difficult to find and retrieve. The thrust of this recommendation is to systematize the storage and access to make the results of the research on education paradigms available for use and scrutiny by the community of engineering educators.

The last of the recommendations recognizes the need to discover ways that make the individual faculty willing to adopt technology-based education. These changes require the investment of time by the faculty and resources by the universities. Students in engineering are oft times reluctant to leave the anonymity that is implicit in the tradition lecture mode and enter into student-centered activities. The faculty by in large are comfortable following the traditional model of "teaching what I know" and need positive encouragement and help to through away an educational model that has been persistent in university education.

The educational delivery mechanisms group offers the following preamble and recommendations:

- Create National Digital Libraries
- Investigate how to create student centered delivery mechanisms
- Promote use of computer tools and faculty training.
- Extend and improve current work on National educational delivery mechanisms
- Study: electronically mediated instruction, team projects, virtual peer learning, design labs cooperative learning environments, self-paced instruction and modular laboratories.

Table 43. Recommendations for Delivery Mechanisms

In order to promote a student centered education by serving more diverse students, attracting and retaining better students, building on new ideas about teaching and learning, and providing mechanisms for life-long learning; to meet economic pressures by reducing delivery costs and increasing faculty effectiveness; to ensure professional relevance by using modern tools, promoting team work, providing hands-on design experiences, developing communications skills, and making clear the organizational context and constraints under which engineers work; and to exploit modern technologies to encourage learning despite barriers in time and space and to provide access to emerging information resources, the engineering education community should:

- develop an infrastructure based on the NEEDS model[4]
- promote experimentation with new educational delivery mechanisms, including particularly
- integrated, electronically mediated instruction
- team projects
- virtual peer-to-peer learning
- design labs
- cooperative learning environments
- self-paced instruction
- modular labs
- and encourage faculty to invest in learning to develop and use the new tools and techniques.

[4]The NEEDS (National Engineering Education Delivery System) model is the delivery system created by the Synthesis Coalition.

V. SUMMARY RECOMMENDATIONS

The recommendations from this study can be summarized in three categories as shown in Table 44. This table shows categories and suggested methods for facilitating achieving progress in these areas.

Category	Recommendations
1. Capturing expertise and making it easily available	Create Electronic Digital Libraries: study access mechanisms, knowledge structures, software, standards
	Collect, store, disseminate expert knowledge
2. Research on how we teach and learn in engineering education	Study cooperative learning environments, team learning methodologies.
	Investigate student centered learning including anchored instruction (= learning design and problem solving).
	Determine how visualization/simulation teaching and learning methods can assist in engineering education.
3. Training/dissemination	Assist faculty in learning about use of technology in engineering education.
	Outreach

Table 44. Summary Recommendations

If one compares these findings with the results of the tradeoffs conducted at the outset of the study one finds that categories one and three were added as a result of the focus group meetings. Recall from the information provided earlier in this part that the recommendations of the first tradeoff were to improve efforts in teaching and learning in "problem solving, teaming and learning facilitation using tools". This result is directly comparable with the second finding in Table 44 above. Adding the category of capturing and distributing knowledge acknowledges the national need for organizing knowledge and making it easily accessible to everyone. Inclusion of a training and dissemination category recognizes the resistance to change of the engineering education culture.

VI. ANALYSIS OF ALTERNATIVES: A ROAD MAP FOR THE FUTURE

Providing knowledge, understanding how to learn and training the practitioners of engineering education to use technology in engineering education are the three focus areas found by the study team to be of most importance in moving from the current lecture/laboratory model of engineering education to a future state. That future state may well be quite different from today if we follow these directions. This section sets forth a road map for change in engineering education based on the above findings. We will first examine the types of National programs that might be promoted by the government, and conclude with a discussion of what can be done now by each engineering college to prepare for the future.

A. RECOMMENDATIONS FOR GOVERNMENT

To stimulate the creation and development of new teaching and learning paradigms in engineering education, the following areas are recommended as potentially useful in assisting in the way faculty teach and students learn.

Creating Knowledge Sources. In 1994, there are current knowledge sources in many academic and recreational areas that can be accessed worldwide via the Internet. In engineering education, notable early work has created and engineering education digital resource that is part of the engineering education Synthesis coalition. This prototypical effort has demonstrated the feasibility of creating digital resources. Efforts in this direction need to be continued and expanded. Examples of expansion would include creation of knowledge taxonomies to classify engineering knowledge so that engineering educators and students could contribute to digital resources in a coherent way and so that digital libraries could be linked in meaningful ways. Another expansion would be in building software that would make it possible for faculty and students to easily access knowledge in digital libraries in more direct and simple ways than is now possible. Standards need to be studies. These include standards for building pedagogical materials and for organizing knowledge. The latter organizational problem is substantial since no current classifications of engineering knowledge exist. Yet, this type of classification is essential if we are to classify world knowledge in engineering and make it available instantaneously to our students.

Retaining, classifying, promulgating, updating and critically appraising expert knowledge should be another focus area. Understanding history and moving past current boundaries of knowledge requires accurate understanding and capture of expert knowledge. Digital libraries of information will require standards in much the same way as the journal and book review process of the past has operated. How we should maintain standards for materials that are disseminated in electronic form should be studied.

Ways to deliver distributed education should be studied. The effectiveness of one-to-many broadcasting of expert lectures, many-to-many learner interactions, and team design learning environments should be investigated. To accomplish such investigation, distant institutions should join together to conduct experiments in this area.

Learning how to effectively teach engineering education is the most significant upcoming challenge. Whereas we understand (more-or-less) how to create digital learning resources and what these resources would be used for, we are much less advanced in understanding just what are the teaching learning methods that are most useful for the future. This study has identified several candidate areas for investigation, notably anchored instruction and team-oriented education, both with a problem solving and design focus. In general, the recommendation is to understand how to move from faculty centered curriculum to a student centered curriculum. The current learning paradigm is thought to be faculty-centered in the sense that faculty deliver lectures and students feedback information on tests. There is little in the way of innovation, invention or exploration. Student-centered education would focus more in the latter areas, presumably by engaging students in team-centered design experiences that are directly relevant to learning to work in the real world so that the problems solved have real meaning to the students.

Finally, to make real progress, faculty should be trained to use new technology. As shown in the tradeoffs above, faculty acceptance always had an influence on the ranking of alternatives. Current practices of providing support for workshops should be continued in this area. Further, efforts should be made to reach out to the K-12 community, community colleges and the community at large. The capabilities of emerging technologically-based informatics infrastructure for this country will permit colleges and universities to play new teaching and learning roles in these areas. They should take advantage of these possibilities!

B. RECOMMENDATIONS FOR ENGINEERING COLLEGES

Change in the teaching and learning paradigms are upon us. Colleges of Engineering can make significant changes now that will assist in adapting to the future as it appears. Several recommendations for how Colleges of Engineering can prepare themselves are listed next.

First, it is important for faculty and administration to note that teaching and learning paradigms can and should change. Faculty who facilitate change should be encouraged. For example, experiments in use of new technology should be encouraged. A ground swell of experimentation could make a significant difference in understanding which things work and which don't.

Second, if computer communications infrastructures are not yet in place in Colleges of engineering, efforts should be made to insure that all students have access to computers and that networking gateways exist to permit access to the world outside the university. Faculty should be encouraged to assist their students use electronic media for learning; for example, including use of international network information in their coursework and encouraging students to us bulletin boards for exchanging information and knowledge.

Finally, computer usage among students and faculty should become ubiquitous.

INDEX

—A—

—B—

—C—

—D—

—E—

—U—

—V—

—W—